USER CENTERED SYSTEM DESIGN

USER CENTERED SYSTEM DESIGN

New Perspectives on Human-Computer Interaction

Edited by

DONALD A. NORMAN
STEPHEN W. DRAPER

University of California, San Diego

LAWRENCE ERLBAUM ASSOCIATES, PUBLISHERS
1986 Hillsdale, New Jersey London

The research conducted at UCSD was supported by Contract N00014-79-C-0323, NR 667-437, and by Contract N00014-85-C-0133, NR 667-541, with the Personnel and Training Research Programs of the Office of Naval Research and by a grant from the System Development Foundation. Edwin Hutchins and James Hollan's research was also sponsored by the Naval Personnel Research and Development Center.

Andrea diSessa's research was supported, in part, by the Advanced Research Projects Agency of the Department of Defense, monitored by the Office of Naval Research under contract N00014-75-C-0661.

Rachel Reichman's work was supported by grant number IST81-200018 from the National Science Foundation.

The Consul system, discussed in William Mark's chapter, and many of the ideas underlying this approach to user interface design, were developed by the Consul group at the University of Southern California Information Sciences Institute under DARPA funding during the years 1979–1984.

Lawrence Erlbaum Associates, Inc., Publishers
365 Broadway
Hillsdale, New Jersey 07642

Library of Congress Cataloging-in-Publication Data
Main entry under title:

User centered system design.

Bibliography: p.
Includes index
1. Interactive computer systems—Addresses, essays, lectures. 2. System design—Addresses, essays, lectures. 3. Human engineering—Addresses, essays, lectures.
I. Norman, Donald A. II. Draper, Stephen W.
QA76.9.I58U73 1986 004'.01'9 85-25207
ISBN Hardback 0-89859-781-1
ISBN Paperback 0-89859-872-9

Printed in the United States of America
10 9 8

Contents

Preface

This book had its genesis in the project on Human–Machine Interaction at the University of California, San Diego. The project started initially from the common interests of the two of us, one starting from Psychology, the other from Artificial Intelligence. We put together a team of researchers—some of them graduate students, some postdoctoral fellows, and some research faculty in the Institute for Cognitive Science, but all selected specifically for the project. All along, our goal was to understand the issues, to raise the important questions, and to seek methods and for a philosophy for approaching the subject. The result has been extensive collaborative effort, culminating finally in this book. Paul Smolensky generated the name of the project alliteratively with the name of the University: So from UCSD, the University, has come UCSD, the project, and, eventually, the name of this book: *User Centered System Design.*

The book represents the combined efforts of a large number of people. All the participants have interacted intensely. As the ideas of the research project took shape, we interacted with other groups of similar-minded researchers across the country, visiting some and inviting others to visit us. Toward the end of 1984 we invited a small number of these people to join us in contributing to the book and to the workshop that played a central role in its creation. The result is a collaborative effort, one that spans considerable time and distance and based on an unusual amount of interaction among authors of an edited book. The amount of interaction and cross-fertilization of ideas is immense, so much so that it is often difficult to recall where any particular idea or concept came from.

The final stages of book production were the most intense periods of collaboration and work. Initial drafts of chapters were circulated among everyone, all chapters were reviewed by from two to five reviewers, with extensive editorial criticism and heated debate among participants. We then met for a four-day retreat at the Asilomar Conference Center, in Pacific Grove, California, where this process was intensified, including a word-by-word review of each chapter. And, finally, there came the last revisions and then the extensive editing and re-editing to cross-check, cross-reference, and elucidate the ideas.

More people participated than wrote chapters. And many people deserve to be thanked for their efforts in putting all of this together. Participants in the research efforts who also attended the Asilomar Conference and aided in the review of chapters, but who did not themselves write a chapter include Jonathan Grudin (Wang Laboratories), Lissa Monty (UCSD and Xerox Palo Alto Research Center), Clark Quinn (UCSD), and Mike Eisenberg (MIT). In addition to the authors and participants, the following people served as "outside" reviewers of chapters: Gerhard Fischer (University of Colorado), Mike Jordan (UCSD), Tom Malone (MIT), and Ben Shneiderman (University of Maryland).

Some of the staff at UCSD who made this possible include Sondra Buffett for editorial supervision and general administration of the entire project (actually, the entire Institute) and Eileen Conway for her help with the illustrations, tables, and for her participation with the initial research efforts. Kathy Farrelly and Judi Turis provided the essential editorial assistance, somehow managing to keep up with the sometimes hourly revisions of chapters and organizational material. Mark Wallen, UNIX Wizard, kept the systems running.

The vicissitudes of academic and corporate funding for this combined theoretical and applied discipline showed itself. Four participants changed companies, another had to drop out due to the intense work schedule in his new company. Two people were originally part of the Atari Research Laboratories, but during the period of this project, Atari changed its character, got sold, and dismantled its research laboratories. The change of company and university affiliations as the book progressed provided a fascinating picture of the social structure of this field. Even the main contributors at UCSD changed; our main sponsor, the System Development Foundation, completed its mission, gave out all its money, and stopped giving grants. Thus, most of the UCSD people are moving on to new locations. Steve Draper has now returned to his starting base at The University of Sussex, Brighton (England).

We owe an immense debt to the wisdom and support of our funding agencies. In one sense, the debt is obvious: Without their money this work could not have been done. But their support went far beyond money. People form our funding agencies; the Office of Naval Research (ONR) and the System Development Foundation (SDF) took personal interest in the work, made positive suggestions, and helped us out with ideas, pointers to parts of the literature, the names of people we should visit, and extra funding. The ONR contractors' meetings have become one of the nation's premier scientific conferences, and many of the people involved in this book participated in these conferences.

Our interests in Human–Computer Interaction started in the Program in Cognitive Science, supported by the Sloan Foundation, in which Steve Draper was a postdoctoral fellow. Formal research started with funding from ONR through a contract from the Personnel and Training Research Program, monitored by Henry Halff and Marshall Farr. Eventually, however, that support terminated because research on Human–Computer Interaction did not fit the mandate of that ONR program. (Norman now has new support from ONR for the joint efforts with Hollan and Hutchins on Direct Manipulation reported in Chapter 7.) Marshall Farr and Henry Halff were visionary, supportive project monitors, but they too have been added to the list of people who changed jobs: Both have now left ONR.

The largest amount of funding, and the support that has made this book possible, was provided by a large grant from SDF, awarded jointly to Donald Norman and David Rumelhart to support our complementary efforts on Human–Computer Interaction and a new theoretical approach, the Parallel Distributed Processing methods. (described in the two volume book edited by Rumelhart and McClelland). Charlie Smith of SDF was extremely supportive and helpful in the conduct of this research, and without SDF support, this project would not have been possible. However, Smith too has moved on: He is now at Schlumberger/CAS.

All in all, this has been a period of intense collaborative effort, a remarkable experience during which more ideas were generated than could be pursued, of frequent interactions with researchers across the United States, Europe, and Japan, and one which has changed the way we think of humans and of computers.

Donald A. Norman
La Jolla, California

Stephen W. Draper
Brighton, Sussex

Authors and Participants

Liam J. Bannon
16, Fortfield Avenue
Terenure
Dublin 6, Ireland

John Seely Brown
Xerox PARC
3333 Coyote Hill Rd.
Palo Alto, California 94304

Sondra Buffett
Institute for Cognitive Science
University of California, San Diego
La Jolla, California 92093

William Buxton
Computer Systems Research Group
University of Toronto
Toronto, Ontario, Canada M5S 1A4

Allen Cypher
Intellicorp
1975 El Camino Real
Mountain View, California 94040

Andrea A. diSessa
Tolman Hall
School of Education
University of California, Berkeley
Berkeley, California 94720

Stephen W. Draper
Experimental Psychology
The University of Sussex
Brighton BN1 9QG England

James D. Hollan
NPRDC-UCSD Intelligent Systems Group
Institute for Cognitive Science
University of California, San Diego
La Jolla, California 92093

Kristina Hooper
Apple Computer Inc.
20525 Mariani Ave.
Cupertino, California 95014

Edwin L. Hutchins
NPRDC-UCSD Intelligent Systems Group
Institute for Cognitive Science
University of California, San Diego
La Jolla, California 92093

Brenda K. Laurel
Activision, Inc.
Drawer No. 7286
Mountain View, California 94039

Clayton Lewis
Department of Computer Science
Campus Box 430
University of Colorado
Boulder, Colorodo 80309

William Mark
Savoir
2219 Main St.
Santa Monica, California 90405

Yoshiro Miyata
Institute for Cognitive Science
University of California, San Diego
La Jolla, California 92093

Donald A. Norman
Institute for Cognitive Science
University of California, San Diego
La Jolla, California 92093

Claire E. O'Malley
Cognitive Studies
The University of Sussex
Brighton BN1 9QN England

David Owen
Institute for Cognitive Science
University of California, San Diego
La Jolla, California 92093

Rachel Reichman (Adar)
Institute for Cognitive Science
University of California, San Diego
La Jolla, California 92093

Mary S. Riley
Institute for Cognitive Science
University of California, San Diego
La Jolla, California 92093

Introduction

STEPHEN W. DRAPER and DONALD A. NORMAN
With a postscript by CLAYTON LEWIS

A PLURALISTIC FIELD, A PLURALISTIC BOOK

What is this field of Human–Computer Interaction? People are quite different from computers. This is hardly a novel observation, but whenever people use computers, there is necessarily a zone of mutual accommodation and this defines our area of interest. People are so adaptable that they are capable of shouldering the entire burden of accommodation to an artifact, but skillful designers make large parts of this burden vanish by adapting the artifact to its users. To understand successful design requires an understanding of the technology, the person, and their mutual interaction, and that is what this book is about.

The computer can be thought of from the perspective of its technology—from the field of computer science. Or it can be thought of as a social tool, a structure that will change social interaction and social policy, for better or for worse. It can be thought of as a personal assistant, where the goals and intentions of the user become of primary concern. It can be viewed from the experience of the user, a view that changes considerably with the task, the person, the design of the system. The field of human–computer interaction needs all these views, all these issues, and more besides. Studying these various perspectives can involve many disciplines: computer science, psychology, artificial intelligence, linguistics, anthropology, and sociology—the cognitive sciences. We are prepared to take on board any discipline, any approach that helps. It is a pluralistic field, so this is a pluralistic book.

A Book of Questions, Not of Answers

This is a book about the design of computers, but from the user's point of view: *User Centered System Design.* The emphasis is on people, rather than technology, although the powers and limits of contemporary machines are considered in order to know how to take that next step from today's limited machines toward more user-centered ones. This book is about the directions in which we must move to get there and does not follow traditional paradigms from any of the contributing disciplines. It is not a book on how to do things. It does not cover techniques and tools. Nor is it a book of fantasies, of possible dream worlds. Instead, it treads an intermediate path, neither detailed science nor flights of fancy. Think of it as a book of ideas, of analytical techniques described for their purpose, not for the details of their methods. Think of it as a book from which to derive the new directions in which we must move. Think of this as a book of questions, not a book of answers.

Some Possible Dimensions

This book is primarily an expression of a pluralistic approach, but if it has a common theme—a unity in its diversity—it is that human-computer interface design is not one small aspect of the main business of software design, nor will it be illuminated (let alone "solved") by a single methodology or technical innovation. To begin with, we do not wish to ask how to improve upon an interface to a program whose function and even implementation has already been decided. We wish to attempt User Centered System Design, to ask what the goals and needs of the users are, what tools they need, what kind of tasks they wish to perform, and what methods they would prefer to use. We would like to start with the users, and to work from there.

Granted the premise of User Centered System Design, though, what follows? The more we study it, the bigger the subject seems to become. Pluralism is the result of the piecemeal recognition of more and more important aspects to the subject. We are at the point (in the mid 1980s) of realizing just how much bigger the problem is than has usually been acknowledged, but we are not within sight of a grand synthesis or a unifying theory. This book offers "perspectives"—pluralistic voices laying claim to your attention. The authors contributing to this book interacted to a considerable extent during its writing. As a result, many mutual connections have been found and are mentioned in the chapters, but nothing like a single synthesis has yet been constructed. The main message remains that pluralism is necessary and appropriate at this stage of the field. The chapters reflect this pluralism implicitly, not by design.

This book has been difficult to organize, for the essence of pluralism is that there is no single *best* approach, no single dominant dimension, no single approach that fully organizes the issues to be faced or the solutions and perspectives that are offered.

One approach, a rather traditional one, is to start with considerations of the person, the study of the human information processing structures, and from this to develop the appropriate dimensions of the user interface. Many of the chapters in this book can be considered to be developments of this approach.

Another approach is to examine the subjective experience of the user and how it might be enhanced. When we read or watch a play or movie, we do not think of ourselves as interpreting light images. We become a part of the action: We imagine ourselves in the scenes being depicted. We have a "first person experience." So too with well-constructed video arcade games and with simulation tools. Well, why not with computers? The ideal associated with this approach is the feeling of "direct engagement," the feeling that the computer is invisible, not even there; but rather, what is present is the world we are exploring, be that world music, art, words, business, mathematics, literature, or whatever your imagination and task provide you.

A different approach would be to focus upon the social context of computing, on the fact that computers are meant to be tools for carrying out tasks, tasks that are normally done in association with (or for the benefit of) other people. This approach focuses upon social interaction, upon the nature of the workplace, about the kinds of assistance people get from one another rather than from manuals or formal instruction, and on the ways that we might aid and assist this process. This is a perspective in which the issue is the use that can be made of interfaces in designing human work, jobs, offices, social interactions—even society itself—i.e., a complete subordination of interface design to social concerns.

All of the above approaches are included in this book. After considering a half dozen ways of organizing the chapters we decided to group the chapters in terms of the scale or grain of analysis adopted. We see approaches that focus on one user pursuing one task on one machine, as representing only one point on a range of possible scales of analysis. On a finer scale we have studies that examine hand motions, keystrokes, and the information-processing structures of actions and perceptions, intentions and thoughts. On a larger scale are the chapters on users' activity structures and on the nature of interaction among people, between people, between people and systems, and systems and society.

These different kinds of analysis are often associated with yet another cause for multiple approaches: differences in methodology, for example in the use of experimental techniques, of implementing trial systems, observing and questioning users, or simply doing theoretical analyses. For instance, Mark's chapter revolves around an implementation, whereas that by Miyata and Norman involves applying existing work in psychology to the concerns of Human–Computer Interaction. But the chapters do not stress methodology, and we rejected the idea of organizing by methodological dimensions.

THE STRUCTURE OF THE BOOK

The chapters are grouped into sections, each with its own short introduction, which typically includes brief chapter summaries. Following Section I that surveys the field as a whole, we jump to what we see as medium scale approaches. This is Section II, "The Interface Experience." The studies in this section are traditional in that they focus on one user, one task, and one machine, but their emphasis on the subjective experience of the user is not conventional. This section deals with a phenomenologically oriented approach that directly asks the ultimately central question "what is the experience like for the user?" In the end, that is the basic question underlying all user-centered design. However, asking it does not necessarily or directly lead either to useful theory or to practical design methods which is why only a few chapters take this approach.

In the next section of the book, Section III on "Users' Understandings," we move to a finer scale. One approach in this section is to start with a design or implementation technique and pursue its consequences, which, of course, one hopes will ultimately have beneficial effects for the user. For instance, in his chapter, Mark describes an aid for designers that ensures a consistent use of concepts in an interface. That this will benefit users is assumed rather than directly investigated in that work. Yet another kind of approach in this section offers various ideas about the nature and content of how a user comes to understand a system. The hope, of course, is that better theories will lead to more informed and hence better interface design, but again that is an indirect consequence. This section of the book in fact shows still further diversity in the kinds of argument found to be interesting:

Whereas some authors (e.g., Riley) concentrate on the kinds of content users' understandings can have, diSessa concentrates on the different origins that understanding can have (e.g., loose analogies with natural language, as opposed to a detailed "functional" model from a different instantiation of the same task domain).

We next move to the larger scale chapters, going from an analysis of user activity structures (Section IV), to the pragmatic aspects of human–machine interaction (Section V), to the considerations of the flow of information among people and between person and systems (Section VI), and lastly to the societal and cultural impact of design (Section VII). This approach has let us zoom progressively outward to larger and larger scales in which first multiple tasks are considered (the section on user activities) and then the interactions between users (some of the chapters on information flow) are considered. Finally, we finish by considering computer tools as they fit into and affect the social structures within which users live and work—the last chapter of Section VI by Bannon and the concluding chapter of the book, the chapter by Brown in Section VII.

Stephen W. Draper
Donald A. Norman

*****

Postscript

A user interface functions at the intersection of many different kinds of things: people, machines, tasks, groups of people, groups of machines, and more. These different things contribute diverse constraints and opportunities to the design process. None of them, not even machines, is captured by adequate theories, that is, theories adequate to tell the designer what's going to happen when a design is put to work. Two results of this state of affairs are that design is very hard, and that there are many ideas about how it should be approached.

One view, argued by Gould and Lewis (1985), is that design must be treated as fundamentally empirical. Designers must work hard to learn as much as possible about the users of the system and the work they will do with it. They must assume that their initial design ideas, even given this background information, will be wrong, and plan for repeated redesign. They must base these redesigns on empirical measurements of the success of the design, made on actual use of an implementation, a prototype, or mockup.

A second view, represented in the influential work of Card, Moran, and Newell (1983), is that the chaos of interface design can be reduced over time to a kind of approximate order, in which quantitative rules are discovered that make useful predictions about the good and bad features of potential design.

A third view, perhaps never stated but implicit in most real design work, is that interface design is driven by technology. Improvements in interfaces stem from improvements in the size and quality of displays, or the invention of new widgets for moving cursors, or new ways to arrange information on the screen. Designers should keep up with the news and think about how to exploit these new possibilities.

Each of these viewpoints has its corresponding design wisdom. Empirical design needs methods: ways to learn about users, ways to detect, measure, and diagnose the problems in a design. Card, Moran, and Newell want those quantitative rules. Widgeteers want to know about new widgets. But this book supplies very little of any of this.

As the title indicates, this book is a collection of a variety of new ideas about human–computer interaction: new interpretations of phenomena, ideas about phenomena in other areas that may be relevant here, and (yes) some ideas about widgets, or at least widgets-in-the-large, such as knowledge-representation technology. We don't think much of this will be of direct use in design, unlike wisdom about methods, rules, and widgets. Rather, we hope it will indicate new directions that design might take, or that experimental designs might explore.

We think this is important. Empirical methods and quantitative rules don't expand the design space in which we work (except when empirical study reveals a problem that we really are forced to solve). Riding the flow of widgets takes us into new territory, but not necessarily the territory we want to be in. We hope some of the ideas we discuss here will lead to new kinds of interfaces, kinds that are new in some way that we want.

These ideas are not intended to supplant the familiar wisdom. With some differences in balance, we all endorse all of those viewpoints on design, and the forms of knowledge that go with them. No one should design an interface without knowing the technology, without thinking about keystroke counts, or without planning for measurement and redesign. So this book is a supplement, not a substitute.

Clayton Lewis

SECTION I

USER CENTERED SYSTEM DESIGN

Chapters 1 and 2 of this section raise some issues concerning the nature of design. The essay by Hooper (Chapter 1) draws a number of interesting parallels between the design of architectural and computational artifacts and searches for some lessons we might learn from the successes and, perhaps more important, the failures of the older design profession. The chapter does not provide a list of principles of design or even a set of guidelines for design that can be adopted by members of the human–computer interaction community, but it does show that "movements" and slogans in design can become overarching, with the result that the ultimate purpose of the design artifact is overlooked. Concern for the needs of the users should be primary.

The short note by Bannon (Chapter 2) expands on this issue, arguing for a socially conscious design process that is embedded in a full understanding of the users' needs. He looks at the interaction between the design and use of technology, stressing that artifacts can and do have an effect on the people who use or inhabit them and that building usable and useful systems requires that attention be paid to user needs at a variety of levels, from ergonomic concerns to organizational issues. The use of design as a social and cultural force is a contentious issue raised in the Bannon and Hooper chapters and returned to in the concluding chapter by Brown.

In the last chapter of this section, Norman discusses the applied side of Cognitive Science, what he calls *Cognitive Engineering*. He is concerned with putting together the parts of the formal science that will make a science of interface design possible. This requires, says Norman, some formal models of people and of interaction, models that need only be approximations of the ultimate theory, but that are precise enough to lead to design rules. The trick is to develop the engineering guidelines in such a way so as not to lose sight of the overall goals of cooperative, enjoyable interaction—of pleasurable engagement, to use the term coined by Laurel in Chapter 5. Norman's chapter introduces a wide range of topics that are dealt with further in subsequent chapters.

CHAPTER 1

Architectural Design: An Analogy

KRISTINA HOOPER

In research on interface design we frequently allude to the *creation of environments for enhanced interaction and problem solving.* Similarly we often distinguish the aesthetics of an interface from its functionality, and we emphasize the importance of the *satisfaction of a human user as a criterion for evaluation* rather than the objective analysis of the technological power of a particular system.

Interestingly, the language we use in these expressions comes quite directly from an examination of our physical environments and of the topics which we consider when we alter their form. They refer to the field of architectural design (of buildings not computers), a well-established field in which controversies concerning these issues have been going on for generations, controversies that have generated a range of different kinds of buildings in our environments as well as a history of the particular ideas.

In the following sections, I outline topics that seem common to architectural and interface design, trying to use architectural examples and experiences as a way to make concrete a number of complex issues in the interface domain. I hope that consideration of this analogous domain might offer insights to individuals working in the design and evaluation of interfaces. The intent of this chapter is to offer shortcuts to our early analyses as well as some time-saving cautions.

In each section I begin by giving a general sense of the arguments involved, pointing directly to architectural analogies. I then describe similarities with interface design problems. In some cases I offer suggestive guidelines for interface design. In most cases I provide cautions about the complexity of a problem, inviting more serious considerations than we have provided to date.

FUNCTIONALITY

A primary consideration in the design of a building or of an interface to a computer system is that it *works*, that it fulfills the purposes for which it was intended. In the architectural domain this assessment is articulated with a *design program* at the initiation of a design, a program that specifies design goals such as square footage, adjacencies, and circulation patterns. In some cases, attributes such as style or general psychological impact are specified, and in some cases the coordination of new and existing materials becomes a principle guideline. Given these specifications, the architect knows the constraints and the client can judge whether the design is adequate. Of course for any design program there are a number of solutions. University dormitories provide a good case in point: Many of those diverse buildings on college campuses were generated from very similar design programs.

Beyond the satisfaction of the initial design program there are a number of attributes of a building that contribute to its *working.* At a very basic level, it is expected that a building functions effectively as a shelter, that the inhabitants will be protected from harsh elements in their environments (e.g., summer sun and winter storms) and that they will be provided with basic services such as electricity and plumbing. Beyond this level the criteria for functionality depend on the individual designer and the social and cultural climate. Early in this century, Western architecture was often judged in terms of the grand statement that it made. Libraries were designed to inspire and make one humble. Churches in the United States were designed to imitate the grandness of European cathedrals. Homes were built to demonstrate the level of prestige of the owner. The Bauhaus movement, begun in Germany but influential through most of the western world, countered this approach to design, arguing that form and function should be one. Form should not be designed to imitate or inspire. Structures should *look functional*, and this often meant that buildings should display the look of a machine. This approach to design was seen in a slightly modified form in the United States in Frank Lloyd Wright's work. As articulated by Wright, the intent in his forms was to display the functionality of a building in an organic way. Rather than *looking functional*, buildings should be appropriate to their function, much like the forms in organic architecture are uniquely suited to their purposes. Hence the form of a building should reflect its function.

In the late 1960s and early 1970s a different level of analysis was applied to the consideration of buildings in the United States. This analysis considered directly the psychological impact of buildings on their users. Rather than considering the *statement* made by a building, or the functionality displayed, or the harmony of functionality and form, the focus shifted from the buildings themselves to the inhabitants and to the interaction between the building and the inhabitants. Issues were phrased in terms of crowding rather than density, as an example, psychological and sociological assessments replaced more objective criteria and concerns. The extension of this approach did not, however, provide practitioners with the kinds of guidelines they wanted in designing buildings; they seemed academic and relevant to the more classical considerations of functionality with which the designers were familiar and comfortable. And so design for psychological functionality has become less and less popular except in a very general way; postmodernistic playfulness and cynicism has focused again on the building as an object rather than as an element in an interaction with humans.

Now consider computer interfaces. We are well behind the architectural domain in considering this range of issues in the consideration of computer system design and the design of interfaces. We don't yet have available widespread general experience in the articulation of a design program. We do have quite a well-developed way to describe hardware specifications and basic software performance in the computer development area, much like the architectural profession is rather proficient in detailing engineering specifications and square footage requirements. We have yet to develop any systematic or even pragmatic way to describe the requirements for an interface, and we have little experience with interfaces that can let us translate our objective criteria with more subjective analyses. We don't yet know what is fast or slow, or high quality or low quality, in the ways that we know a 200 square-foot house will be cramped. The development of such a methodology and set of intuitions is clearly important to the progress of the field of interface design and evaluation.

We are currently developing a general sense of basic requirements for a computer system. A memory restricted to 64K is judged to be inadequate for most serious personal computing tasks, as are 8-bit machines. Yet in terms of even the very basic level of interfaces, that of screen display for example, we are only beginning to be precise. We generally assume that 80 bits per inch can generate legible text and that a 640 x 480 pixel screen can provide reasonable quality images. Yet we have little knowledge of the accuracy of these guesses on a range of systems or anticipated uses. And though we can always fall back on the basic criterion that a system works if it yields the desired product (e.g., a printed manuscript) given appropriate inputs (e.g., keystrokes), we are only now beginning to develop a vocabulary with which we can describe the basic criteria for the process involved in using a system, or of the interface which provides access to available computing power.

Relative to the general articulation of functionality and form, the field of interface design has only begun to consider the range of ways these can relate, and the alternative results of different design philosophies in these regards. Many of our current interfaces do seem designed to make statements, much as buildings have done in the past. Some seem purposely designed to inspire awe, like our early churches and public institutions. Others are designed to look friendly and approachable, regardless of their actual functionality. Many seem to imitate other things we are familiar with, like paper and desktops. It seems impossible at this stage of development to display the unique computing characteristics of a system while maintaining comprehensibility. But architecture suggests that this too will probably change with time: American architecture no longer imitates European architecture, as it has now developed its own style.

Interestingly, the machine design aesthetic has already faded from prominence in the design of personal computers, as the nonobtrusive incorporation of computing resources into furniture and buildings has already become popular. We tried so hard to have our buildings look like machines in the middle of the 20th century, but we seem content to hide our new computing machines in nonmechanistic surrounds.

Of course, part of the reason we have not considered the relationship between the functionality and form of our systems is that we only recently have had access to computing resources which enable the consideration of a range of alternatives. Windowing systems provide a great deviation from other systems, for example, as do systems that focus on graphical as opposed to textual interactions. In addition, the notion that different forms may represent the same functionality, depending on the characteristics of a user, makes it imperative that we quickly familiarize ourselves with historical arguments about the relationships between form and function in order to use these new capabilities to approach the issues with some knowledge.

An analysis of the psychological characteristics of a particular interface is clearly appropriate given the highly cognitive nature of computer processing tasks. In this sense the kind of deliberate ignoring of this aspect in the manner done within the architectural domain seems unlikely. However, it is important that we consider the importance of articulating our basic principles clearly, and in a practically applicable form, if we expect them to be taken seriously. And we should acknowledge that this articulation is very complicated, conceptually as well as pragmatically, because the description of an interaction—which is necessary if we are to have a sophisticated view of the nature of an interface—is a very complex task.

INTERFACE AS FACADE?

Many analyses of buildings emphasize the facade as a principle element. This is what most people experience directly; it is the introduction to a building. In fact, if one examines architectural magazines like Progressive Architecture or Architectural Digest one becomes convinced that many prize-winning homes were designed solely for their striking two dimensional images of the facade or of interior spaces. There are a number of challenges to this approach to building design. Many theorists emphasize that the facade is a membrane between the inside and the outside, and that its purpose is to articulate the relationship between the two. Others argue that a facade should contain information about the structure of the building, hence suggesting other viewpoints. Current postmodernistic designs focus on false facades as *jokes*, delighting in the lack of necessary relationship between exterior facades and interior designs.

In many of our discussions of interfaces we talk about them as though they were facades, the face that a computer systems *presents to the world.* Given this context, we discuss *well-designed* interfaces that present orderly and attractive views to the user. And we consider carefully how graphical and textual elements relate on a two-dimensional screen, arguing about the importance of different elements in providing an approachable interface.

Yet, a brief consideration of the architectural examples cautions against such a literal approach, suggesting that although an interface is first experienced as a screen display, an examination of interfaces requires much more analysis. For although screen display is certainly a critical aspect of any interface—it is typically the window through which we experience our interactions with a system—it is only the surface aspect of complex behaviors. It is the stage on which our interactions with computer systems are shown, but it is the nature of the play that is of relevance, not just the physical actions of the players.

The notion of a false facade—an interface that does not reflect the workings of a computer system—makes this issue concrete. For no matter how beautiful a screen display is, an interface will not be effective unless the functionality of a system is revealed preferably directly. In contrast to the architectural domain where one can walk into a building and then evaluate the informativeness of the facade, one is stuck at a screen, using the tracings presented there to infer the structure of the computer system.

Consideration of an interface as an entranceway is somewhat more informative than interface as facade. In the architectural domain, entrances have been traditionally finely crafted to control the presentation of a place to a visitor. Consider European cathedrals and formal Japanese gardens: The viewpoint of a visitor is carefully planned to reveal the whole of a place in a very systematic way. In the design of interfaces one must also consider carefully how one selectively informs a user about a particular system, providing well-chosen bits and pieces that can constitute a general understanding of a system. Some of the metaphors of entranceways may be useful in addressing this issue of selective disclosure to new users of a particular computer system.

FLEXIBILITY AND ADAPTABILITY

In vernacular architecture, buildings built by or for individuals, there has always been an emphasis on flexibility and adaptability. Buildings change: They are enlarged to contain larger families, or they are changed to improve upon the earlier effort. Buildings grow and contract to suit the residents and their ideas about their own behaviors.

In more formal architecture this has not typically been the case. A general assumption is that the designer is well-informed to provide a total design concept, and that changes in this by nonexperts will interfere with the coherence of the design. In addition, many architectural designs are at a scale such that changes are extremely complicated; unless they were anticipated initially, they become impossible.

Many architects have argued that buildings should be adaptable, that they should be designed to grow and change. Such a design philosophy is witnessed in the development of housing that is to be personalized by users. Some dormitories at UC Santa Cruz were designed this way, for example, where students worked together to fill in a shell to suit their own tastes. The design of modular furniture for particular places to encourage interior changeability provides another example of how some personalization of places can be encouraged by a designer.

An outgrowth of this emphasis on flexibility and adaptability has been seen in the school landscape in the last few decades. General purpose auditoriums have been designed to fulfill a range of activities. Open classrooms have been built to encourage a number of different styles of teaching. Experiments with flexibility have been judged failures in these examples. For by providing spaces that do lots of things, many of the attributes which make a single space do one thing well (such as maintain sound control in a classroom) are lost. Flexibility does not necessarily imply adaptability for any single activity.

Critical to the relationship between flexibility and adaptability seems to be the mechanisms for change. An intriguing architectural example is evident in the new office buildings that have rediscovered the *window* as a means of temperature control. In the 1970s office buildings installed elaborate temperature control systems that were designed to minimize the effect of weather patterns on individuals' comfort. This meant no windows that could be opened. Now it has been found that individual comfort can be maximized by including local controls (e.g., windows) in the complex systems.

There are important cautions offered by this particular architectural analogy to our design of interfaces. For in our design of interfaces, we do currently emphasize the personalization of interfaces. Many systems now provide ways to provide idiosyncratic protocols for particular tasks, like reading mail, and they let the user select the timing of particular controllers (such as how much time is allowed to do a double click on the Macintosh). Yet, the architectural example shows us that flexibility in personalization may not necessarily provide adaptable systems. One may want to rely on expert judgments of a *best* system as a first approximation, making changes available from this base level. One might want to prevent the *moving of walls*, for example, but encourage the *rearrangement of furniture*.

CULTURAL IMPACT

Members of the German Bauhaus Movement proposed that the introduction of new buildings into an environment directly imposed a set of cultural values on the people who would inhabit them, whether or not this was done self-consciously. From this starting point they proceeded to articulate a coherent set of principles of design that considered the range of ways designs would impact people. They then deliberately chose amongst alternative designs based on their judgment of appropriate impact.

The relationship between a design and cultural values was demonstrated dramatically by the failure of the Pruitt-Igoe housing complex in St. Louis. This complex, a public housing project designed with great care by a talented group, turned out to be a total disaster in facilitating the behaviors of the individuals who lived in it. (Bannon, Chapter 2, gives more detail.)

Given this experience, and others like it, a large emphasis was introduced in architectural design to attempt a match between designs and inhabitants. *User-centered design* and *community participation in design* became major attributes of design processes, particularly of publicly funded projects.

Yet the activity of incorporating the values of users proved to be more complicated than it had seemed initially. In addition to logistical and methodological issues, a number of conceptual issues emerged. For one thing it was difficult to define users and their needs. The people who were initial residents of a building were often very different than those who were residents later; it soon became clear that buildings had to be designed to anticipate future users and uses. In addition, a contrast was often noted between users' current desires and behaviors and those which a designer might want to encourage or which might be anticipated, given residence in a new living situation.

The parallels of this situation in the interface design field are many. Once we acknowledge that importing particular interfaces and computer systems into different environments fundamentally alter these environments, in terms of social interactions as well as personal development, we realize that a consideration of the nature of these environments in the preparation of our systems is imperative. And yet, as in the architectural examples described, the definition of the environments we are entering and the users we are providing for, is a complicated endeavor. Quite clearly, for example, users of computer systems will increase considerably in sophistication in the near future. Problems we now must address in terms of *computer phobias* and novice users will not be central to our concerns. In addition, it is most probably the case that the cognitive processing of computer users will change given exposure to these new systems. Instead of doing things that are done without computers in more efficient ways, there will be new things that will be done with computers, and people will probably think quite differently about what they are doing (see Brown, Chapter 22).

Given this, just who is the user we should be studying in designing our interfaces? Is it the current novice or the current sophisticate? Or is it our guesses about future sophisticates? And do we design for a majority of people, or for an elite whom we judge to be good models of what the majority will be in the future? Moreover, how much of our effort do we put into adapting our machines to people, and how much in adapting people to our machines (e.g., in providing good tutorials)? And just how will we deal with issues of future changes? Where in computer design are the green-belts which urban planners have provided to insure comfortable physical environments in the future?

Our intent to consider the users of our systems seriously is clearly a good first instinct, as it is the case in the architectural domain.

However, it is clear that we have not solved any problems in this acknowledgment, but instead have just begun to articulate them.

DESIGN METHODOLOGIES

In the architectural domain there is a great emphasis on a *design philosophy*, a coherent set of principles that guide a designer on individual projects. Typically these philosophies are most visibly demonstrated in movements which characterize particular styles of presentation. Often these movements are reactions to earlier movements, hence the neoclassic revival which imitated classical forms, or the current postmodern movement which can be generally characterized as a "take-off" (often humorous) on modernistic principles (where Modernism was an earlier movement). Occasionally they are initiated by single individuals or firms, hence the impact of Frank Lloyd Wright or Skidmore, Owens, and Merrill on modern design. More infrequently than one would expect, a design philosophy is articulated which is based on a set of logical and consistent principles that are carefully and systematically presented. The Bauhaus provided us with an example of this general approach to design. This movement did not rest with general concepts like *aesthetics* or *style,* nor did it only provide examples without analysis. Instead, members of the Bauhaus attempted to articulate their concepts in a complete way, and then to show the results of their design philosophy in a range of media, including weavings, paintings, sculptures, and plays as well as buildings.

It is this structured approach to design that seems favored by most individuals interested in *interface as science,* where one systematically builds knowledge which becomes more complete and coherent. Studies of the Bauhaus approach to this problem are then relevant. However, it is important to acknowledge that there are other mechanisms that will affect the evolution of interfaces. Already, we see the computer industry working in a reactionary mode, building interfaces where the main criterion is either that they are just like something else (either in the computer domain or the physical world) or that they represent reactions to existing approaches. And we have started to see classic interfaces be developed which are innovations, and hence copied. Already, for example, the phrase *Macintosh-like* has been incorporated into marketing vocabularies as well as design analyses. The parallels to design philosophies in the architectural domain might therefore be useful to interface designers as they anticipate future issues.

In addition to a philosophy which generates a design, there needs also to be criteria for the evaluation of this design if learning is to take place. In the field of architecture, evaluation has typically been based on an *artistic model,* where issues of aesthetics and general acceptance have been central in judging the impact of a new building. Most cities have architectural critics who review major buildings with great fervor, for example, much like movie critics review new films. The criteria are typically cultural and personal to the expert involved, and the biases of the critics are purposely made quite obvious for informed interpretation.

There was a major shift in the 1960s and 1970s in architectural criticism, as a social science perspective was introduced that began systematic analyses of the effectiveness of buildings from a user's point of view. New fields of investigation were introduced by these efforts. The field of environmental psychology emerged as an important one in the field of psychology, for example, contributing new information on the relationship between people and their environments as well as new experimental methodologies and journals. However, most of these studies did not yield any applicable results for the designer involved in a particular building project, and they were criticized by designers for this reason. Also, studies of any complexity were not typically available to designers during the time frame of a particular project. Research directed to particular problems on particular projects served typically as post hoc evaluations rather than as formative evaluations.

In the evaluation of interfaces and/or of software packages in general, we are addressing many of the same issues considered in the architectural domain. One alternative is to rely on expert critics in our assessment of particular instances, even though this class of analysis is not systematic nor predictive of *how to do things better.* This is the approach taken, for example, in the *Whole Earth Software Review Catalog* for the evaluation of software.

This approach will probably become widespread in the analysis of interfaces, as it provides consumers with the kinds of information they require in making selections from available materials. Yet it does not greatly assist the developer in improving interfaces in any fundamental way. We must introduce some systematic analyses into the arena of interface evaluation. In doing this, there are a couple of things that can be learned from the architecture experience. One technique that has proven very successful in the architectural area, for example, is simulation. The motivation of this approach is that *if one could only experience a building before it is built, one could assess the basic nature of this building, and make required alterations.* Such an approach is key to the development of Computer Aided Design (CAD) stations that enable a designer to explore an environment that is being designed from a number of physical viewing locations, as well as a number of systems viewpoints (e.g., electrical or plumbing). The attraction of such systems is that the designer can quickly examine subsolutions and make rapid alterations to plans. The Environmental Simulation Project at UC Berkeley is another example of an approach to simulations in the architectural community. The emphasis of this project is on the human view of an environment, and great pains have been taken to generate realistic still and moving pictures of environments. Similarly, patterns of observer movement have been designed that correspond to human movement in environments (e.g., eye-level views rather than helicopter views). Rapid prototyping laboratories provide the equivalent kind of experience for the designer of interfaces, so that ranges of different classes of interfaces can be *tried out* well before entire systems are complete. In addition, candidate interfaces can be evaluated experimentally in laboratory settings by a number of individuals besides the designers, something that is possible in the architectural case, but that is typically not pursued.

A rapid prototyping environment does not, however, guarantee that we will end up with results that are useful for designers, even if they contribute magnificently to our understanding of the general field of interface design. We should take some caution in our goals in this area, then, taking advantage of the ambiguities in the experiences of the social scientists in the architectural domain.

More critically, it is important to acknowledge that, just as in architectural simulation, a rapid prototyping environment does not take care of the major problems of systematic analysis. It does not solve the problems of defining the basic elements of an interface or of choosing criteria for evaluation. Just as the environmental psychologist struggled to define the central attributes of buildings only to find that simple parameters were not obvious, and then worked to develop descriptions of interactions between buildings and people that seemed even less obvious, interface researchers have a number of problems to solve. If their work is to be of much value, they must develop descriptions of interfaces that go beyond a consideration of input and output methods, instead emphasizing descriptions which focus on protocols of human–machine interactions and all the conversational activities that this implies.

In theory, the analyses that are done in the evaluation of interfaces will have greater impact on designs than do evaluations of buildings. For one thing, they can be done concurrently with the development of a system and hence provide formative evaluation, evaluation that can be incorporated into designs. In addition, the similarities between interfaces are currently quite extensive, so that the results of evaluations can be thought to transfer from one system to another in a more natural way than is possible in the diverse architectural domain.

However, it is the case that interface evaluators can benefit from the experiences of social scientists in the architectural domain, realizing the contexts in which their evaluations will exist.

POLITICS OF DESIGN

Designers and evaluators in both architectural and interface design are not typically *true scientists*, as they are generally interested in making designs better as well as understanding them. Environmental psychologists were, for example, typically advocates of a human centered view, of providing people with better living conditions. Similarly, many interface researchers are interested in the humanizing of computer environments as well as in understanding human–machine interactions. Given this, it is important that one consider problems of politics as well as science in each of these endeavors, for there is clearly a difference between knowing *the truth* and having any effect on the development of new systems.

The impact of social scientists on architectural design has been disappointingly tiny given the scale of effort and apparent success in the 60s and 70s. Most architecture schools did indeed hire a few social scientists on their staff during these years, but there seem to be few new positions offered and no substantial resources set forth to develop full-scale programs in these areas (or to maintain those initiated in the past). In addition, few social scientists have been hired into architects' offices, even where there is sympathy toward a human-centered view. More profoundly, those social scientists who have been involved in the architectural design problems have typically had little influence on the overall design process and have hence been generally unsuccessful in influencing designs.

Interestingly, the main effect of all the work on human-centered design seems to have been accomplished by a few books on the subject that have been read by a large number of designers. *Design Awareness* by Robert Sommer (1971) was, for example, a very influential book that articulated the basic tenets of human-oriented environments, providing succinct anecdotal examples as well as convincing arguments for the importance of considering human experience. The *Image of the City* by Kevin Lynch (1960) set forth an equally successful argument for the importance of considering the experiences of humans in the interpretation of their environments. Introducing the concept of a *cognitive map* to the design community as a fundamental gauge of the satisfaction of inhabitants, using this to propose the importance of *urban design* as well as building design, Lynch provided both a methodology and a set of practicable directives to the architectural and planning professions.

Designers who were influenced by these books, finding in them a way to articulate things they already suspected or as a new view on the topic of architectural design, have quite clearly introduced some carefully crafted human oriented buildings in the environment. In addition, they have been influential in the formulations of policies for planning which address issues of human aesthetics and livability. Design review boards at a local level now routinely provide for community input to architectural design proposals, for example, something that was quite unheard of a few decades ago. Many urban design plans emphasize livability issues, including the amount of light available on urban streets and the open areas available in dense urban centers for general access (e.g., plazas for eating lunch in sunny areas beneath high rises). In addition, there are a number of laws that require extensive community participation in the design of public design projects. In fact some design theorists are now arguing that, given this orientation toward community participation as a method to incorporate human values, the interesting theoretical issue is to address the architectural design process from the perspective of community involvement as opposed to the attempt to articulate basic design principles. Christopher Alexander, for example, has shifted his approach to architectural design from the provision of sets of systematic rules to generate good designs to the articulation of patterns in the environment which can be useful to a community involved in a design process. Hence while his early book, *Notes on the Synthesis of Form* (1964), focused on the mathematical expression of the design process, his later book, *A Pattern Language* (1977), provides a set of descriptions of elements in the physical environment (e.g., the farmhouse kitchen, window seats) which might be useful to individuals and communities involved in the design of their own environments.

The field of interface design has fared somewhat better than that of human-oriented architectural design in that there are a number of professionals in the computer design profession addressing these topics. And, though the results of this work are only recently moving from research laboratories to commercially available products, the field of interface design and evaluation has been established in industry. Yet it is still an open question as to how this group of professionals might be best incorporated into the computer design area. Should they be partners in computer design, providing a particular perspective on the design problem? Or should groups of interface designers work separately from other designers (e.g., hardware design, general software design), setting forth alternative proposals for systems and analyses of current candidate solutions? Or should they work with marketing departments that consider human needs and desires in the computer area? Clearly each of these is a reasonable approach, and an approach taken on different projects. It will be interesting to see how these different approaches are differentiated in the future.

Interface designers and researchers also seem to have had an impact in academic institutions, as classes in these topics are starting to appear on major campuses. It is important to learn from the architectural experience in this area, however, and to keep track of how the commitment is sustained.

Unlike architecture, there are no current examples of classic books in the human interface area. For though there have been a large number of books published recently on this topic, and though there are a number of journals that provide a conduit for information about this research area, it does not seem that any single book or set of articles has caught peoples' imaginations. It will be interesting to see if there is some small set of examples in this area which will influence the field in the future. Clearly there will be a set of idiosyncratic events (like the introduction of a book or a new machine) which will influence the development of interfaces in ways that can never be anticipated.

Of course one hopes that there will be some influence of theory on design. One possible result will be the development of standards for interfaces (required protocols for particular classes of tasks), or of a set of standard interfaces. Yet, unlike the field of architectural design, these standards and standard interfaces probably will not be supported by legislation, but instead by their adoption by large computer manufacturers. An alternative to standardization, one which seems compelling, is to set forth a limited number of design principles which are so well-documented, or so compelling, that people will rush to adopt them. This would provide for quality as well as diversity in design. An acknowledgment of the cost of purchasing and developing complex interfaces for computer systems makes this, unfortunately, an unlikely possibility. Yet experiences in the architectural domain suggest that just the provision of examples which provide new models does in fact influence future developments. So the strategy of clear articulation, documentation, and example-making does then seem like a promising one.

CONCLUSIONS

Although there are a number of differences between the architectural design domain and that of human–computer interface design, I have tried to show that serious consideration of the architectural domain can offer some insights on the problems of the interface designer and researcher. Decisions about interface design influence a huge number of people because there are numerous replicas of each design (in contrast to most examples in architecture). It is, therefore, especially important for us to avoid naivete when building human–computer interfaces.

CHAPTER 2

Issues in Design: Some Notes

LIAM J. BANNON

The chapters in this book are concerned with the design of computer systems that are tailored to the needs of users. Many of the chapters discuss the need for a better understanding of the task domains that users are working in, and the need for a more complete psychology of users. The ultimate function of design and the role of design in society are more fundamental issues underlying this work that are rarely articulated within the field of human–machine interaction, yet they certainly influence the "What?" and "How?" of design in this field as in others.

These notes discuss aspects of the design process, starting on topics that are relatively straightforward and accepted within the design community, and moving on to more controversial issues. As computers become more accepted as a part and parcel of everyday life for people, both at work and at home, the importance of the design of computer systems becomes more evident.

A THEORY OF THE USER

The emphasis within the human–machine interaction community on the need to "study the user" (Gould & Lewis, 1985) attests to the gulf in understanding that exists between system designers and users. The need is not simply for more detailed psychological models of how people think and communicate, although such models are of course fundamental to the building of more usable systems, but for a more comprehensive, more enlightened view of people that recognizes their need for variety and challenge in the tasks that they perform. Let us take the issue of building "idiot-proof" systems, a topic that has concerned members of the human factors and human engineering disciplines. The term is used to refer to the design of human–computer interfaces that are easy to use. This goal is quite legitimate, but the phrase "idiot-proof" has other connotations that are inappropriate and misleading. One interpretation of the term is that somehow we must build a system that is proof against the bunglings of people who might wreak havoc with the system. Here, the emphasis is clearly on protecting the machine from the errant humans, a surprising viewpoint for those professionals involved in "helping the user." A more important critique is that any artifact that we design embodies a theory not only of the domain for which it is applicable, but also a theory of the human user at the other end. If we start out with a theory of users that assumes mass idiocy, the likely result is an artifact that is suitable for idiots.[1]

A THEORY OF THE TASK DOMAIN

This is again a topic that is acknowledged within the field as being of key importance in the design of usable systems. This can in some instances override problems with the user interface per se. The initial success of the VISICALC spreadsheet package is an example of this phenomenon, which provided an extremely good model of the task domain for many accounting purposes, even though the interface was poor in other respects (see Norman, Chapter 3). Conversely, one can have an adequate model of the user, but if the understanding of the task domain is incomplete then the utility of the resulting system will be low. We would have a toy, not a tool.

THE ORGANIZATION OF WORK AROUND TECHNOLOGY

Having better theories of users and task domains will allow us to build more usable human–computer interfaces, but it is important to realize that the ultimate usability of a system in a given environment is governed as much by the organization of work around the system as it is by fundamental characteristics of the technology itself. Boddy and

[1] I would like to thank Edwin Hutchins for this observation.

Buchanan (1982), in a report on case studies of new technology applications, note how the technology can be used to complement or replace human capabilities, and that both the level of worker satisfaction and the overall use of the technology were sensitive to the organization of work as a whole rather than to specific aspects of systems. Bikson and Gutek (1983) found that a key feature that predicts how well users integrate information technology into their work and how happy they are involves the amount of variety they have in their work. This does not imply that lower-level issues of user interface design are unimportant, rather it points out that the design of interfaces should include a concern for how features of the interface might accommodate different kinds of work organization, allowing for flexible tailoring of the system to accommodate particular user groups.

Although the technology does not determine the work organization, it can bias things so that certain kinds of procedures are more likely to be adopted than others. It is at this level that designers can, perhaps inadvertently, influence how work gets done. This argues for a more thorough analysis of how work currently gets done that goes beyond traditional task analysis and examines the social and organizational context that influences the operation of work activities. Restricting oneself to abstract work procedures and designing systems at this level can lead to unworkable systems.

THE ROLE OF DESIGN IN SOCIETY

Designers should be concerned about the intended purpose of the artifacts they create and their ultimate effects on society for good or ill. It might be useful to ponder the utility of products and to ask whether they satisfy the real needs of people. The artifacts in society are both a product and a reflection of specific economic, political, and cultural history. They in turn help shape society and affect the quality of people's lives. Designers of artifacts thus play a critical role especially in a society as self-conscious as ours about design (Alexander, 1964). The designer is placed in a peculiar tension between serving the expressed needs of the client and yet attempting to influence aspects of the society through the creation of artifacts. The "push–pull" forces that meet in design are very evident in the interaction between urban planning and architectural design. (See Hooper, Chapter 1, for further commentary on the lessons that can be learned from architecture.)

Much of the modern movement in architecture has taken seriously the special role that design can play in shaping society, and many designs are explicitly noted as "making a statement" about societal problems or providing possible solutions to these issues. Le Corbusier's plans for cities are a clear example of this. While applauding the concern with the role of values in design and the need for social responsibility among designers, this must be grounded in a thorough understanding of the needs of the people we are designing for, and the larger context within which the design must fit. Failure to pay attention to these more basic issues has led to a number of celebrated disasters. The best intentions in the world do not, of themselves, necessarily lead to better design. Conversely, the quality of people's lives is not enhanced by well-engineered, yet useless artifacts that are, unfortunately, not uncommon in our society today.

In the field of architecture, there are many examples of designs that, although accepted and even praised by other designers, do not meet the real needs of people for shelter and community. Perhaps the classic example of this mismatch is the story of the Pruitt-Igoe low-income housing complex built in St. Louis in the early fifties. The design, by Minoru Yamasaki, was in the "modern" style as exemplified in the designs of Le Corbusier—large apartment complexes surrounded by open spaces. This new urban development did away with traditional streets, gardens, and semiprivate spaces. It won an award from the American Institute of Architects. As Jencks (1984) notes:

> Its Purist style, its clean salubrious hospital metaphor, was meant to instill, by good example, corresponding virtues in the inhabitants. Good form was to lead to good content, or at least good conduct; the intelligent planning of abstract space was to promote healthy behavior. (p. 9)

The problem was that the design was totally inappropriate for the needs of the occupants, many of whom (Southern migrants) had no experience of living in such densely packed living compartments which gave little scope for the expression of individuality. With few places available for traditional social activity, and those that were available not integrated coherently into the pattern of their daily activities, the covered walkways on the floors of the complex—Le Corbusier's "streets in the air"—became the site of vandalism, drug abuse, and crime. The place became a nightmare for the residents, and people moved out. Finally, after several attempts to rehabilitate the complex, a commission asked the few remaining tenants what they thought should be done with the buildings. The residents replied that the authorities should blow them up, which they ultimately did in 1972.[2]

---

[2] In the interest of historical accuracy, it should be added that not all of the blocks were blown up, and that some of the blocks continue to be inhabited to some degree.

Here was a case where the "vision" of the designer and the real needs of the people being designed for were dramatically different. Part of the problem was that the architect was working for a client whose values were not representative of those of the ultimate inhabitants of the buildings. Related to this is the tendency for designers, or more often, design teams, to perform their work in a location remote from the actual site of the building, thus losing both the physical and cultural context for their work. Obviously, fault does not lie at the door of the designer alone, as that person is but one element in a complex socioeconomic system that makes decisions about the features of artifacts, but it behooves the designer to display a thorough understanding of the needs of the people that will ultimately live and work around these artifacts, be they houses or computers.

CONCLUSIONS

Willingly or not, designers, by the choice of how artifacts are designed, can affect society.[3] To this extent, even without forethought, the designer makes choices, and these are affected by an underlying set of values. It is important to make this fact more salient. I am suggesting a value-laden approach to design that emphasizes an understanding and indeed the active involvement of the people who will be affected by the design. Our designs can shape society even inadvertently, and as a consequence we need to pay more attention to the longer-term effects our designs have on society.

ACKNOWLEDGMENTS

I wish to thank Sondra Buffett, Edwin Hutchins, Donald Norman, and Claire O'Malley for discussions on the issues raised in this note.

[3] The organizational and social dimensions of computer system design—specifically in software application systems—has been investigated most comprehensively by Kling (see, for example, Kling, 1980, 1984; Kling & Scacchi, 1982).

CHAPTER 3

Cognitive Engineering

DONALD A. NORMAN

PROLOGUE

Cognitive Engineering, a term invented to reflect the enterprise I find myself engaged in: neither Cognitive Psychology, nor Cognitive Science, nor Human Factors. It is a type of applied Cognitive Science, trying to apply what is known from science to the design and construction of machines. It is a surprising business. On the one hand, there actually is quite a lot known in Cognitive Science that can be applied. But on the other hand, our lack of knowledge is appalling. On the one hand, computers are ridiculously difficult to use. On the other hand, many devices are difficult to use—the problem is not restricted to computers, there are fundamental difficulties in understanding and using most complex devices. So the goal of Cognitive Engineering is to come to understand the issues, to show how to make better choices when they exist, and to show what the tradeoffs are when, as is the usual case, an improvement in one domain leads to deficits in another.

In this chapter I address some of the problems of applications that have been of primary concern to me over the past few years and that have guided the selection of contributors and themes of this book. The chapter is not intended to be a coherent discourse on Cognitive Engineering. Instead, I discuss a few issues that seem central to the way that people interact with machines. The goal is to determine what are the critical phenomena: The details can come later. Overall, I have two major goals:

1. To understand the fundamental principles behind human action and performance that are relevant for the development of engineering principles of design.

2. To devise systems that are pleasant to use—the goal is neither efficiency nor ease nor power, although these are all to be desired, but rather systems that are pleasant, even fun: to produce what Laurel calls "pleasurable engagement" (Chapter 4).

AN ANALYSIS OF TASK COMPLEXITY

Start with an elementary example: how a person performs a simple task. Suppose there are two variables to be controlled. How should we build a device to control these variables? The control question seems trivial: If there are two variables to be controlled, why not simply have two controls, one for each? What is the problem? It turns out that there is more to be considered than is obvious at first thought. Even the task of controlling a single variable by means of a single control mechanism raises a score of interesting issues.

One has only to watch a novice sailor attempt to steer a small boat to a compass course to appreciate how difficult it can be to use a single control mechanism (the tiller) to affect a single outcome (boat direction). The mapping from tiller motion to boat direction is the opposite of what novice sailors sometimes expect. And the mapping of compass movement to boat movement is similarly confusing. If the sailor attempts to control the boat by examining the compass, determining in which direction to move the boat, and only then moving the tiller, the task can be extremely difficult.

Experienced sailors will point out that this formulation puts the problem in its clumsiest, most difficult form: With the right formulation, or the right conceptual model, the task is not complex. That comment makes two points. First, the description I gave is a reasonable one for many novice sailors: The task is quite difficult for them. The point is not that there are simpler ways of viewing the task, but that even a task that has but a single mechanism to control a single variable can be difficult to understand, to learn, and to do. Second, the comment reveals the power of the proper conceptual model of the situation: The correct conceptual model can transform confusing, difficult tasks into simple, straightforward ones. This is an important point that forms the theme of a later section.

Psychological Variables Differ From Physical Variables

There is a discrepancy between the person's *psychologically* expressed goals and the *physical* controls and variables of the task. The person starts with goals and intentions. These are *psychological* variables. They exist in the mind of the person and they relate directly to the needs and concerns of the person. However, the task is to be performed on a *physical* system, with physical mechanisms to be manipulated, resulting in changes to the physical variables and system state. Thus, the person must interpret the physical variables into terms relevant to the psychological goals and must translate the psychological intentions into physical actions upon the mechanisms. This means that there must be a stage of interpretation that relates physical and psychological variables, as well as functions that relate the manipulation of the physical variables to the resulting change in physical state.

In many situations the variables that can easily be controlled are not those that the person cares about. Consider the example of bathtub water control. The person wants to control rate of total water flow and temperature. But water arrives through two pipes: hot and cold. The easiest system to build has two faucets and two spouts. As a result, the physical mechanisms control rate of hot water and rate of cold water. Thus, the variables of interest to the user interact with the two physical variables: Rate of total flow is the sum of the two physical variables; temperature is a function of their difference (or ratio). The problems come from several sources:

1. *Mapping problems.* Which control is hot, which is cold? Which way should each control be turned to increase or decrease the flow? (Despite the appearance of universal standards for these mappings, there are sufficient variations in the standards, idiosyncratic layouts, and violations of expectations, that each new faucet poses potential problems.)

2. *Ease of control.* To make the water hotter while maintaining total rate constant requires simultaneous manipulation of both faucets.

3. *Evaluation.* With two spouts, it is sometimes difficult to determine if the correct outcome has been reached.

Faucet technology evolved to solve the problem. First, mixing spouts were devised that aided the evaluation problem. Then, "single control" faucets were devised that varied the psychological factors directly: One dimension of movement of the control affects rate of flow, another orthogonal dimension affects temperature. These controls are clearly superior to use. They still do have a mapping problem—knowing what kind of movement to which part of the mechanism controls which variable—and because the mechanism is no longer as visible as in the two-faucet case, they are not quite so easy to understand for the first-time user. Still, faucet design can be used as a positive example of how technology has responded to provide control over the variables of psychological interest rather than over the physical variables that are easier and more obvious.

It is surprisingly easy to find other examples of the two-variable—two-control task. The water faucets is one example. The loudness and balance controls on some audio sets is another. The temperature controls of some refrigerator–freezer units is another. Let me examine this latter example, for it illustrates a few more issues that need to be considered, including the invisibility of the control mechanisms and a long time delay between adjustment of the control and the resulting change of temperature.

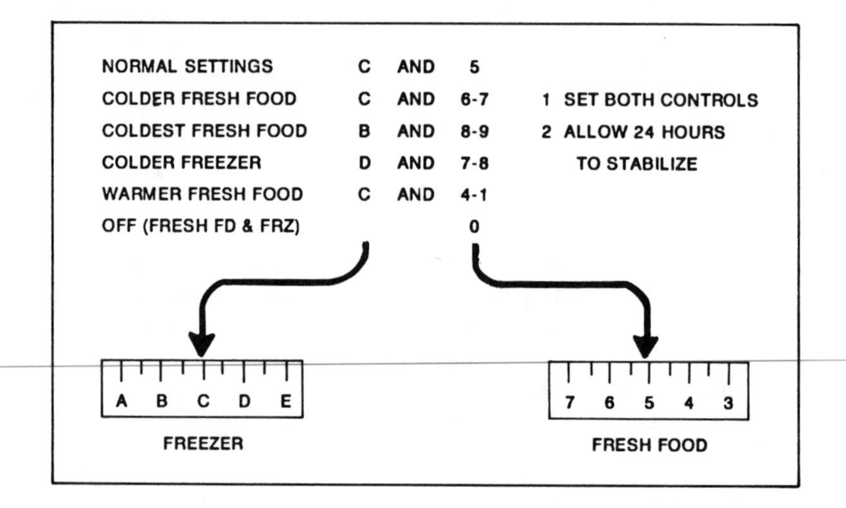

There are two variables of concern to the user: the temperature of the freezer compartment and the temperature of the regular "fresh food" compartment. At first, this seems just like the water control example, but there is a difference. Consider the refrigerator that I own. It has two compartments, a freezer and a fresh foods one, and two controls, both located in the fresh foods section. One control is labeled "freezer," the other "fresh food," and there is an associated instruction plate (see the illustration). But what does each control do? What is the mapping between their settings and my goal? The labels seem clear enough, but if you read the "instructions" confusion can rapidly set in. Experience suggests that the action is not as labeled: The two controls interact with one another. The problems introduced by this example seem to exist at almost every level:

1. Matching the psychological variables of interest to the physical variables being controlled. Although the labels on the control mechanisms indicate some relationship to the desired psychological variables, in fact, they do not control those variables directly.

2. The mapping relationships. There is clearly strong interaction between the two controls, making simple mapping between control and function or control and outcome difficult.

3. Feedback. Very slow, so that by the time one is able to determine the result of an action, so much time has passed that the action is no longer remembered, making "correction" of the action difficult.

4. Conceptual model. None. The instructions seem deliberately opaque and nondescriptive of the actual operations.

I suspect that this problem results from the way this refrigerator's cooling mechanism is constructed. The two variables of psychological interest cannot be controlled directly. Instead, there is only one cooling mechanism and one thermostat, which therefore, must be located in either the "fresh food" section or in the freezer, but not both. A good description of this mechanism, stating which control affected which function would probably make matters workable. If one mechanism were clearly shown to control the thermostat and the other to control the relative proportion of cold air directed toward the freezer and fresh foods section, the task would be much easier. The user would be able to get a clear conceptual model of the operation. Without a conceptual model, with a 24-hour delay between setting the controls and determining the results, it is almost impossible to determine how to operate the controls. Two variables: two controls. Who could believe that it would be so difficult?

Even Simple Tasks Involve a Large Number of Aspects

The conclusion to draw from these examples is that even with two variables, the number of aspects that must be considered is surprisingly large. Thus, suppose the person has two psychological goals, G_1 and G_2. These give rise to two intentions, I_1 and I_2, to satisfy the goals. The system has some physical state, S, realized through the values of its variables: For convenience, let there be two variables of interest, V_1 and V_2. And let there be two mechanisms that control the system, M_1 and M_2. So we have the psychological goals and intentions (G and I) and the physical state, mechanisms, and variables (S, M, and V). First, the person must examine the current system state, S, and evaluate it with respect to the goals, G. This requires translating the physical state of the system into a form consistent with the psychological goal. Thus, in the case of steering a boat, the goal is to reach some target, but the physical state is the numerical compass heading. In writing a paper, the goal may be a particular appearance of the manuscript, but the physical state may be the presence of formatting commands in the midst of the text. The difference between desired goal and current state gives rise to an intention, again stated in psychological terms. This must get translated into an action sequence, the specification of what physical acts will be performed upon the mechanisms of the system. To go from intention to action specification requires consideration of the mapping between physical mechanisms and system state, and between system state and the resulting psychological interpretation. There may not be a simple mapping between the mechanisms and the resulting physical variables, nor between the physical variables and the resulting psychological states. Thus, each physical variable might be affected by an interaction of the control mechanisms: $V_1 = f(M_1, M_2)$ and $V_2 = g(M_1, M_2)$. In turn, the system state, S is a function of all its variables: $S = h(V_1, V_2)$. And finally, the mapping between system state and psychological interpretation is complex. All in all, the two variable–two mechanism situation can involve a surprising number of aspects. The list of aspects is shown and defined in Table 3.1.

TABLE 3.1
ASPECTS OF A TASK

Aspect	Description
Goals and intentions.	A goal is the state the person wishes to achieve; an intention is the decision to act so as to achieve the goal.
Specification of the action sequence.	The psychological process of determining the psychological representation of the actions that are to be executed by the user on the mechanisms of the system.
Mapping from psychological goals and intentions to action sequence.	In order to specify the action sequence, the user must translate the psychological goals and intentions into the desired system state, then determine what settings of the control mechanisms will yield that state, and then determine what physical manipulations of the mechanisms are required. The result is the internal, mental specification of the actions that are to be executed.
Physical state of the system.	The physical state of the system, determined by the values of all its physical variables.
Control mechanisms.	The physical devices that control the physical variables.
Mapping between the physical mechanisms and system state.	The relationship between the settings of the mechanisms of the system and the system state.
Interpretation of system state.	The relationship between the physical state of the system and the psychological goals of the user can only be determined by first translating the physical state into psychological states (perception), then interpreting the perceived system state in terms of the psychological variables of interest.
Evaluating the outcome.	Evaluation of the system state requires comparing the interpretation of the perceived system state with the desired goals. This often leads to a new set of goals and intentions.

TOWARD A THEORY OF ACTION

It seems clear that we need to develop theoretical tools to understand what the user is doing. We need to know more about how people actually do things, which means a theory of action. There isn't any realistic hope of getting *the* theory of action, at least for a long time, but certainly we should be able to develop approximate theories.[1] And that is what follows: an approximate theory for action which distinguishes among different stages of activities, not necessarily always used nor applied in that order, but different kinds of activities that appear to capture the critical aspects of doing things. The stages have proved to be useful in analyzing systems and in guiding design. The essential components of the theory have already been introduced in Table 3.1.

In the theory of action to be considered here, a person interacts with a system, in this case a computer. Recall that the person's goals are expressed in terms relevant to the person—in psychological terms—and the system's mechanisms and states are expressed in terms relative to it—in physical terms. The discrepancy between psychological and physical variables creates the major issues that must be addressed in the design, analysis, and use of systems. I represent the discrepancies as two gulfs that must be bridged: the *Gulf of Execution* and the *Gulf of Evaluation*, both shown in Figure 3.1.[2]

The Gulfs of Execution and Evaluation

The user of the system starts off with goals expressed in psychological terms. The system, however, presents its current state in physical terms. Goals and system state differ significantly in form and content, creating the Gulfs that need to be bridged if the system can be used (Figure 3.1). The Gulfs can be bridged by starting in either direction. The designer can bridge the Gulfs by starting at the system side and moving closer to the person by constructing the input and output characteristics of the interface so as to make better matches to the

---

[1] There is little prior work in psychology that can act as a guide. Some of the principles come from the study of servomechanisms and cybernetics. The first study known to me in psychology—and in many ways still the most important analysis—is the book *Plans and the Structure of Behavior* by Miller, Galanter, and Pribram (1960) early in the history of information processing psychology. Powers (1973) applied concepts from control theory to cognitive concerns. In the work most relevant to the study of Human-Computer Interaction, Card, Moran, and Newell (1983), analyzed the cycle of activities from Goal through Selection: the GOMS model (*Goal, Operator, Methods, Selection*). Their work is closely related to the approach given here. This is an issue that has concerned me for some time, so some of my own work is relevant: the analysis of errors, of typing, and of the attentional control of actions (Norman, 1981a, 1984b, 1986; Norman & Shallice, 1985; Rumelhart & Norman, 1982).

[2] The emphasis on the the discrepancy between the user and the system, and the suggestion that we should conceive of the discrepancy as a Gulf that must be bridged by the user and the system designer, came from Jim Hollan and Ed Hutchins during one of the many revisions of the Direct Manipulation chapter (Chapter 5).

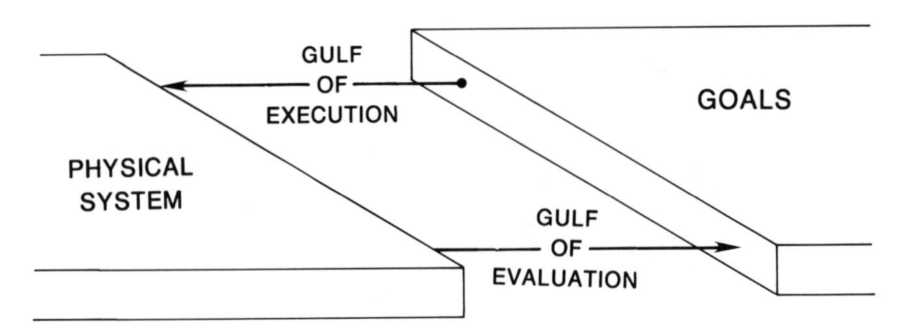

FIGURE 3.1. The Gulfs of Execution and Evaluation. Each Gulf is unidirectional: The Gulf of Execution goes from Goals to Physical System; the Gulf of Evaluation goes from Physical System to Goals.

psychological needs of the user. The user can bridge the Gulfs by creating plans, action sequences, and interpretations that move the normal description of the goals and intentions closer to the description required by the physical system (Figure 3.2).

Bridging the Gulf of Execution. The gap from goals to physical system is bridged in four segments: intention formation, specifying the action sequence, executing the action, and, finally, making contact with the input mechanisms of the interface. The intention is the first step, and it starts to bridge the gulf, in part because the interaction language demanded by the physical system comes to color the thoughts of the person, a point expanded upon in Chapter 5 by Hutchins, Hollan, and Norman. Specifying the action sequence is a nontrivial exercise in planning (see Riley & O'Malley, 1985). It is what Moran calls matching the internal specification to the external (Moran, 1983). In the terms of the aspects listed in Table 3.1, specifying the action requires translating the psychological goals of the intention into the changes to be made to the physical variables actually under control of the system. This, in turn, requires following the mapping between the psychological intentions and the physical actions permitted on the mechanisms of the system, as well as the mapping between the physical mechanisms and the resulting physical state variables, and between the physical state of the system and the psychological goals and intentions.

After an appropriate action sequence is determined, the actions must be executed. Execution is the first physical action in this sequence: Forming the goals and intentions and specifying the action sequence were all mental events. Execution of an action means to do something, whether it is just to say something or to perform a complex motor

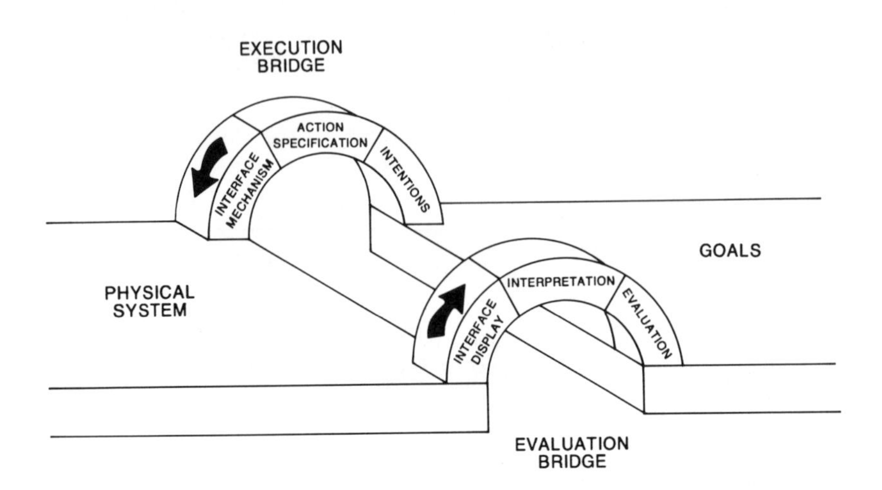

FIGURE 3.2. Bridging the Gulfs of Execution and Evaluation. The Gulf of *Execution* is bridged from the psychology side by the user's formation of intentions relevant to the system and the determination of an action sequence. It is bridged from the system side when the designer of the system builds the input characteristics of the interface. The Gulf of *Evaluation* is bridged from the psychology side by the user's perception of the system state and the interpretation placed on that perception, which is then evaluated by comparing it with the original goals and intentions. It is bridged from the system side when the designer builds the output characteristics of the interface.

sequence. Just what physical actions are required is determined by the choice of input devices on the system, and this can make a major difference in the usability of the system. Because some physical actions are more difficult than others, the choice of input devices can affect the selection of actions, which in turn affects how well the system matches with intentions. On the whole, theorists in this business tend to ignore the input devices, but in fact, the choice of input device can often make an important impact on the usability of a system. (See Chapter 15 by Buxton for a discussion of this frequently overlooked point.)

Bridging the Gulf of Evaluation. Evaluation requires comparing the interpretation of system state with the original goals and intentions. One problem is to determine what the system state is, a task that can be assisted by appropriate output displays by the system itself. The outcomes are likely to be expressed in terms of physical variables that bear complex relationships to the psychological variables of concern to the user and in which the intentions were formulated. The gap from system to user is bridged in four segments: starting with the output displays of the interface, moving to the perceptual processing of those displays, to its interpretation, and finally, to the evaluation—the comparison of the interpretation of system state with the original goals and intention. But in doing all this, there is one more problem, one just beginning to be understood, and one not assisted by the usual forms of displays: the problem of level. There may be many levels of outcomes that must be matched with different levels of intentions (see Norman, 1981a; Rasmussen in press; Rasmussen & Lind, 1981). And, finally, if the change in system state does not occur immediately following the execution of the action sequence, the resulting delay can severely impede the process of evaluation, for the user may no longer remember the details of the intentions or the action sequence.

Stages of User Activities

A convenient summary of the analysis of tasks is is that the process of performing and evaluating an action can be approximated by seven stages of user activity[3] (Figure 3.3):

- Establishing the Goal
- Forming the Intention
- Specifying the Action Sequence
- Executing the Action
- Perceiving the System State
- Interpreting the State
- Evaluating the System State with respect to the Goals and Intentions

[3] The last two times I spoke of an approximate theory of action (Norman, 1984a, 1985) I spoke of four stages. Now I speak of seven. An explanation seems to be in order. The answer really is simple. The full theory of action is not yet in existence, but whatever its form, it involves a continuum of stages on both the action/execution side and the perception/evaluation side. The notion of stages is a simplification of the underlying theory: I do not believe that there really are clean, separable stages. However, for practical application, approximating the activity into stages seems reasonable and useful. Just what division of stages should be made, however, seems less clear. In my original formulations, I suggested four stages: intention, action sequence, execution, and evaluation. In this chapter I separated goals and intentions and expanded the analysis of evaluation by adding perception and interpretation, thus making the stages of evaluation correspond better with the stages of execution: Perception is the evaluatory equivalent of execution, interpretation the equivalent of the action sequence, and evaluation the equivalent of forming the intention. The present formulation seems a richer, more satisfactory analysis.

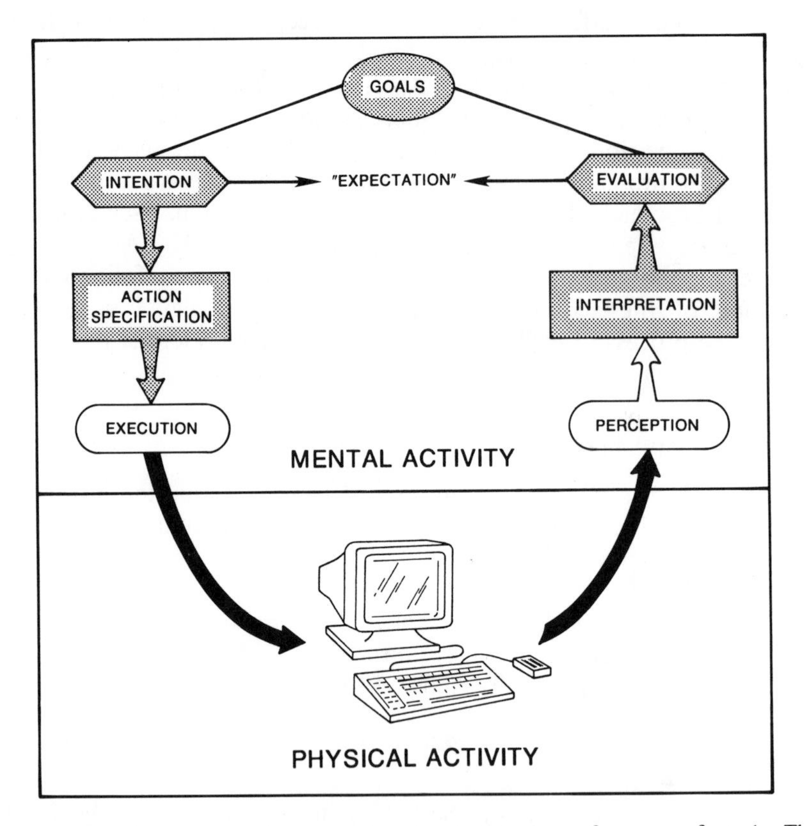

FIGURE 3.3. Seven stages of user activities involved in the performance of a task. The primary, central stage is the establishment of the goal. Then, to carry out an action requires three stages: forming the intention, specifying the action sequence, and executing the action. To assess the effect of the action also requires three stages, each in some sense complementary to the three stages of carrying out the action: perceiving the system state, interpreting the state, and evaluating the interpreted state with respect to the original goals and intentions.

Real activity does not progress as a simple sequence of stages. Stages appear out of order, some may be skipped, some repeated. Even the analysis of relatively simple tasks demonstrates the complexities. Moreover, in some situations, the person is reactive—event or data driven—responding to events, as opposed to starting with goals and intentions. Consider the task of monitoring a complex, ongoing operation. The person's task is to respond to observations about the state of the system. Thus, when an indicator starts to move a bit out of range, or when something goes wrong and an alarm is triggered, the operator must diagnose the situation and respond appropriately. The diagnosis leads to the formation of goals and intentions: Evaluation includes not only checking on whether the intended actions were executed properly and intentions satisfied, but whether the original diagnosis was appropriate. Thus, although the stage analysis is relevant, it must be used in ways appropriate to the situation.

Consider the example of someone who has written a letter on a computer word-processing system. The overall goal is to convey a message to the intended recipient. Along the way, the person prints a draft of the letter. Suppose the person decides that the draft, shown in Figure 3.4A, doesn't look right: The person, therefore, establishes the intention "Improve the appearance of the letter." Call this first intention $intention_1$. Note that this intention gives little hint of how the task is to be accomplished. As a result, some problem solving is required, perhaps ending with $intention_2$: "Change the indented paragraphs to block paragraphs." To do this requires $intention_3$: "Change the occurrences of *.pp* in the source code for the letter to *.sp*." This in turn requires the person to generate an action sequence appropriate for the text editor, and then, finally, to execute the actions on the computer keyboard. Now, to evaluate the results of the operation requires still further operations, including generation of a fourth intention, $intention_4$: "Format the file" (in order to see whether $intention_2$ and $intention_1$ were satisfied). The entire sequence of stages is shown in Figure 3.4B. The final product, the reformatted letter, is shown in Figure 3.4C. Even intentions that appear to be quite simple (e.g., $intention_1$: "Approve the appearance of the letter") lead to numerous subintentions. The intermediary stages may require generating some new subintentions.

Practical Implications

The existence of the two gulfs points out a critical requirement for the design of the interface: to bridge the gap between goals and system. Moreover, as we have seen, there are only two ways to do this: move the system closer to the user; move the user closer to the system. Moving from the system to the user means providing an interface that matches the user's needs, in a form that can be readily interpreted and manipulated. This confronts the designer with a large number of issues. Not only do users differ in their knowledge, skills, and needs, but for even a single user the requirements for one stage of activity can conflict with the requirements for another. Thus, menus can be thought of as information to assist in the stages of intention formation and action specification, but they frequently make execution more

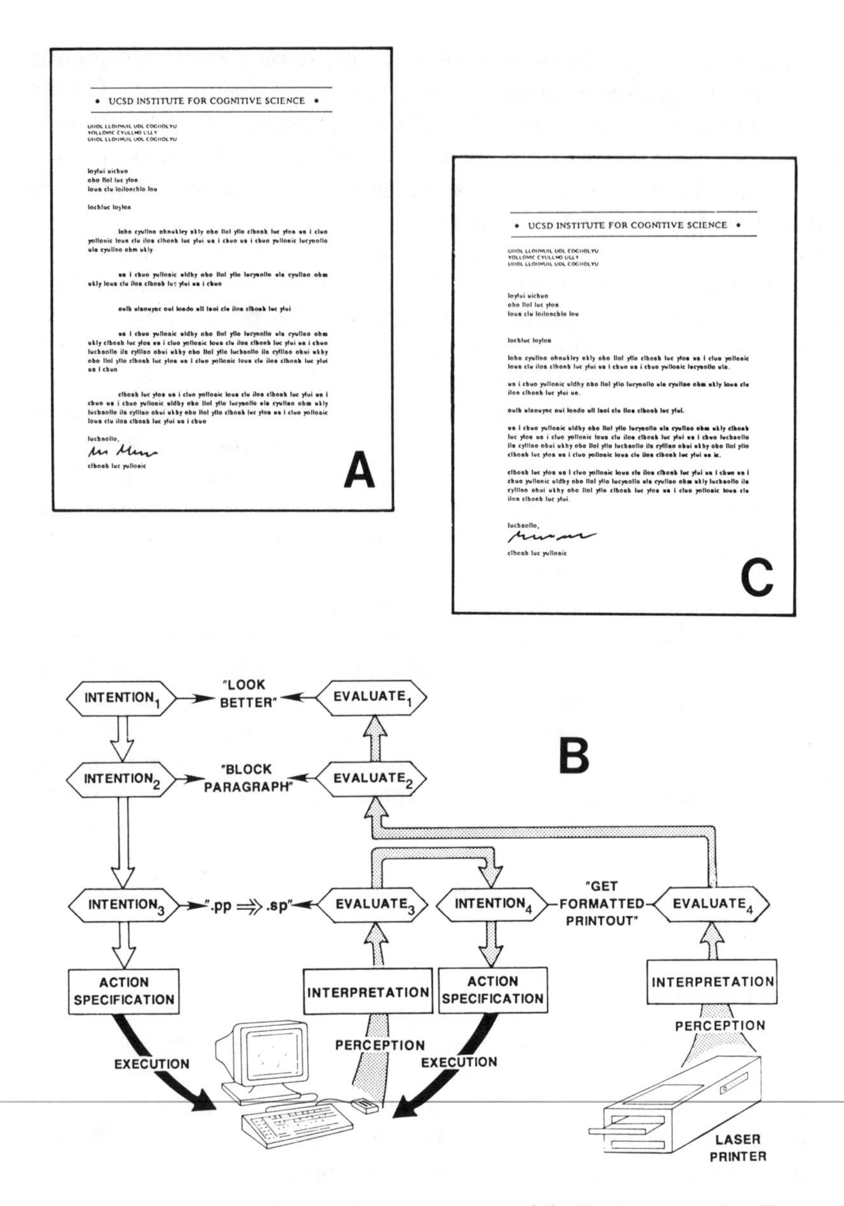

FIGURE 3.4. Sequence of stages in a typical task. **(A)** The starting point. The letter doesn't "look right," so the initial intention is "improve the appearance of the letter." **(B)** The sequence of stages necessary to make the appropriate changes to the source file of the manuscript, then to get a printed, formatted copy of the letter, and finally, to evaluate the outcome against the several levels of intentions. **(C)** The final product, the reformatted letter.

difficult. The attempt to aid evaluation by presenting extra information can impair intention selection, in part by providing distractions. On the other hand, failure to provide information can make life more complex for the user, making it harder to get the job done and adding to the frustrations with the system if the user is left bewildered, not knowing what options are available or what is happening.

Many systems can be characterized by how well they support the different stages. The argument over whether action specification should be done by command language or by pointing at menu options or icons turns out to be an argument over the relative merits of support for the stages of *Execution* and *Action Specification*.

Visual presence can aid the various stages of activity. Thus, we give support to the generation of intentions by reminding the user of what is possible. We support action selection because the visible items act as a direct translation into possible actions. We aid execution, especially if execution by pointing (throwing switches) is possible. And we aid evaluation by making it possible to provide visual reminders of what was done. Visual structure can aid in the interpretation. Thus, for some purposes, graphs, pictures, and moving images will be superior to words: In other situations words will be superior.

Moving from psychological variables to physical variables can take effort. The user must translate goals conceived in psychological terms to actions suitable for the system. Then, when the system responds, the user must interpret the output, translating the physical display of the interface back into psychological terms. The major responsibility should rest with the system designer to assist the user in understanding the system. This means providing a good, coherent design model and a consistent, relevant system image.

CONCEPTUAL MODELS AND THE SYSTEM IMAGE

There are two sides to the interface: the system side and the human side. The stages of execution and perception mediate between psychological and physical representations. And the input mechanism and output displays of the system mediate between the psychological and physical representations. We change the interface at the system side through proper design. We change the interface at the human side through training and experience. In the ideal case, no psychological effort is required to bridge the gulfs. But such a situation occurs only either with simple situations or with experienced, expert users. With complex tasks or with nonexpert users, the user must engage in a planning process to go from intentions to action sequence. This planning process, oftentimes involving active problem solving, is aided when the person has a good conceptual understanding of the physical system, an argument developed more fully by Riley in Chapter 7.

Think of a conceptual model of the system as providing a scaffolding upon which to build the bridges across the gulfs. The scaffoldings provided by these conceptual models are probably only important during learning and trouble-shooting. But for these situations they are essential. Expert users can usually do without them. They allow the user to derive possible courses of action and possible system responses. The problem is to design the system so that, first, it follows a consistent, coherent conceptualization—a design model—and, second, so that the user can develop a mental model of that system—a user model—consistent with the design model.

Mental models seem a pervasive property of humans. I believe that people form internal, mental models of themselves and of the things and people with whom they interact. These models provide predictive and explanatory power for understanding the interaction. Mental models evolve naturally through interaction with the world and with the particular system under consideration (see Owen's description in Chapter 9 and the discussion by Riley, Chapter 7). These models are highly affected by the nature of the interaction, coupled with the person's prior knowledge and understanding. The models are neither complete nor accurate (see Norman, 1983c), but nonetheless they function to guide much human behavior.

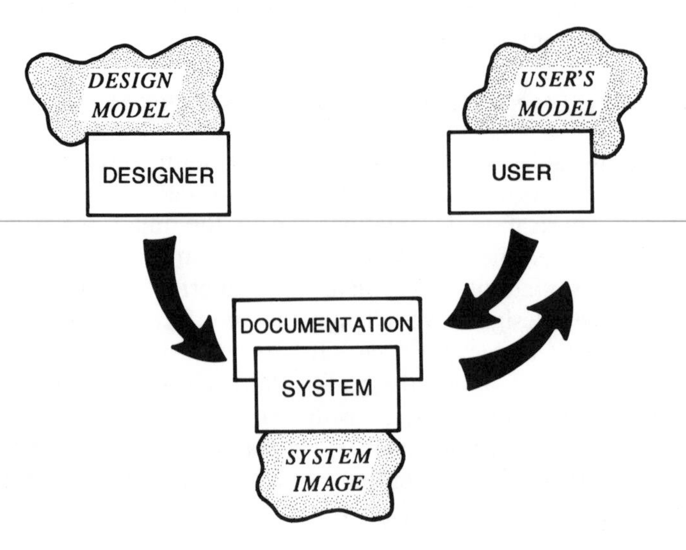

There really are three different concepts to be considered: two mental, one physical. First, there is the conceptualization of the system held by designer; second, there is the conceptual model constructed by the user; and third, there is the physical image of the system from which the users develop their conceptual models. Both of the conceptual models are what have been called "mental models," but to separate the several different meanings of that term, I refer to these two aspects by different terms. I call the conceptual model held by the designer the *Design Model*, and the conceptual model formed by the user the *User's Model*. The third concept is the image resulting from the physical structure that has been built (including the documentation and instructions): I call that the *System Image*.

The Design Model is the conceptual model of the system to be built. Ideally, this conceptualization is based on the user's task, requirements, and capabilities. The conceptualization must also consider the user's background, experience, and the powers and limitations of the user's information processing mechanisms, most especially processing resources and short-term memory limits.

The user develops a mental model of the system—the *User's Model*. Note that the user model is not formed from the Design Model: It results from the way the user interprets the *System Image*. Thus, in many ways, the primary task of the designer is to construct an appropriate System Image, realizing that everything the user interacts with helps to form that image: the physical knobs, dials, keyboards, and displays, and the documentation, including instruction manuals, help facilities, text input and output, and error messages. The designer should want the User's Model to be compatible with the underlying conceptual model, the Design Model. And this can only happen through interaction with the System Image. These comments place a severe burden on the designer. If one hopes for the user to understand a system, to use it properly, and to enjoy using it, then it is up to the designer to make the System Image explicit, intelligible, consistent. And this goes for everything associated with the system. Remember too that people do not always read documentation, and so the major (perhaps entire) burden is placed on the image that the system projects.[4]

---

[4] The story is actually more complex. The "user's model" can refer to two distinctive things: the individual user's own personal, idiosyncratic model (which is the meaning I intended); or the generalized "typical user" model that is what the designer develops to help in the formulation of the "Design Model." I jumped between these two different meanings in this paragraph. Finally, there is yet another model to worry about: the model that an intelligent program might construct of the person with which it is interacting. This too has been called a user model and is discussed by Mark in Chapter 11.

There do exist good examples of systems that present a System Image to the user in a clear, consistent fashion, following a carefully chosen conceptual model in such a way that the User's Model matches the Design Model. One example is the spreadsheet programs (starting with VISICALC), systems that match the conceptualizations of the targeted user, the accountant or budget planner. Another good example is the stack calculator, especially the early designs from Hewlett Packard. And a third example is the "office desk" metaphor followed in the Xerox *Star*, Apple *Lisa* and *Macintosh* workstations.

> *It is easier to design consistent Design Models for some things than for others. In general, the more specialized the tool, the higher the level at which a system operates, the easier the task. Spreadsheets are relatively straightforward. General purpose operating systems or programming languages are not. Whenever there is one single task and one set of users, the task of developing the conceptual model is much simplified. When the system is general purpose, with a relatively unlimited set of users and power, then the task becomes complex, perhaps undoable. In this case, it may be necessary to have conceptualizations that depend on the use to which the system is being put.*

This discussion is meant to introduce the importance and the difficulties of conceptual models.[5] Further discussion of these issues occurs throughout this book, but most especially in the chapters by diSessa (Chapter 10), Mark (Chapter 11), Owen (Chapter 9), and Riley (Chapter 7).

ON THE QUALITY OF HUMAN-COMPUTER INTERACTION

The theme of quality of the interaction and "conviviality" of the interface is important, a theme worth speaking of with force. So for the moment, let me move from a discussion of theories of action and

[5] There has been a lot said, but little accomplished, on the nature and importance of mental models in the use of complex systems. The book, *Mental Models,* edited by Gentner and Stevens (1983) is perhaps the first attempt to spell out some of the issues. And Johnson-Laird's book (1983), with the same title, gets at one possible theoretical understanding of the mental models that people create and use in everyday life. At the time this is being written, the best publication on the role of a mental model in learning and using a complex system is the paper by Kieras and Bovair (1984).

conceptual models and speak of the qualitative nature of human–computer interaction. The details of the interaction matter, ease of use matters, but I want more than correct details, more than a system that is easy to learn or to use: I want a system that is enjoyable to use.

This is an important, dominating design philosophy, easier to say than to do. It implies developing systems that provide a strong sense of understanding and control. This means tools that reveal their underlying conceptual model and allow for interaction, tools that emphasize comfort, ease, and pleasure of use: for what Illich (1973) has called *convivial tools.* A major factor in this debate is the feeling of control that the user has over the operations that are being performed. A "powerful," "intelligent" system can lead to the well documented problems of "overautomation," causing the user to be a passive observer of operations, no longer in control of either what operations take place, or of how they are done. On the other hand, systems that are not sufficiently powerful or intelligent can leave too large a gap in the mappings from intention to action execution and from system state to psychological interpretation. The result is that operation and interpretation are complex and difficult, and the user again feels out of control, distanced from the system.

Laurel approaches this issue of control over one's activities from the perspective of drama in her chapter, *Interface as Mimesis* (Chapter 4). To Laurel, the critical aspect is "pleasurable engagement," by which she means the complete and full engagement of the person in pursuit of the "end cause" of the activity. The computer should be invisible to the user, acting as the means by which the person enters into the engagement, but avoiding intrusion into the ongoing thoughts and activities.

The Power of Tools

When I look around at instances of good system design—systems that I think have had profound influence upon the users, I find that what seems more important than anything else is that they are viewed as tools. That is, the system is deemed useful because it offers powerful tools that the user is able to apply constructively and creatively, with understanding. Here is a partial list of system innovations that follow these principles:

- *Smalltalk.* This language—and more importantly, the design philosophy used in getting there—emphasize the development of tools at an appropriate conceptual level, with object-oriented, message-passing software, where new instances or procedures are derived from old instances, with derived (inherited) conditions and values, and with the operations visible as graphic objects, if you so want them to be (Goldberg, 1984; Tesler, 1981).

- *The Xerox* Star *computer.* A carefully done, psychologically motivated approach to the user interface, emphasizing a consistent, well-thought-through user model (Smith, Irby, Kimball, Verplank, & Harslem, 1982). The implementation has changed how we think of interfaces. The Star was heavily influenced by Smalltalk and it, in turn, led to the Apple *Lisa* and *Macintosh.*

- *UNIX.* The underlying philosophy is to provide a number of small, carefully crafted operations that can be combined in a flexible manner under the control of the user to do the task at hand. It is something like a construction set of computational procedures. The mechanisms that make this possible are a consistent data structure and the ability to concatenate programs (via "pipes" and input–output redirection). The interface suffers multiple flaws and is easily made the subject of much ridicule. But the interface has good ideas: aliases, shell scripts, pipes, terminal independence, and an emphasis on shared files and learning by browsing. Elsewhere I have scolded it for its shortcomings (Compton, 1984; Norman, 1981b), but we should not overlook its strengths.

- *Interlisp (and the Lisp machines).* Providing a powerful environment for Lisp program development, integrating editor, debugger, compiler, and interpreter, nowadays coupled with graphics and windows. To say nothing of DWIM — *Do What I Mean* (See Teitelman & Masinter, 1981).

- *Spreadsheets.* Merging the computational power of the computer with a clean, useful conceptual model, allowing the interface to drive the entire system, providing just the right tools for a surprising variety of applications.

- *Steamer.* A teaching system based on the concept of intelligent graphics that make visible to the student the operations of an otherwise abstract and complex steam generator system for large ships. (Hollan, Hutchins, & Weizman, 1984).

- *Bill Budge's Pinball Construction Set* (Budge, 1983). A game, but one that illustrates the toolkit notion of interface, for the user can manipulate the structures at will to create the game of choice. It is easy to learn, easy to use, yet powerful. There is no such thing as an illegal operation, there are no error messages—and no need for any. Errors are simply situations where the operation is not what is desired. No new concepts are in this game over those illustrated by the other items on this list, but the other examples require powerful computers, whereas this works on home machines such as the Apple II, thus bringing the concept to the home.

This list is idiosyncratic. It leaves out some important examples in favor of ones of lesser importance. Nonetheless, these are the items that have affected me the most. The major thing all these systems offer is a set of powerful tools to the user.

The Problem With Tools

The *Pinball Construction Set* illustrates some of the conflicts that tools present, especially conflict over how much intelligence should be present. Much as I enjoy manipulating the parts of the pinball sets, much as my 4-year-old son could learn to work it with almost no training or bother, neither of us are any good at constructing pinball sets. I can't quite get the parts in the right places: When I stretch a part to change its shape, I usually end up with an unworkable part. Balls get stuck in weird corners. The action is either too fast or too slow. Yes, it is easy to change each problem as it is discovered, but the number seems endless. I wish the tools were more intelligent—do as I am intending, not as I am doing. (This point is examined in more detail in Chapter 5 by Hutchins, Hollan, and Norman.)

Simple tools have problems because they can require too much skill from the user. Intelligent tools can have problems if they fail to give any indication of how they operate and of what they are doing. The user can feel like a bystander, watching while unexplained operations take place. The result is a feeling of lack of control over events. This is a serious problem, one that is well known to students of social psychology. It is a problem whether it occurs to the individual while interacting with colleagues, while a passenger in a runaway vehicle, or while using a computer. If we take the notion of "conviviality" seriously, we will develop tools that make visible their operations and assumptions. The argument really comes down to presenting an appropriate system image to the user, to assist the user's understanding of what is going on: to keep the user in control. These are topics discussed in Mark's chapter (Chapter 11). They require, among other things, developing a good model of the user. In addition, the user must have a good user's model of the system.

> *When systems take too much control of the environment, they can cause serious social problems. Many observers have commented on the dehumanizing results of automation in the workplace. In part, this automatically results from the systems that take control away from the users. As Ehn and Kyng (1984) put it, such a result follows naturally when the office or workplace is thought of as a system, so that the computer reduces "the jobs of the workers to algorithmic procedures" minimizing the need for skill or control, and thereby the attractiveness of the workplace. The alternative view, that of tools, offers more control to the worker. For Eng and Kyng, tools "are under complete and continuous manual control of the worker, are fashioned for the use of the skilled worker to create products of good use quality, and are extensions of the accumulated knowledge of tools and materials of a given labour process." The problem arises over and over again as various workplaces become automated, whether it is the factory, the office, or the aviation cockpit. I believe the difficulties arise from the tension between the natural desire to want intelligent systems that can compensate for our inadequacies and the desire to feel in control of the outcome. Proponents of automatic systems do not wish to make the workplace less pleasant. On the contrary, they wish to improve it. And proponents of tools often wish for the power of the automated systems. (See Chapters 2, 19, and 21 by Bannon for further discussion of these issues.)*

The Gulfs of Execution and Evaluation, Revisited

The stages of action play important roles in the analysis of the interface, for they define the psychological stages that need support from the interface. Moreover, the quality of the interaction probably depends heavily upon the "directness" of the relationship between the psychological and physical variables: just how the Gulfs of Figure 3.1 are bridged. The theory suggests that two of the mappings of Table 3.1 play critical roles: (a) the mapping from the psychological variables in which the goals are stated to the physical variables upon which the control is actually exerted; (b) the mapping from the physical variables of the system to psychological variables. The easier and more direct these two mappings, the easier and more pleasant the learning and use of the interface, at least so goes the theory.[6] In many ways, the design efforts must focus upon the mappings much more than the stages. This issue forms the focus of much of the discussion in the chapter by Hutchins, Hollan, and Norman (Chapter 5), where it is the mappings that are discussed explicitly as helping bridge the gulf between the demands of the machine and the thought processes and actions of the user. In that chapter the discussion soon turns to the qualitative feeling of control that can develop when one perceives that manipulation is directly operating upon the objects of concern to the user: The actions and the results occur instantaneously upon the same object. That chapter provides a start toward a more formal analysis of these qualitative feelings of "conviviality" or what Hutchins, Hollan, and Norman call "direct engagement" with the task.

> **The problem of level.** *A major issue in the development of tools is to determine the proper level. Tools that are too primitive, no matter how much their power, are difficult to work with. The primitive commands of a Turing machine are of sufficient power to do any task doable on a computer, but who would ever want to program any real task with them? This is the "Turing tarpit" discussed in Chapter 5 by Hutchins, Hollan, and Norman. When I program a computer, I want a language that matches my level of thought or action. A programming language is precisely in the spirit of a tool: It is a set of operations and construction procedures that allows a machine to do anything doable, unrestricted by conventions or preconceived notions. The power of computers comes about in part because their languages do follow the tool formulation. But not everyone should do this kind of programming. Most people need higher-level tools, tools where the components are already closely matched to the task. On the other hand, tools that are at too high a level are too specialized. An apple-peeler is well matched to its purpose, but it has a restricted set of uses. Spelling checkers are powerful tools, but of little aid outside their domain. Specialized tools are invaluable when*

[6] Streitz (1985) has expressed a similar view, stating that "An interactive computer system (ICS) is the more user-oriented the less discrepancies do exist between the relevant knowledge representations on the user's side and on the side of the ICS."

they match the level and intentions of the user, frustrating when they do not.

How do we determine the proper level of a tool? That is a topic that needs more study. There are strong and legitimate arguments against systems that are too specialized. Equally, there are strong arguments against tools that are too primitive, that operate at too low a level. We want higher-level tools that are crafted to the task. We need lower-level tools in order to create and modify higher-level ones. The level of the tool has to match the level of the intention. Again, easier to say than to do.

DESIGN ISSUES

Designing computer systems for people is especially difficult for a number of reasons. First, the number of variables and potential actions is large, possibly in the thousands. Second, the technology available today is limited: limited in the nature of what kinds of input mechanisms exist; limited in the form and variety of output; limited in the amount of affordable memory and computational power. This means that the various mappings (see Table 3.1) are particularly arbitrary. On the other hand, the computer has the potential to make visible much more of the operation of the system and, more importantly, to translate the system's operations into psychologically meaningful variables and displays than any other machine. But, as the opening sections of this chapter attempted to demonstrate, the problem is intrinsically difficult: It isn't just computers that are difficult to use, interaction with any complex device is difficult.

Any real system is the result of a series of tradeoffs that balance one design decision against another, that take into account time, effort, and expense. Almost always the benefits of a design decision along one dimension lead to deficits along some other dimension. The designer must consider the wide class of users, the physical limitations, the constraints caused by time and economics, and the limitations of the technology. Moreover, the science and engineering disciplines necessary for a proper design of the interface do not yet exist. So what is the designer to do? What do those of us who are developing the design principles need to do? In this section I review some of the issues, starting with a discussion of the need for approximate theory, moving to a discussion of the general nature of tradeoffs, and then to an exhortation to attend first to the first-order issues. In all of this, the goal is a

User-Centered Interface, which means providing intelligent, understandable, tools that bridge the gap between people and systems: convivial tools.

What Is It We Want in Computer Design?

Approximate science. In part we need a combined science and engineering discipline that guides the design, construction, and use of systems. An important point to realize is that *approximate methods suffice*, at least for most applications. This is true of most applied disciplines, from the linear model of transistor circuits to the stress analysis of bridges and buildings: The engineering models are only approximations to reality, but the answers are precise enough for the purpose. Note, of course, that the designer must know both the approximate model and its limits.

Consider an example from Psychology: the nature of short-term memory (STM). Even though there is still not an agreed upon theory of memory, and even though the exact nature of STM is still in doubt, quite a bit is known about the phenomena of STM. The following approximation captures a large portion of the phenomena of STM and is, therefore, a valuable tool for many purposes:

> **The five-slot approximate model of STM.** *Short-term memory consists of 5 slots, each capable of holding one item (which might be a pointer to a complex memory structure). Each item decays with a half-life of 15 seconds. Most information is lost from STM as a result of interference, new information that takes up the available slots.*

Although the approximate model is clearly wrong in all its details, in most practical applications the details of STM do not matter: This approximate model can be very valuable. Other approximate models are easy to find. The time to find something can be approximated by assuming that one object can be examined within the fovea at any one time, and that saccades take place at approximately 5 per second. Reaction and decision times can be approximated by cycles of 100 milliseconds. The book by Card, Moran, and Newell (1983) provides sophisticated examples of the power of approximate models of human cognition. All these models can be criticized at the theoretical level. But they all provide numerical assessment of behavior that will be accurate enough for almost all applications.

Tradeoffs

Design is a series of tradeoffs: Assistance for one stage is apt to interfere with another. Any single design technique is apt to have its virtues along one dimension compensated by deficiencies along another. Each technique provides a set of tradeoffs. The lesson applies to almost any aspect of design. Add extra help for the unskilled user and you run the risk of frustrating the experienced user. Make the display screen larger and some tasks get better, but others get more confused. Display more information, and the time to paint the display goes up, the memory requirement goes up, programs become larger, bulkier, slower. It is well known that different tasks and classes of users have different needs and requirements.

The design choices depend on the technology being used, the class of users, and the goals of the design. The designers must decide which aspects of the interface should gain, which can be left wanting. This focus on the tradeoffs emphasizes that the design problem must be looked at as a whole, not in isolated pieces, for the optimal choice for one part of the problem will probably not be optimal for another. According to this view, there are no correct answers, only tradeoffs among alternatives.

> *It might be useful to point out that although there may not be any best solution to a problem in which the needs of different parts conflict, there is a worst solution. And even if no design is "best" along all dimensions, some designs are clearly better than others—along all dimensions. It clearly is possible to design a bad system. Equally, it is possible to avoid bad design.*

The prototypical tradeoff: information versus time. One basic tradeoff pervades many design issues: *Factors that increase informativeness tend to decrease the amount of available workspace and system responsiveness.* On the one hand, the more informative and complete the display, the more useful when the user has doubts or lacks understanding. On the other hand, the more complete the display, the longer it takes to be displayed and the more space it must occupy physically. This tradeoff of amount of information versus space and time appears in many guises and is one of the major interface issues that must be handled (Norman, 1983a). To appreciate its importance, one has only to examine a few recent commercial offerings, highly touted for their innovative (and impressive) human factors design that were intended to make the system easy and pleasurable to use, but which so degraded system response time that serious user complaints resulted. The term "user friendly" has taken on a negative meaning as a result of badly engineered tradeoffs, sacrificing utility, efficiency, and ease of use for the benefit of some hypothetical, ill-informed, first-time user.

It is often stated that current computer systems do not provide beginning users with sufficient information. However, the long, informative displays or sequence of questions, options, or menus that may make a system usable by the beginner are disruptive to the experienced user who knows exactly what action is to be specified and wishes to minimize the time and mental effort required to do the specification. The tradeoff here is not only between different needs, but between different stages of activity. After all, the extra information required by the beginner would not bother the experienced users if they could ignore it. However, this information usually cannot be ignored. It is apt to take excess time to be displayed or to use up valuable space on the display, in either case impeding the experienced users in executing and evaluating their actions. We pit the experienced user's requirement for ease of specification against the beginner's requirement for knowledge.

First- and second-order issues. One major tradeoff concerns just which aspects of the system will be worked on. With limited time and people, the design team has to make decisions: Some parts of the system will receive careful attention, others will not. Each different aspect of the design takes time, energy, and resources, none of which is apt to be readily available. Therefore, it is important to be able to distinguish the first order effects from the secondary effects—the big issues from the little issues.

I argue that it is the conceptual models that are of primary importance: the design model, the system image, the user's model. If you don't have the right design model, then all else fades to insignificance. Get the major issue right first—the Design Model and the System Image. Then, and only then, worry about the second order issues.

Example: VISICALC. At the time VISICALC was introduced, it represented a significant breakthrough in design. Bookkeepers and accountants were often wary of computers, especially those who were involved in small and medium size enterprises where they had to work alone, without the assistance of corps of programmers and computer specialists. VISICALC changed all this. It let the users work on their own terms, putting together a "spreadsheet" of figures, readily changing the numbers and watching the implications appear in the relevant spots.

It would be useful to explore the various design issues involved in the construction of VISICALC. The designers not only were faced with the creation of a conceptualization unlike anything else that existed, but they chose to do it on a relatively small and limited machine, one in which the two major languages available were BASIC and Assembler code, which could only display 24 rows of 40 columns worth of upper-case letters and digits. Yet, spreadsheets require matrices with hundreds of rows and columns of numerals. The success of VISICALC was due both to the power of the original conceptualization and the clever use of design techniques to overcome the limitations of the machine. Probably an important key to its sucess was that the design team consisted of just two people, one a user (at the time, he was a student in the Harvard Business School who needed a tool to do business analyses and projections), the other a programmer.

But look at the command structure used in VISICALC: cryptic, obscure, and unmeaningful. It is easy to make errors, difficult to remember the appropriate operations. The choice of command names could be used as an exercise in how not to do things, for they appear to be the typical conventions chosen by computer programmers, for computer programmers. The point of this is to note that VISICALC was a success story, despite the poor choice of command structure. Yes, VISICALC would have been much improved had the commands been better. People would have liked it better, users would have been happier. But the commands were a second-order issue. The designers of VISICALC were working with limited time, manpower, and budget: They were wise in concentrating on the important conceptualizations and letting the problems of command names go for later. I certainly do not wish to advocate the use of poor commands, but the names are second-order issues.

Why was the command structure less important than the overall conceptual structure? Two factors helped:

- The system was self-contained.
- The typical user was a frequent user.

First, VISICALC was a self-contained system. That is, many users of VISICALC, especially the first wave of users, used only VISICALC. They put the floppy disk containing VISICALC into the computer, turned it on, did their work, and then turned off the computer. Therefore, there were no conflicts between the command choices used by VISICALC and other programs. This eliminated one major source of difficulty. Second, most users of VISICALC were practiced, experienced users of the system. The prime audience of the system was the professional who worked with spreadsheet computations on a regular basis. Therefore, the commands would be expected to be used frequently. And whenever there is much experience and practice, lack of meaning and consistency is not so important. Yes, the learning time might be long, but it only need take place once and then, once the commands have been learned well, they become automatic, causing no further difficulty. Choices of command names are especially critical when many different systems are to be used, each with its own cryptic, idiosyncratic choice of names. Problems arise when different systems are involved, oftentimes with similar functions that have different names and conventions, and with similar names that have different meanings. When a system is heavily used by beginners or casual users, then command names take on added significance.

Prescriptions for Design Principles

What is it that we need to do? What should we accomplish? What is the function of *Cognitive Engineering*? The list of things is long, for here we speak of creating an entirely new discipline, one moreover that combines two already complex fields: psychology and computer science. Moreover, it requires breaking new ground, for our knowledge of what fosters good interactions among people and between people and devices is young, without a well-developed foundation. We are going to need a good, solid technical grounding in the principles of human processing. In addition, we need to understand the more global issues that determine the essence of interaction. We need to understand the way that hardware affects the interaction: As Chapter 15 by Buxton points out, even subtle changes in hardware can make large changes in the usability of a system. And we need to explore the technology into far richer and more expressive domains than has so far been done.

On the one hand, we do need to go deeper into the details of the design. On the other hand, we need to determine some of the higher, overriding principles. The analysis of the stages of interaction moves us in the former direction, into the details of interaction. In this chapter I have raised a number of the issues relevant to the second issue: the higher, more global concerns of human–machine interaction. The general ideas and the global framework lead to a set of overriding design guidelines, not for guiding specific details of the design, but for structuring how the design process might proceed. Here are some prescriptions for design:

- *Create a science of user-centered design.* For this, we need principles that can be applied at the time of the design, principles that get the design to a pretty good state the first time around.

This requires sufficient design principles and simulation tools for establishing the design of an interface *before* constructing it. There will still have to be continual iterations, testing, and refinement of the interface—all areas of design need that—but the first pass ought to be close.

- *Take interface design seriously as an independent and important problem.* It takes at least three kinds of special knowledge to design an interface: first, knowledge of design, of programming and of the technology; second, knowledge of people, of the principles of mental computation, of communication, and of interaction; and third, expert knowledge of the task that is to be accomplished. Most programmers and designers of computer systems have the first kind of knowledge, but not the second or third. Most psychologists have the second, but not the first or third. And the potential user is apt to have the third, but not the first or second. As a result, if a computer system is to be constructed with a truly user-centered design, it will have to be done in collaboration with people trained in all these areas. We need either especially trained interface specialists or teams of designers, some members expert in the topic domain of the device, some expert in the mechanics of the device, and some expert about people. (This procedure is already in use by a number of companies: often those with the best interfaces, I might add.)

- *Separate the design of the interface from the design of the system.* This is the principle of modularization in design. It allows the previous point to work. Today, in most systems, everyone has access to control of the screen or mouse. This means that even the deepest, darkest, most technical systems programmer can send a message to the user when trouble arises: Hence arises my favorite mystical error message: "longjmp botch, core dump" or du Boulay's favorite compiler error message: "Fatal error in pass zero" (Draper & Norman, 1984; du Boulay & Matthew, 1984). It is only the interface module that should be in communication with the user, for it is only this module that can know which messages to give, which to defer, to know where on the screen messages should go without interfering with the main task, or to know the associated information that should be provided. Messages are interruptions (and sometimes reminders), in the sense described in the chapters by Cypher (Chapter 12) and Miyata and Norman (Chapter 13).

Because they affect the ongoing task, they have to be presented at the right time, at the right level of specification.

Modularity also allows for change: The system can change without affecting the interface; the interface can change without affecting the system. Different users may need different interfaces, even for the same task and the same system. Evaluations of the usability of the interface may lead to changes—the principle of iterative, interactive design—and this should be possible without disruption to the rest of the system. This is not possible if user interaction is scattered throughout the system: It is possible if the interface is a separate, independent module.

- *Do user-centered system design: Start with the needs of the user.* From the point of view of the user, the interface *is* the system. Concern for the nature of the interaction and for the user—these are the things that should force the design. Let the requirements for the interaction drive the design of the interface, let ideas about the interface drive the technology. The final design is a collaborative effort among many different disciplines, trading off the virtues and deficits of many different design approaches. But user-centered design emphasizes that the purpose of the system is to serve the user, not to use a specific technology, not to be an elegant piece of programming. The needs of the users should dominate the design of the interface, and the needs of the interface should dominate the design of the rest of the system.

ACKNOWLEDGMENTS

The chapter has been much aided by the comments of numerous people. I thank Eileen Conway for her aid with the illustrations. Julie Norman and Sondra Buffett provided extensive editorial comments for each of the numerous revisions. Liam Bannon, Steve Draper, and Dave Owen provided a number of useful comments and suggestions. Jonathan Grudin was most savage of the lot, and therefore the most helpful. And the Asilomar Workshop group provided a thorough reading, followed by two hours of intensive commentary. All this effort on the part of the critics led to major revision and reorganization. For all this assistance, I am grateful.

SECTION II

THE INTERFACE EXPERIENCE

This section of the book contains chapters that get directly at the question of the quality of the user's experience. This is of course the ultimate criterion of User Centered System Design, but most workers approach it obliquely in various ways such as exploring the implementation techniques, or applying existing cognitive concepts. These chapters attempt more direct analyses. The spirit of the enterprise is expressed by the image of computer usage described by Laurel in her chapter: *Interface as Mimesis.* Laurel suggests we think of the computer as a theater stage, capable of letting the user experience a world. What we want is to figure out how to design things so that we, as users, can enter the world. When we watch a play or movie, we do not think of ourselves as interpreting light images: We become a part of the action: We imagine ourselves in the scenes being depicted. We have a "first person experience." So too with well-constructed video arcade games and with simulation tools. Well, why not with computers?

In this section we start with Laurel's notion of *Mimesis,* a concept borrowed from that under-credited developer of computational concepts, Aristotle. Aristotle was a dreamer, but like proper dreamers, he provided deep analyses of his ideas in ways that are fruitful to consider today, a mere 2000 years later. Hutchins, Hollan, and Norman follow-up with an analysis of a new form of interface, one that Ben Shneiderman has called "direct manipulation,"— an interface idea made practical by the folks at the Xerox Palo Alto Research Center, and made popular by the folks at Apple, in the Macintosh computer. Direct manipulation offers great promise of being a route toward Laurel's "first person experience," but it also provides traps for the unwary. The Hutchins, Hollan, and Norman chapter offers detailed analyses of what might really be going on. After that, diSessa, speaking from his experience in the design of languages for children of all ages—first Logo, then Boxer—discusses the future of programming, and the directions in which the programming world might move.

There have been major changes in interaction in computers in general and with programming in particular, with the advent of highly interactive, personal computers. One change came about naturally with the introduction of screen editors (text editors that provide automatic and full display of the contents of the text being manipulated). These editors instantly reveal the changes to the text, thus giving the illusion of working directly with the objects on the screen, an illusion that is more difficult to sustain with earlier generations of line or character-oriented editors. A similar revolution is occurring in the development of computer systems and in programming, perhaps most prominently with the programming conventions used in spreadsheets. Again, the hallmark is continual visual presence, a self-revealing display showing at all times the state of the system. Not very many programming systems have all these properties. Important examples are the Smalltalk programming system (Goldberg, 1984) and Boxer, the system described by diSessa.

We somehow have to separate the concepts of immediate visual feedback from other components of interactive computing. Visual programming and immediate representation are of crucial importance, but this is not all there is to the experience of directness. The nature of the interaction between one's thoughts and the structures available in the computing environment are also crucial. Hutchins, Hollan, and Norman speak of the Gulfs of Execution and Evaluation, gulfs comprised of the distances between mental concepts and the physical requirements and displays of the computational system. DiSessa shows how different programming environments offer quite different tools and powers to the user. Surface appearance is only a small part of the story, although it is an essential component.

The phrase "direct manipulation" has become a catchword, connoting great ease of use, visual depiction of the events upon the screen, and specification of computer operation by moving icons around on a screen, perhaps supplemented by pop-up menus. But both chapters in the section that analyze the concept—Chapter 5 by Hutchins, Hollan, and Norman and Chapter 6 by diSessa—show this view to be simplistic. The concept of directness is complex, involving different concepts for the directness of specifying the action to be performed and the process of evaluating the result. Moreover, there is a directness between a person's intended meaning and the commands and displays available on the system ("semantic directness") and a directness of the actions that are to to be performed and the input actions required or the depiction of the result on the displays ("articulatory directness"). Finally, the feeling of "direct engagement," the concept of which Laurel speaks and which, implicitly, is what is believed by many to be the hallmark of "direct manipulation," occurs only under special conditions, when both the Gulfs of Execution and Evaluation are bridged, and when there is inter-referential constancy between the objects manipulated at the input and viewed at the output, i.e., input and output representations are unified.

The nature of programming environments is further explored in the chapter by diSessa, who examines in detail the underlying principles of programming, from conventional languages to "device programming" and "direct manipulation." diSessa illustrates the way that programming languages borrow from and count upon analogies with natural language and experience, illustrating his point with reference to the language, Boxer, he has helped develop at MIT.

CHAPTER 4

Interface as Mimesis

BRENDA K. LAUREL

AN INITIAL ASSERTION

In a recent article entitled "The Importance of Interaction," theorist Robert M. Dunn observes: "When we compound the 14 categories of interaction with the 25 different possible states of discourse in any dialog, we see that there are 350 theoretically distinct forms of interaction. This is the primary reason why building interactive systems is still an art" (Dunn, 1984).

What? All those categories, and it's not science yet?

The design of the human interface to interactive systems is "still an art," and likely to remain one, not because design "science" has not yet come of age, but because an interface is by nature a form of artistic imitation: a *mimesis.* It need not be a dark art, fraught with mysterious rituals and ethereal constructs. If designing interfaces feels like painting on cave walls by flickering torchlight, it is only because we, the designers, have not availed ourselves of better illumination: the science of the mimetic arts, poetics.

Many of us in the computer industry have been struck by the thrashing about of software designers and analysts as they attempt to identify usable criteria for the design of good interfaces. Since the birth of the consumer software business in the 70s, I have watched software engineers grapple with the issues of engagement and ease of use (designers, as we know, remain a "luxury item" in the consumer end of the industry). Good aesthetic intuitions and crisp logic have marked those engineers who have created successful products. Yet "artistic" concerns—indeed, the very idea of an artistic approach to design problems—has been blushed about as "fluff," while the "harder" disciplines of psychology and computer science are assiduously mined for guiding principles.

As a student of theatre and dramatic criticism, as well as a software designer and researcher, I am impressed again and again by the applicability of dramatic theory to the problems of interface design. I am also impressed by the availability and comprehensiveness of that theory. The dramatic theory of Aristotle, for instance, has been around for 2000 years. It can guide us at levels far deeper than the "fluff" usually associated with art ("What color should the border of my menu be?"). The purpose of this chapter is to advocate an approach to the problems of interface design—both theoretical and practical—that utilizes dramatic theory.

THE PURPOSE OF AN INTERFACE

According to Aristotle, the *end cause* of a thing is the function that it is intended to serve; that is, what it is supposed to *do*. The end cause of a play, for instance, is "the arousal and catharsis of emotion." By emotional arousal, we mean that the audience becomes emotionally involved in the action of the play. The term *catharsis* is best understood as a kind of closure: All the questions and issues that have been raised are brought to a satisfactory conclusion, so that the audience is not troubled by a bunch of loose ends. For catharsis to occur, the audience's experience must have a rational component; that is, the emotional content of the play is orchestrated so that the events represented are, on some level, both probable and believable. Thus, the end cause of a play is the emotional and rational *engagement* of its audience. It is Aristotle's premise that such engagement, when successfully achieved, is intrinsically pleasurable.

A specific play may be intended by its author to teach the audience something, to incite the audience to action, or to provide an object for aesthetic appreciation. But these second-order functions can only be fulfilled if the play succeeds in achieving its primary end: the *engagement* afforded by the dramatic form. If the audience loses interest, falls asleep, or cannot follow the action, none of the other desired effects can occur.

Likewise, an interactive computer program may be intended to enable its user to do a variety of different things—find information, compose and format a document, play a game, or explore a virtual world. The user's goals for a given application may be recreational, utilitarian, or some combination of both, but it is only through *engagement* at the level of the interface that those goals can be met. An interface, like a play, must represent a comprehensible world comprehensibly. That representation must have qualities which enable a person to become engaged, rationally and emotionally, in its unique context.

> *Naturally, there is disagreement on this point. The critics point to the all-too-common case in which computer users are willing to submit themselves to heinous interfaces and have dreadful experiences, as long as they are able to get the job done (my own experiences with UNIX Emacs come to mind). So might drama buffs endure perfectly awful plays, to the end of being able to chat about them later. But in seeking design principles for good interfaces, we must, it seems to me, concern ourselves with the best case, and ask, not what the users are willing to endure, but what the ideal user experience might be, and what sort of interface might provide it.*

A QUESTION OF FORM

While the end cause of a thing is that which it is intended to do, the *formal cause* of that thing is the shape it must take in order to do it. Form provides the means whereby the end cause can be realized, through the exercise of its powers to engage, to evoke emotional and rational response, and to provide *catharsis*—emotional and intellectual closure.

To persuade ourselves that it is form and not content alone that has these powers, we need only consider the differences between a good play and a "day in a life." During a normal day, a lot of time is wasted with the mundane (scratching, staring out windows, sitting in traffic); the relations between causes and effects are often murky; things rarely build to a climax and then come to a satisfactory conclusion. Selecting incidents from that day and arranging them in a dramatic *form* will create a universe in which our thoughts and feelings make sense and assume a coherent pattern. It is the form of our "day-play" which orchestrates our experience, the form which creates an organic whole from the chaos—a whole in which we can engage with a special kind of pleasure.

The form of a play and that of an interface are similar in a fundamental way: Both are *mimetic*. A *mimesis* is a particular kind of representation. It is a made thing, not an accidental or arbitrary one: using a pebble to represent a person is not mimetic; using a doll is. The object of a mimesis (e.g., that which it is intended to represent) may be a "real" thing or a "virtual" one. A painting, for instance, might be a mimesis of a real landscape or an imaginary one; a play may be a mimesis of events (literally, a series of actions) that are taken from history or that are entirely "made up."

A mimesis is a closed system; it has finite potential and is limited in some way. A play like *Hamlet* can be seen as a closed system in several ways. The most obvious is that it has a clear beginning, middle, and end: When the play is finished, all the potential for action that was present in the beginning has been exhausted. The play is also closed in the sense that new potential cannot be introduced. Laertes cannot threaten Hamlet with a Jedi's light sabre, for instance, and King Claudius cannot suddenly sprout wings to escape his fate (although in other mimetic worlds the potential for such actions exists; e.g., *Star Wars*). An interactive example of the destructive effects of introducing new potential in midstream would be a spreadsheet program that suddenly invites the user to play a few rounds of "Space Invaders."

This is not to say that a "closed" system is not extensible. The world of "Star Trek," for instance, is the host for countless dramas and narratives. But it is precisely the closed nature of the "Star Trek" world that allows new stories and dramas to be created and recognized as part of a larger whole. When new characters, things, or actions are introduced (e.g., the appearance of Kirk's son or the invention of a new star drive), they must conform to the principles and probabilities already established in the closed mimetic world; in other words, such "new" materials do not represent the introduction of new potential, but are rather new formulations of existing potential within the closed mimetic system. The trilogies of ancient Greek drama, as well as contemporary soap operas,[1] are based on mimetic contexts that are "closed" in the same sense.

Because a mimesis is a closed system, it is theoretically completely knowable. Aspects of the mimetic context that are not immediately apparent may be deduced from the principles and probabilities that are

[1] Michael Lebowitz of Columbia, in his work on an AI system that will generate serial-like narratives, provides operational means for maintaining internal consistency in the creation of new characters and events in his soap opera world, UNIVERSE. (See Lebowitz, 1984.)

built into it. Finally, a mimesis is internally consistent; e.g., a character in a play will not exhibit a radical mood shift without some identifiable and plausible reason (even if that reason is apparent only retrospectively).

In the *Poetics*, Aristotle describes and analyzes those forms of mimesis that have what we have called "pleasurable engagement" as their end cause—the various species of poetry, including narratives and epics, dance, musical performances, and drama. These species are differentiated from one another in terms of the *means*, *manner*, and *object* that each employs in its representations. In this analysis, more similarities between plays and interfaces emerge:

1. In both forms, the *means* consist of signs—linguistic, imagistic, etc.—that are arranged in harmonious, or pleasing, ways (the pleasure afforded by such harmony has both emotional and rational components). While other mimetic forms use subsets of these means (i.e., music employs rhythm and harmony but not language), dramatic and interactive forms (at least potentially) utilize all of them.

2. The *manner* of representation in both plays and interfaces may be termed "enactment"; that is, things are meant to be acted out rather than described. It is worthwhile here to emphasize the distinction between *narrative* and *dramatic* forms. In literature, narrative forms include things like novels and stories, while plays are dramatic in form. The difference lies in the *manner* in which each form presents its representation. Stories and novels are intended to be read, while plays are intended to be acted out. A similar distinction can be observed between certain kinds of computer applications and the human interface to them. When conceptually divorced from its interface, a text-only database or adventure game may indeed be "narrative" in form—it is a descriptive representation. However, the interface to such an application is primarily concerned, not with description, but with *action*. Novels and databases describe things; plays and interfaces act them out. In the process of *enacting* the transactions between user and system, interfaces, like plays, are represented in a medium that includes elements like music and spectacle as well as language.

3. The *object* of both kinds of representation (plays and interfaces) is a whole action—or interaction—with a beginning, middle, and end.

   This observation is not meant to preclude interactive representations that take the form of "extended," re-usable environments (as opposed to stand-alone applications). But each interactive encounter or session, like the separate plays in a trilogy or single episodes of a soap opera, must offer sufficient closure (catharsis) to make the experience pleasurable. Many of those who hated the movie "The Empire Strikes Back," for instance, felt that the piece did not have an adequate ending, even though they knew it to be part of a trilogy (knowing that we would have to wait at least three years for the next installment didn't help).

   Aristotle emphasizes that plays represent *actions* as their primary object, and that characters are represented only to serve the action. Likewise, *interaction* is the primary object of an interface, with other objects—agents, environmental elements, etc.—represented only as called for by that primary object.

What does all this do for our understanding of a good interface? It provides a basic way of thinking about things. The consideration of means gives us grounds to believe (besides our good intuitions) that good interfaces are ultimately multimodal. The lawful relations among such "modes" (visual, auditory, lexical, etc.) are described in Aristotle's theory, as well as principles for their orchestration. "Enactment" as the manner of representation steers us away from interface structures that depend on description and supplication and toward those that focus on enabling the user to *act*. The primacy of action as the object of representation gives us (inter-)action and its component parts as the primary structural elements of the interface, as opposed to objects, menus, "screens," command modes, or other secondary sorts of entities. "The thing itself" emerges as a particular, real-time interaction of a particular shape, with parts and their relations defined by the nature of that interaction—an organic whole.

MIMESIS AND INTERACTION

The most important distinction between a play and an interface is that an interface is *interactive*, while a play is not. The audience in a theatre can have no substantive effect upon the action of a play. Their applause, laughter, or inattention may influence individual performers to be louder or slower or to make more faces, but nothing an audience can do will change one word of dialogue or one event in the plot—least of all its outcome. The action of the play is predetermined. The plot, with its characters and environments, has been painstakingly crafted by the playwright to be an organic whole, an instance of dramatic form, with all its powers and pleasures.

An interface, on the other hand, is literally co-created by its human user every time it is used. We have said that the interface represents a *whole interaction*, just as a play represents a whole action, with beginning, middle, and end. The potential for that interaction exists within the system, as well as within the user who structures goals in relation to it. Between them, those two fields of potential must contain the predispositions and constraints necessary to guarantee that the interaction, when it occurs, will be an organic whole. That whole is collaboratively formulated in real time by user and system. The user functions as an agent (or character, if you like) within the mimetic context—the context offered by the representation. Just as changing the goals and actions of a character in a play will change its outcome (and thus the shape of the whole action), so the user, through actions, has a profound effect on the whole interaction.

Few would disagree that the function of the interface is to present a representation—a context—which enables a person to interact coherently with the underlying application. The idea that the user must therefore be engaged in the mimetic context is an obvious corollary. But obvious as it seems, the leading traditions ("metaphors," if you like) in interface design contradict this notion. Consider the following examples:

1. A young woman is playing a popular adventure game on her home computer. She types a command: "Go north." Words appear on the TV screen: "YOU ARE IN A DARK CAVE." "Damn," she types. "I DON'T KNOW THAT WORD," replies the screen, unperturbed. "A NASTY TROLL IS APPROACHING FROM THE EAST." "Kill the troll," she types frantically. There is a suspenseful pause. "THE TROLL IS DEAD. YOUR

SWORD HAS BEEN BROKEN." She queries the system, "How?" "I DON'T KNOW THAT WORD."

2. A few hours later, the same young woman is trying to edit her "recipe" database. Having discovered that it is vile, she is attempting to remove the recipe for "Fish Bisque." She consults her user manual to find the command that invokes the file system editor. Then she must select the command for displaying files, then select the file to be deleted from the list of filenames, then return to the command menu and select the "delete" command. "Why can't I just delete Fish Bisque?" she moans.

These two interfaces have one important thing in common: They create a situation in which the user is not doing what she wants to be doing. In the first case, she wants to have an adventure in a fantasy world. In the second, she wants to yank a rotten recipe from her collection. In both cases, however, the interface insists that *what she is doing is using a computer*. This fixation springs from the idea that the computer is a *tool*.

The notion of the interface exemplified above uses programming as its model for human–machine interaction. The logic behind the "tool metaphor" goes like this: Regardless of what they *think* they are doing (e.g., playing a game, searching a database, or designing a cathedral), end users are *actually* using the computer as a tool to carry out commands, just like programmers. It follows, then, that what the user is interacting with is the computer itself, with outcomes like game-playing and database management as secondary consequences of that interaction.

Of course, that is silly. Just as the behind-the-scenes workings of a theatre are distinct from the action of a play produced in it, so a computer is distinct from any representation that it presents. End users are not interested in *making* a representation (like a programmer); they want to move around *inside* one. The context in which they wish to operate is the *mimetic context*.

It seems to me that the "computer as tool" convention is an artifact of the rather haphazard evolution of interface design, as well as of the evolution of computer usage. The notion of dedicated applications with end users who were not programmers did not spring crisply from the thigh of Zeus. When things were far enough along for somebody to notice the dissonance created for users by the dueling contexts present in most interfaces (namely, the mimetic context and the context of the computer as tool), a cognitive patch was developed to bridge the gap: the convention of the *intermediary*.

The intermediary is an ill-formed presence or persona that belongs wholly to neither context, but attempts to mediate between them for the user. The user tells the intermediary (read "interface") what is wanted and the intermediary takes care of making it happen. In the adventure game example, the intermediary seems to act as an agent for both the program in its evaluation of user input ("I DON'T UNDERSTAND THAT WORD") and as an agent for the user in its performance of functions like "inventory" and "look." It also "stands in" for the user: It swings the sword, takes the lumps, and reports what happens. In the file management example, the intermediary takes the form of command menus that are invoked in order to activate processes in the program that will create the desired results. The user does not have the experience of pushing files around, stowing them and grabbing them, or blowing them away. Instead, the user has the experience of communicating with the file management intermediary.[2]

The fundamental problem with both the tool metaphor and the intermediary convention is that they rob users of the experience of direct agency—the ability to *act*—within the mimetic context. What users get to do is persuade the system, via the interface, to take the action they would like to take themselves. But isn't it sufficient to enable users to affect a process or outcome, by whatever means? What is our justification for insisting upon direct agency for users *within* the mimetic context?

Here I return to that "obvious" assertion made a while ago. I insist upon full participation in the mimetic context because it is quite simply the *experience* we desire. It is through such direct participation that the full pleasure of the mimetic form is available to us. By becoming *agents* in the mimetic context, our participation is active rather than descriptive, in keeping with the nature of interactive form (the *manner* of the mimesis is enactment, as we have noted, and its *object* is action itself).

---

[2] "Intermediaries" as I describe them here are fundamentally different from what others (Brennan, 1984) have referred to as "agents." In Brennan's formulation, for instance, an "agent" is an explicitly represented persona that is intended as part of the mimesis that constitutes the interface, while the intermediary is a vague presence that attempts to bridge the gap between a "computer as tool" interface and the mimetic context of the application. An interface agent, when represented mimetically as an interface convention, can successfully assist the user in performing complicated, odious, or overly detailed tasks *without* violating the mimetic context. Such an agent must be constructed according to mimetic principles delineated in the section on selection criteria, below. When I speak of the user as an agent, I mean that the user is an initiator of action. In this context, the opposite of "agent" is "patient"—one who sits passively while action is performed by some other entity.

Immersion in the mimetic context of a play by an audience member is possible through an act that Samuel Taylor Coleridge called "the willing suspension of disbelief." As an audience member, I know that the people on the stage are actors and that the castle parapets are cardboard, but I choose, in order to have the pleasure of unencumbered emotional and rational participation, to suspend that knowledge for the duration of the play. My experience will be clobbered if I am constantly reminded of what I have chosen to disregard—I do not wish to watch the stage manager pulling the curtain and calling the cues. If I wanted to invent a form of mimesis that was more participatory—more interactive—than a play, why in the world would I choose to make myself even more aware of the suspended facts by becoming the stage manager? On the contrary; I would want to walk into the dramatic world, participate in the action, and change the world by being in it.

One can imagine such an experience as the ultimate adventure game, but can something like it be possible within the humble world of a spreadsheet or a database manager? Aren't we stretching things a little here? It is my position that the principles of mimetic interaction remain the same in both kinds of worlds: The labyrinth of the adventure game of Zork is no more or less real than the terrain of spreadsheets of Lotus; both are representations enacted on the same "stage." Like Dorothy in *The Wizard of Oz*, in neither case do we want to deal with the man behind the curtain; that would be extraneous to both the form and the experience.

The conclusion can only be that a particular kind of relationship between the user and the context is appropriate to the nature and form of an interactive mimesis. The term I have chosen to describe it is *first-personness*. The first-person metaphor is grammatical: The personness of pronouns reflects where one stands *in relation to* others and to the world. Most movies and novels, for example, are third-person experiences; the viewer or reader is "outside" the action, and would describe what goes on using third-person pronouns: "First he did this, then they did that." Most instructional documents are second-person affairs: "Place the diskette in drive B"; "Honor your father and mother." Operating a computer program with an intermediary interface is a second-person experience: The user makes imperative statements to the system and asks it questions; the system tells the user what to do and what it has done ("File access denied, please try again"). Walking through the woods is a first-person experience; so is playing cowboys and Indians, writing a letter, or wielding a hammer.

Based on an understanding of interface as mimesis and of the necessary "first-person" relationship between user and context, design principles can be formulated and tested. Such principles are intended to indicate how the materials and structure of a mimetic world can be orchestrated to create the experience we desire.

SOME DESIGN PRINCIPLES

Representational Aspects of First-Personness

Personness is affected by the representational aspects of the interface; that is, how the user's choices and actions are introduced into the system, and how the activities of the system are represented to the user. The notion of a first-person interface seems to be more intuitively obvious in applications which, like video games, have the experience itself as the objective of the user.

In the Atari game *Pole Position*, for example, the user "drives" a simulated race car down a track. The user controls the speed of the car by pressure on a pedal that is analogous to an automobile accelerator. The ability of the user to participate in the race-driver fantasy in a first-person way would be significantly lessened if the only means of controlling the speed of the car were to specify speed numerically from a keyboard. Likewise, first-personness would be diminished if the effect of the user's pressure on the pedal were reflected, not by an animated representation of a race car, but by a numerical display of its speed.

In a task-oriented activity like file management, the representational aspect of first-personness can also be explored. In our Fish Bisque example, the functions of file management are represented by a series of command menus. At each level, the selected command tells the system something about the file manipulations desired by the user. But what is really going on? The user wants to get rid of a file. She might, via a touch-sensitive screen, draw a line through the region of text or the name of the file she is trying to remove. She might, via a natural-language interface with speech recognition, simply say, "I want to delete the recipe for Fish Bisque."

The underlying principle here is *mimetic*; that is, first-personness is enhanced by an interface that enables inputs and outputs that are more nearly like their real-world referents, in all relevant sensory modalities. The intuitive correctness of this notion is witnessed by the direction of technical evolution in the areas of simulators and games—toward higher resolution graphics and faster animation, greater sound capabilities, motion platforms, and mimetic input devices like force-feedback controllers. In product-driven applications, new technologies are allowing researchers to replace indirect or symbolic representations and manipulations with direct, concrete ones; e.g., physically pointing or speaking as opposed to typing, spatial and graphical representation of data as opposed to textual representation, etc. Likewise, the evolution of natural-language interfaces is beginning to replace the elaborate conventions of menu- and command-based systems with systems that employ language in ways that are mimetic of real-world activities like conversation and question-and-answer dialogues.

Interactive Aspects of First-Personness

First-personness is affected by the kinds of choices a user may make and the patterns of choice that emerge from the interaction. Three interactive aspects of first-personness are frequency, range, and significance.

The opportunity for users to take action may occur more or less frequently in an interactive work. *Interactive frequency* is a measure of how often user input is enabled. Near one end of the frequency continuum is a program with only a few clearly delimited interactive nodes. The other extreme is exemplified by action games like *Pole Position*, in which users make apparently continuous, real-time inputs.

We must be careful to distinguish between how often users actually make input, and how often they feel able to. Especially in product-driven applications, the frequency criterion of first-personness can be achieved simply by making the users aware that they can express themselves at any time.

Interactive range describes the range of choices available to users at a given moment in the interaction. That "interactive moment" may be a distinct node or a slice of "continuous" interaction. A binary choice, such as the use of a "fire button" in an action game or a yes-or-no question in a file management session, has the narrowest interactive range, even though the *consequences* of that choice may be of great significance to the whole. In natural-language-like interfaces that interpret user input using keywords, the interactive range is determined by the number of words that are both recognized and functional at a given interactive moment. Interactive range is also affected by the users' perceptions of the choices available to them. When there is a discrepancy between the number of choices actually recognized by the system and those perceived by the users, the users' assessment takes precedence.

Interactive significance is a measure of the impact of the users' choices and actions upon the *whole*. In a product-driven activity, that whole is the whole *outcome*, or result, of an interactive session. In a process-driven activity, the whole refers to the whole *experience*—the result of collaboration between user and system. In an interactive fantasy game, interactive significance measures the user's effect on the whole *plot*. In the CyberVision interactive story program *Rumplestiltskin*, for instance, users are allowed to choose which of three characters will serve as the queen's messenger. The choice is of relatively little significance, however, in that the actions performed by the messenger in the story remain the same regardless of the users' choices, hence the choice has little impact on the plot.

In the menu-driven approach to file management portrayed in our example, a significant action—deleting a file—is broken into chunks, none of which may be interpreted as a whole (and potentially "dangerous") action. The design will not permit the user to directly express her knowledge of the state of her files or what she wants to do with them. Thus, the interface prevents the user from making choices with high interactive significance in the name of protecting her from herself.

The user's experience can be enhanced by life-likeness in the realm of agency—the experience of one's ability to *act*. First-personness is most completely realized at the extreme end of each of the interactive variables' continuum: Frequency is continuous; range is infinite; significance is maximal.

Designing User Constraints

Constraints are limitations. They may be expressed as anything from gentle suggestions to stringent rules. People are always operating under some set of constraints: the physical limitations of survival (air to breathe, food, and water); the constraints of language on verbal expression; the limitations of social acceptability in public situations (e.g., wearing clothes). The ability to act without any such constraints is the stuff of fantasy and myth—the power of flight, for instance, or the appeal of immortality. Yet even such fantasy powers can be lost by the failure to comply with other, albeit magical, constraints (witness Prometheus). It is difficult to imagine life, even a fantasy life, in the absence of any constraints at all.

The user of an interactive system is subject to some special kinds of constraints. Some constraints arise from the technical capabilities of the system itself: If the system has no speech processing capability, for instance, the user must employ the keyboard for verbal input, and is constrained by its vicissitudes—the "QWERTY" layout, the presence or absence of function keys, etc. Other constraints arise from the nature of the activity that is appropriate to the application. The user of an adventure game cannot perform calculations with the computer while the game is running on it, even though the computer is capable of complex mathematical operations.

Are constraints simply a necessary evil, or do they perform some positive function for users? In idle fantasies about interactive systems, people tend to imagine magical spaces where they can and do whatever they wish—like gods. Even if such a system were technically feasible, the experience of using it might be more like an existential nightmare than a dream of freedom. When a person is asked to "be creative" with no direction or constraints whatever, the result is often a sense of powerlessness or even complete paralysis of the imagination (May, 1975). Constraints provide the security net that enables people to take imaginative leaps.

Constraints exist to prevent the user from doing things that would cause the system to blow up or resort to second-person intervention. They keep the user's activities within the bounds of the mimetic world. In exchange for his complicity, the user experiences increased potential for effective agency, in a world in which the causal relations among events are not obscured by the randomness and noise characteristic of open systems (like "real life").

How should constraints be determined and expressed? The standard techniques for introducing user constraints—second-person transactions like error messages, or delimiters of interactive frequency and range like explicit nodes with choice menus, for example—are almost always destructive of first-personness.

User constraints can be either explicit or implicit. Explicit constraints, as in the case of menus or command languages, are undisguised and directly available to the user. When the user is in doubt about the "legality" of a certain choice or action, it should be possible to find the rules and protocols of the system straightforwardly expressed, either in the manual, or in an on-line "help" facility. Implicit constraints, on the other hand, are inferred by the user from the behavior of the system. In *Zork*, for example, the user is not given a list of words that the language parser understands, but is informed via an error message ("I DON'T UNDERSTAND THAT WORD") when unfamiliar words are used. Implicit constraints may also be identified by the user when the system fails to allow certain kinds of choices. There is no way, for example, to negotiate with the Zylons in a game of *Star Raiders*, or to get *Word Star* to revise your latest paper.

Explicit constraints can be used without damage to first-personness if they are presented before the action begins. A good example is the determination and expression of rules in child's play, which occurs before play actually begins and creates a contract binding the participants to behave within certain constraints. Once the action has started, however, explicit constraints prove disruptive—an argument about the rules can ruin a perfectly good session of "cowboys and Indians" ("Wait a minute—who says Indians can only be killed with silver bullets?"). Implicit constraints are preferable during the course of the action, simply because the means for expressing them are usually less intrusive than those used for explicit constraints.

Constraints may also be characterized as extrinsic or intrinsic to the mimetic action. Extrinsic constraints have to do, not with the mimetic context, but with the context of the user as operator of the system. Avoiding the "reset" and "escape" keys during play of a game has nothing to do with the game world and everything to do with the behavior of the computer. Frequent breaks in a text editing session to protect your files from power failures and system crashes have nothing to do with the process of composing or editing text.

Extrinsic constraints, when they cannot be handled invisibly, should be expressed in terms of the mimetic context. If the "escape" key is defined as a self-destruct mechanism, for instance, the constraint against pressing it in the course of flying one's mimetic spaceship is intrinsic to the action. The user need not "shift gears" to consider the effect of the key on the computer that is running the game program. Expressing constraints in this manner preserves the contextual aspect of first-personness.

Constraints should be applied without shrinking interactive range or significance as experienced by the user: They should limit, not what the user can do, but what the user is likely to think of doing. Context is the most effective medium for presenting such constraints. The user's ability to recognize and comply with implicit, context-based constraints is a common human skill, exercised automatically in most situations, and not requiring concentrated effort or explicit attention. It is the same skill that a person uses to determine what to say and do when interacting with a group of unfamiliar people—at a cocktail party, for instance. The limitations on behavior are not likely to be consciously known or mulled over; they arise naturally from growing knowledge of the context in which one finds oneself. In the *Phone Slave* (an intelligent phone-answering and message-taking system developed at MIT), for example, users are successfully constrained to provide appropriate and recognizable natural-language input simply by the conversational questions that the system asks (Schmandt & Arons, 1984).[3]

[3] Ed Hutchins has suggested street theatre as a good example of the process of designing context-based constraints. The improvisational actor in the "interactive" context of street theatre knows enough about the audience to set up a context that will elicit a predictable response. This is essentially the same technique that Schmandt employs with the Phone Slave: "When you ask people questions," says Schmandt, "they tend to give you answers."

Selection Criteria

The notion of the interactive representation (interface) as an organic whole allows us to articulate some principles for the selection of materials to be included in that representation. Those materials are contributed by both the system and the user during interaction, and are subject to the same criteria. In the *Poetics*, Aristotle provides a general principle for inclusion: "That which makes no perceptible difference by its presence or absence is no real part of the whole." From this principle he derives selection criteria for the inclusion of materials in dramatic works which can also be applied to interactive representations.

Aristotle describes the plot as the central action of a play—the object of its representation. Incidents which have no direct bearing on the plot should not be represented; e.g., there is no reason to include a scene where Hamlet brushes his teeth. We have all been annoyed by such gratuitous incidents in films or TV shows. My guess is that most of us have been annoyed by the same kinds of incidents in interactive works. A convention in the world of Computer Assisted Instruction, for instance, is to ask the student to enter his name at the beginning of an interactive session. The most that is usually made of this "incident" is a message that replies, "Hello there, Jimmy," before proceeding with the "meat" of the lesson. The name-entering incident (Jimmy and the computer say "hi") has nothing to do with the plot (Jimmy learns his multiplication tables).

Gratuitous incidents most often occur in the presentation of extraneous information. The appearance of menus, prompts, or "helps" whether you want them or not is one common annoyance; the appearance of "garbage text" is another. The mail facility on most large systems appends a header to displayed messages which contains a cacophony of information that should really not clutter up the user's screen—or mind. Information about the path that a message took between nodes in a network, for instance, may be of functional use in delivering a response automatically, but its presentation to the user for perusal is gratuitous in the context of reading and responding to electronic mail.

Another example of gratuitous incidents are bungled efforts to provide "intrinsic motivation" in educational programs (often based on gross misinterpretations of the seminal paper by Malone, 1981) by interspersing problem-solving or tutorial segments with pieces of games: If Jimmy solves three arithmetic problems successfully, he gets to spend 20 seconds playing a version of something like *Asteroids*. Either the math problems or the game segments are gratuitous, depending upon Jimmy's point of view. The proper solution is either to eliminate one of the activities, or to re-shape the context so that it includes both; e.g., a starfighter simulation in which Jimmy must naturally solve math problems in order to operate the ship.

A second selection criterion in the world of drama involves the presentation of character traits. A trait should not be included in the representation unless it either eventuates in some action or sets up an important line of probability. We need never know, for instance, that Dr. Frankenstein is a great dancer (unless, perhaps, a dance with the monster shows up later in the plot, as in Gene Wilder's version).

Knowledge about a trait sets up expectations (probabilities) for the subsequent behavior of the possessor of that trait. Traits are rarely asserted explicitly; they are more often inferred from some action or contextual feature. The user of a text editing system who notices that there are commands that will position a cursor at the end of a word, a line, and a sentence, may infer that a trait of the system is the ability to "go to the end of" various logical units. If then the user discovers that there is no command to go to the end of a paragraph, the inference must be discarded. The trait should never have been implied in the partially consistent behavior of the system. System traits of a similar nature are identified by users on syntactic, visual, and kinesthetic levels (Hulteen, 1984). In poorly designed systems, they lead users to expect the system to perform inaccessible or nonexistent functions.

The converse is also true: Actions should not be represented for which the appropriate traits are not apparent. Peter Pan has traits which make it reasonable for him to fly; Perry Mason does not. Dials and counters that whirl away in the displays of many video games represent actions for which no meaningful traits exist. A game programmer confessed to me recently that he displayed the contents of an address register to "beef up" a video game "control panel." John Seely Brown proposed a mail system that would surreptitiously insert spelling errors in messages to create a climate of equality and informality in electronic communication.[4] An action so sophisticated is not likely to spring from the humble traits of an electronic mail system (in fact, its users would quickly wonder whose traits were at work in the apparent sabotage).

Previously, I distinguished between a representation and its object. I have noted that the object of an interface is a whole interaction, just as the object of a play is a whole action. In both cases, only those traits

---

[4] Unpublished talk. Conference on Human Factors in Computer Systems, 1984, Boston, MA.

that are necessary to represent the object should be included. This seems quite obvious, until one encounters a play in which the author or actor has attempted to create a complete human personality where a dramatic character with a limited number of traits is all that is required. As a student actor, I once auditioned for a character with one line—a maid who said something like "Yes, your lordship." I asked the director at the audition what kind of person the character was. His response was simply, "She is the kind of person who would say 'Yes, your lordship.'" I received a sound drubbing from the director when, in performance, my "maid" delivered the line with such adoring sincerity that the audience whispered conjectures about her love affair with the master of the house. Various dramatic theorists have noted that an audience will find such a character believable in the presence of very few traits (Schwamberger, 1980). Indeed, the audience is distracted and misled when a dramatic character is too "noisy."

A case in point is the design of an electronic "interface agent" (Brennan, 1984). One is tempted to design such an agent as a model of a human personality, with human-like knowledge and thought processes. Yet persuasive "computational personalities" like Eliza and Parry have been represented using a remarkably simple set of traits and memory capabilities (Boden, 1977). The Artificial Intelligence dogma of utilizing human-process models tempts us to misunderstand the nature of the design problem and the techniques appropriate to solving it. Years of blue-sky research and awesome computing resources have already been applied to the elusive goal of creating a simulation of a "real" human personality without success—the problem is too enormous, too elusive, and too complex. Creating an interface agent, on the other hand, is a relatively straightforward problem in artistic representation. The key tasks are specification of the characteristics of the action in which the agent will participate, and identification of the traits necessary to represent that action mimetically.

A FANCIFUL EXAMPLE

In my other life as a student of dramatic theory and criticism, I have been exploring ways to employ computer technology to apply certain dramatic principles prescriptively in the design of a system that will create interactive dramas (Laurel, 1983, 1985a, 1985b). The notions of first-personness and interactive mimesis presented in this chapter have their origins in that project. The Interactive Fantasy (IF) system is a hypothetical beast that is intended to provide the user with an experience very much like becoming a character in a play, co-creating the plot by making choices and performing actions in a fantasy world. IF sprang from my impatience with "dumb" computer games, a fascination with the idea of "interactive movies," and a perverse desire to become Captain James T. Kirk.

What would it really take to create such a "you-are-there" experience? The system would have to be able to figure out what the user-character was doing as it moved, spoke, and "lived" inside the fantasy world. In order to do that, the system would have to be implemented in a multimodal interface environment, with user-sensing capabilities like speech recognition, body-tracking, and eye-tracking. It should be able to parse and understand gestures as well as utterances. It would have to know enough about playwriting to be able to weave a good plot on the fly from materials contributed by the user and those generated by the system itself (the behavior of "system characters," for instance). It would have to be able to represent the action in real time in the appropriate sensory modalities—doors that slam and characters that move and speak. As I began to investigate the "state of the art" in all of these areas—user-sensing, understanding and inference, story generation, "intelligent" animation, etc.—I discovered that "the technology" is not the limiting factor in the design of such a system. Most of the necessary capabilities exist, at least in rudimentary form. All that is needed, it seems, is a reason to put it all together—a reason compelling enough to attract the brains and funding for such an enterprise.

Then it occurred to me that there's a way in which the IF system is *all interface*. Such a system could provide interactive representations for worlds as diverse as Hamlet's Denmark and the *Encyclopedia Brittanica* entry on ancient Egypt. Right now, we usually define an encounter with an "on-line database" as "searching for information." But what if we defined the activity as *finding* the information, or better still, *experiencing* it? We would no longer be "looking up information about the pyramids"—we would be climbing them, looking around their musty innards, reading hieroglyphs, or reincarnating pharaohs. Our notion of a "database interface" would be radically transformed.

The idea of interface as mimesis is based on the primacy of experience. It requires us to focus, not on what we can deliver within the constraints of current technology and convention, but what kinds of experiences we want to have with interactive representations. It provides a means for analyzing the resources from which such experiences can be built and some principles that can govern the orchestration of those resources. It takes as its model a theory that employs logic and aesthetics to create representations that *engage humans in pleasurable ways*. That is, after all, what we're trying to do.

CHAPTER 5

Direct Manipulation Interfaces

EDWIN L. HUTCHINS, JAMES D. HOLLAN,
and DONALD A. NORMAN

DIRECT MANIPULATION: ITS NATURE, FORM, AND HISTORY

The best way to describe a *Direct Manipulation* interface is by example. Suppose we have a set of data to be analyzed with the numbers stored in matrix form. Their source and meaning are not important for this example: The numbers could be the output of a spreadsheet, a matrix of numerical values from the computations of a conventional programming language, or the results of an experiment. Our goal is to analyze the numbers, to see what relations exist among the rows and columns of the matrix. The matrix of numbers is represented on a computer display screen by an icon. To plot one column against another, we simply get a copy of a graph icon, then draw one line from the output of one column to the x-axis of the graph icon and another line from the output of the second column to the y-axis input of the graph icon: See Figure 5.1A. Now, what was wanted? Erase the lines and reconnect them. Want to see other graphs? Make more copies of the graph icons and connect them. Need a logarithmic transformation of one of the axes? Move up a function icon, type in the algebraic function that is desired ($y = log\ x$, in this case) and connect it in the desired data stream. Want the analysis of variance of the logarithm of the data?

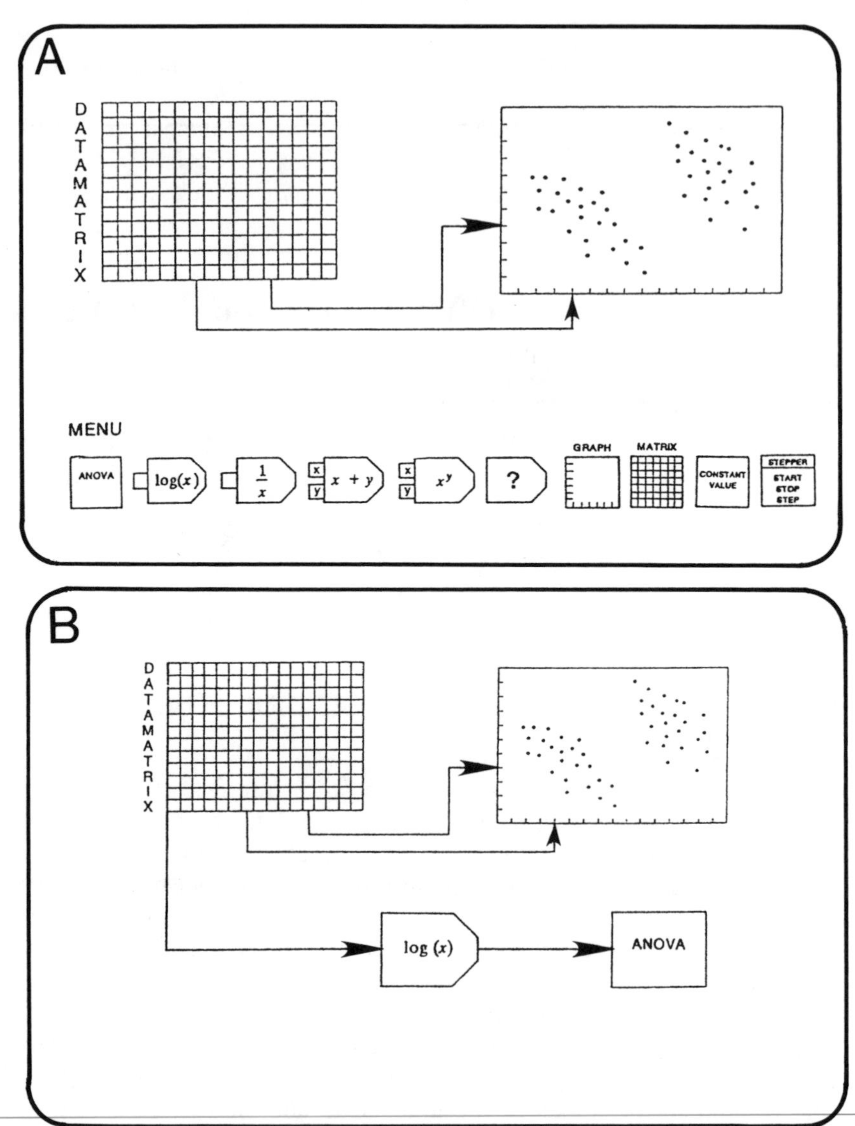

FIGURE 5.1. An elementary example of doing simple statistical computations by Direct Manipulation. **(A)** The basic components: The data are contained in the matrix, represented by the icon in the upper left corner of the screen. At the bottom of the screen are basic icons that represent possible functions. To use one, a copy of the desired icon is moved to the screen and connected up, much as is shown for the graph. **(B)** More complex interconnections, including the use of a logarithmic transformation of the data, a basic statistical package (for means and standard deviations), and an Analysis of Variance Package (ANOVA).

Connect the matrix to the appropriate statistical icons. These examples are illustrated in Figure 5.1B.

Now consider how we could partition the data. Suppose one result of our analysis was the scatter diagram shown in Figure 5.2A. The straight line that has been fitted through the points is clearly inappropriate. The data fall into two quite different clusters and it would be best to analyze those clusters separately. In the actual data matrix, the points that form the two clusters might be scattered randomly throughout the data set. The regularities are apparent only when we plot them. How do we pull out the clusters? Suppose we could simply circle the points of interest and use each circled set as if it were a new matrix of values, each of which could be analyzed in standard ways, as shown in Figure 5.2B.

The examples of Figures 5.1 and 5.2 are partially implemented,

A

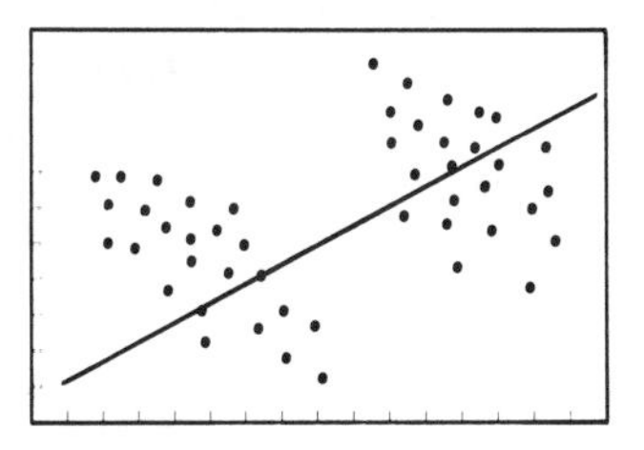

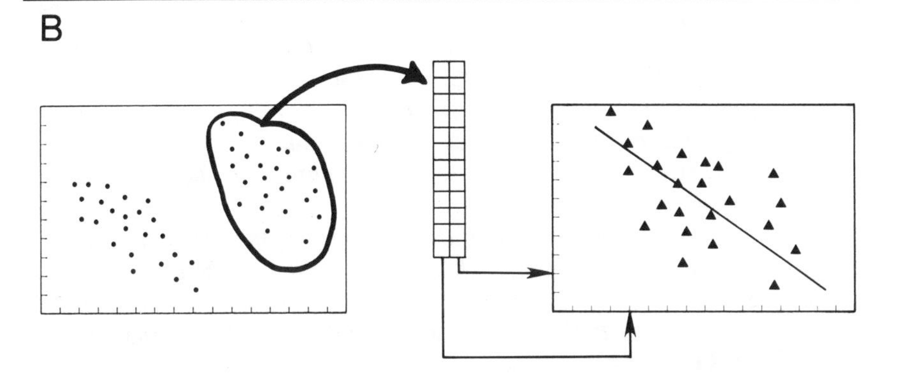

FIGURE 5.2. **(A)** The scatter plot formed in Figure 5.1, along with the best fitting regression line to the data. It is clear that the data really fall into two quite distinct clusters and that it would be best to look at these clusters independently. **(B)** The clusters are analyzed by circling the desired data, then treating the circled data as if they were a new matrix of values, which can be treated as a data source and analyzed in standard ways.

partially hypothetical: But they illustrate a powerful manipulation medium for computation.[1] The promise of Direct Manipulation is that instead of an abstract computational medium, all the "programming" is done graphically, in a form that matches the way one thinks about the problem. The desired operations are done simply by moving the appropriate icons onto the screen and connecting them together. Connecting the icons is the equivalent of writing a program or calling on a set of statistical subroutines, but with the advantage of being able to directly manipulate and interact with the data and the connections. There are no hidden operations, no syntax or command names to learn. What you see is what you get. Some classes of syntax errors are eliminated. For example, you can't point at a nonexistent object. The system requires expertise in the task domain, but only minimal knowledge of the computer or of computing.

> *A number of direct manipulation systems already exist. One was described briefly in Norman's chapter on "Cognitive Engineering": Bill Budge's Pinball Construction Set (Budge, 1983). It allows for the construction of game environments simply by moving the desired pieces onto the game, then stretching, coloring, and otherwise manipulating the environment.*

Direct manipulation interfaces seem remarkably powerful. Shneiderman has suggested that direct manipulation systems have the following virtues:

> 1. *Novices can learn basic functionality quickly, usually through a demonstration by a more experienced user.*
> 2. *Experts can work extremely rapidly to carry out a wide range of tasks, even defining new functions and features.*
> 3. *Knowledgeable intermittent users can retain operational concepts.*
> 4. *Error messages are rarely needed.*
> 5. *Users can see immediately if their actions are furthering their goals, and if not, they can simply change the direction of their activity.*
> 6. *Users have reduced anxiety because the system is comprehensible and because actions are so easily reversible.* (Shneiderman, 1982, p. 251)

[1] The examples are part of a design specification for our research project and at the time this chapter was written, part of the system had been implemented, including much of what is shown in Figure 5.1.

Can this really be true? Certainly there must be problems as well as benefits. It turns out that the concept of Direct Manipulation is complex. A checklist of surface features is unlikely to capture the real sources of power in direct manipulation interfaces. Moreover, although there are important benefits, there are also deficits. There are limitations: Like everything else, direct manipulation systems trade off one set of virtues and vices against another. It is important that we understand all the tradeoffs.

A Brief History of Direct Manipulation

The term "Direct Manipulation" was coined by Shneiderman (see Shneiderman, 1974, 1982, 1983; the 1983 review provides the best treatment of the topic), to refer to interfaces having the following properties:

> 1. *Continuous representation of the object of interest.*
> 2. *Physical actions or labeled button presses instead of complex syntax.*
> 3. *Rapid incremental reversible operations whose impact on the object of interest is immediately visible.* (Shneiderman, 1982, p. 251)

Hints of direct manipulation programming environments have been around for quite some time. The first major landmark is Sutherland's *Sketchpad,* a graphical design program (Sutherland, 1963). Sutherland's goal was to devise a program that would make it possible for a person and a computer "to converse rapidly through the medium of line drawings." Sutherland's work is a landmark not only because of historical priority but because of the ideas that he helped develop: He was one of the first to discuss the power of graphical interfaces, the conception of a display as "sheets of paper," the use of pointing devices, the virtues of constraint representations, and the importance of depicting abstractions graphically.

Sutherland's ideas took 20 years to have widespread impact. The lag is perhaps due more to hardware limitations than anything else. Highly interactive, graphical programming requires the ready availability of considerable computational power and it is only recently that machines capable of supporting this type of computational environment have become inexpensive enough to be generally available. But now we see these ideas in many of the computer-aided design and computer-aided manufacturing systems; many of which can trace their heritage directly to Sutherland's work. Borning's *ThingLab* program (1979) explored a general programming environment, building upon many of Sutherland's ideas within the Smalltalk programming environment. More recently, Direct Manipulation systems have been appearing with reasonable frequency. We have already discussed *Bill Budge's Pinball Construction Set* (Budge, 1983). Other examples exist in the area of intelligent training systems (e.g., Steamer: Hollan, Hutchins, & Weitzman, 1984, Hollan, Stevens, & Williams, 1980:) Steamer makes use of similar techniques and also provide tools for the construction of interactive graphical interfaces. Finally, the spreadsheet program, first available on the Apple II computer and now commonplace, incorporates many of the essential features of Direct Manipulation. In the lead article of *Scientific American*'s special issue on computer software, Kay (1984) claims that the development of dynamic spreadsheet systems gives strong hints that programming styles are in the offing that will make programming as it has been done for the past 40 years obsolete.

The Goal: A Cognitive Account of Direct Manipulation

We see promise in the notion of direct manipulation, but as yet we see no explanation of it. There are systems with attractive features, and claims for the benefits of systems that give the user a certain sort of feeling, and even lists of properties that seem to be shared by systems that provide that feeling, but no account of how particular properties might produce the feeling of directness. The purpose of this chapter is to examine the underlying basis for Direct Manipulation systems. On the one hand, what is it that provides the feeling of "directness"? Why do the examples feel so natural? What is so compelling about the notion? On the other hand, why does it seem so painful at times? Certainly you wouldn't want to do basic programming this way, or would you? What about Norman's complaint (in Chapter 3):

> *Much as I enjoy manipulating the parts of the pinball sets, much as my 4-year-old son could learn to work it with almost no training or bother, neither of us are any good at constructing pinball sets. I can't quite get the parts in the right places: When I stretch a part to change its shape, I usually end up with an unworkable part. Balls get stuck in weird corners. The action is either too fast or too slow. Yes, it is easy to change each problem as it is discovered, but the number seems endless. I wish the tools were more intelligent—do as I am intending, not as I am doing.*

For us, the notion of *Direct Manipulation* is not a unitary concept nor even something that can be quantified in itself. It is an orienting notion. "Directness" is an impression or a feeling about an interface. What we seek to do here is to characterize the space of interfaces and see where within that picture the range of phenomena that contribute to the feeling of directness might reside. The goal is to give cognitive accounts of these phenomena. At the root of our approach is the assumption that the feeling of directness results from the commitment of fewer cognitive resources. Or put the other way round, the need to commit additional cognitive resources in the use of an interface leads to the feeling of indirectness. As we shall see, some of the production of the feeling of directness is due to adaptation by the user, so that the designer can neither completely control the process, nor take full credit for the feeling of directness that may be experienced by the user.

There is a need to provide a characterization of how interfaces work, both in the computer science and the psychological senses, that will allow us to examine their strengths, weaknesses, and limitations. The computer science sense of "how they do it" concerns the computer-based technologies that contribute to interface power: object-oriented programming, constraint-based systems, logic programming, pointing devices, window systems, etc. In this chapter we focus primarily on the psychological sense of "how they do it" because we feel that this is where the power lies. We believe that while many technological advances may be helpful (new types of input devices, for example), even the existing technology is not used to full advantage because of a lack of understanding of the psychological side of interface design.

We will not attempt to set down hard and fast criteria under which an interface can be classified as direct or not direct. The sensation of directness is always relative. It is often due to the interaction of a number of factors. There are costs associated with every factor that increases the sensation of directness. At present we know of no way to measure the tradeoff values, but we will attempt to provide a framework within which one can say what is being traded off against what.

TWO ASPECTS OF DIRECTNESS: DISTANCE AND ENGAGEMENT

There are two separate and distinct aspects of the feeling of directness. One involves a notion of the distance between one's thoughts and the physical requirements of the system under use. A short distance means that the translation is simple and straightforward, that thoughts are readily translated into the physical actions required by the system and that the system output is in a form readily interpreted in terms of the goals of interest to the user. We call this aspect "*distance*" to emphasize the fact that directness is never a property of the interface alone, but involves a relationship between the task the user has in mind and the way that task can be accomplished via the interface. Here the critical issues involve minimizing the effort required to bridge the gulf between the user's goals and the way they must be specified to the system.

The second aspect of directness concerns the qualitative feeling of engagement, the feeling that one is directly manipulating the objects of interest. There are two major metaphors for the nature of human-computer interaction, a conversation metaphor and a model world metaphor. In a system built on the conversation metaphor, the interface is a language medium in which the user and system have a conversation about an assumed, but not explicitly represented world. In this case, the interface is an implied intermediary between the user and the world about which things are said. In a system built on the model world metaphor, the interface is itself a world where the user can act, and that changes state in response to user actions. The world of interest is explicitly represented and there is no intermediary between user and world. Appropriate use of the model world metaphor can create the sensation in the user of acting upon the objects of the task domain themselves. We call this aspect "*direct engagement*."

Distance

An interface introduces distance to the extent there are gulfs between a person's goals and knowledge and the level of description provided by the systems with which the person must deal. These are the *Gulf of Execution* and the *Gulf of Evaluation* (Figure 5.3) that were incorporated

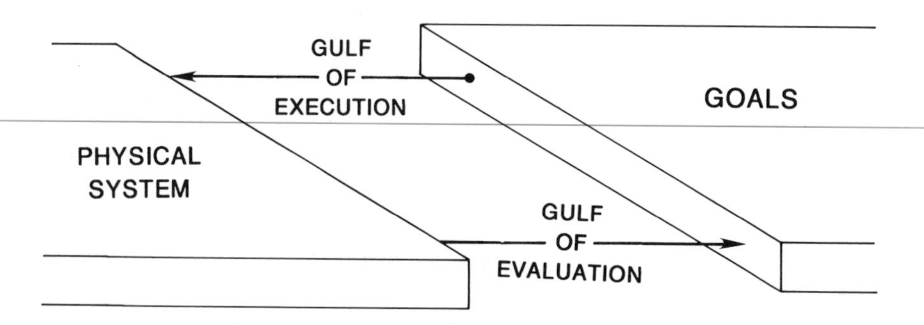

FIGURE 5.3. The Gulfs of Execution and Evaluation. Each gulf is unidirectional: the Gulf of Execution goes from Goals to Physical System; the Gulf of Evaluation goes from Physical System to Goals (from Chapter 3).

by Norman in his chapter on Cognitive Engineering (Chapter 3). The Gulf of Execution is bridged by making the commands and mechanisms of the system match the thoughts and goals of the user as much as possible. The Gulf of Evaluation is bridged by making the output displays present a good Conceptual Model of the system that is readily perceived, interpreted, and evaluated. The goal in both cases is to minimize cognitive effort.

We suggest that the feeling of directness is inversely proportional to the amount of cognitive effort it takes to manipulate and evaluate a system and, moreover, that cognitive effort is a direct result of the Gulfs of Execution and Evaluation. The better the interfaces to the system help bridge the gulfs, the less cognitive effort needed, the more direct the resulting feeling of interaction.

In Chapter 3, Norman suggests that it is useful to consider an approximate theory of action, and in particular, that it is useful to consider separately the seven different stages of activity shown in Figure 5.4 that a person might go through in the performance of a task. For the current discussion, the important point is that the interface must try to match the goals of the person. The two gulfs are bridged in two directions; from the system side by the system interface and from the user side by mental structures and interpretation. The bridges from the user's side are aided through the development of appropriate conceptual models—the design model and the user model discussed in Chapter 3, which are in turn aided by the system image presented by the interface. But a major burden should be carried by the interface itself. The more of the gulf spanned by the interface, the less distance need be bridged by the efforts of the user.

Direct Engagement

The description of the nature of interaction to this point begins to suggest how to make a system less difficult to use, but it misses an important point, a point that is the essence of Direct Manipulation. The analysis of the execution and evaluation process explains why there is difficulty in using a system and it says something about what must be done to minimize the mental effort required to use a system. But there is more to it than that. The systems that best exemplify Direct Manipulation all give us the qualitative feeling that we are directly engaged with control of the objects—not with the programs, not with the computer, but with the semantic objects of our goals and intentions. This is the feeling that Laurel discusses in Chapter 4: a feeling of first-personness, of direct engagement with the objects that concern us. Are we analyzing data? Then we should be manipulating the data

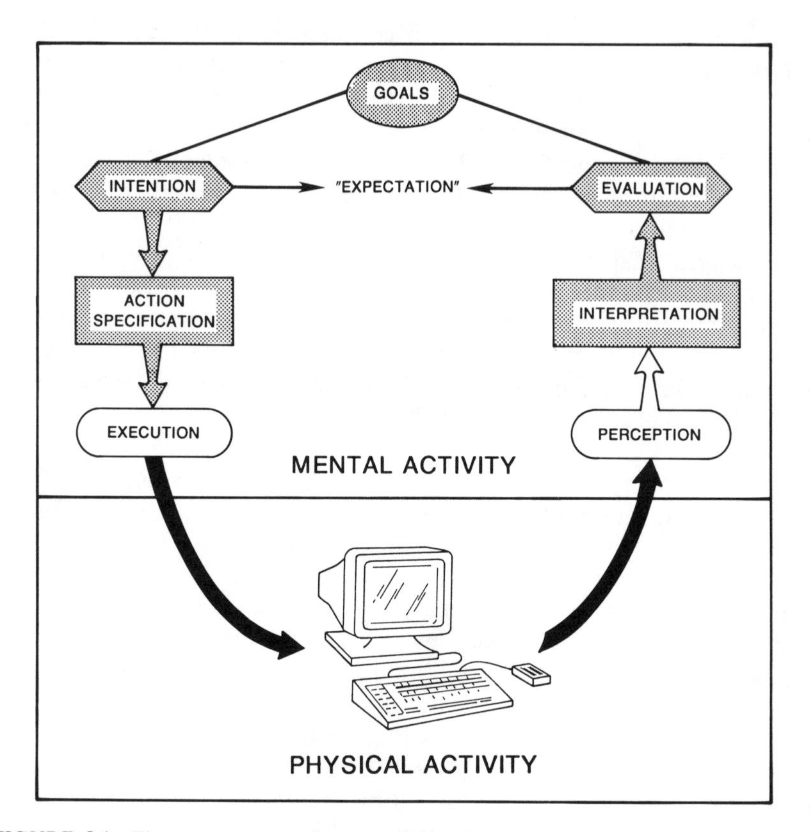

FIGURE 5.4. The seven stages of user activities involved in the performance of a task (from Chapter 3).

themselves or if we are designing an analysis of data, we should be manipulating the analytic structures themselves. Are we playing a game? Then we should be manipulating directly the game world, touching and controlling the objects in that world, with the output of the system responding directly to our actions, and in a form compatible with them.

Historically, most interfaces have been built on the conversational metaphor. There is power in the abstractions that language provides (we discuss some of this later), but the implicit role of interface as an intermediary to a hidden world denies the user direct engagement with the objects of interest. Instead, the user is in direct contact with linguistic structures, structures that can be interpreted as referring to the objects of interest, but that are not those objects themselves.

Making the central metaphor of the interface that of the model world supports the sensation of directness: Instead of describing the actions of interest, the user performs those actions. In the conventional interface, the system describes the results of the actions: In the model world the system would present directly the actions taken upon the objects. This change in central metaphor is made possible by relatively recent advances in technology. One of the exciting prospects for the study of Direct Manipulation is the exploration of the properties of systems that provide for direct engagement.

Building interfaces on the model world metaphor requires a special sort of relationship between the input interface language and the output interface language. In particular, the output language must represent its subject of discourse in a way that natural language does not normally do. The expressions of a Direct Manipulation output language must behave in such a way that the user can assume that they, in some sense, *are* the things they refer to. Furthermore, the nature of the relationship between input and output language must be such that an output expression can serve as a component of an input expression. When these conditions are met, it is as if we are directly manipulating the things that the system represents.

Thus, consider a system in which a file is represented by an image on the screen and actions are done by pointing to and manipulating the image. In this case, if we can specify a file by pointing at the screen representation, we have met the goal that an expression in the output "language" (in this case an image) be allowed as a component of the input expression (in this case, by pointing at the screen representation). If we ask for a listing of files, we would want the result to be a physical display that can, in turn, be used directly to specify the further operations to be done. Notice that this is not how a conversation works. In conversation, one may refer to what has previously been said, but one cannot operate upon what has been said. This kind of requirement does not require an interface of pictures, diagrams, or icons. It can be done with words and descriptions. The key properties are that the objects, whatever their form, have behaviors, can be referred to by other objects, and that referring to an object causes it to behave. In the file listing example, we must be able to use the output expression that represents the file in question as a part of the input expression calling for whatever operation we desire upon that file, and the output expression that represents the file must change as a result of being referred to in this way. The goal is to permit the user to act as if the representation is the thing itself.

These conditions are met in many screen editors when the task is the arrangement of strings of characters. The characters appear as they are typed. They are then available for further operations. We treat them (but not their meanings!) as though they are the things we are manipulating. These conditions are also met in the statistics example with which we opened this chapter (Figure 5.1), in Steamer, and by the *Pin-Ball Construction Set*. The special conditions are not met in file listing commands on most systems, the commands that allow one to display the names and attributes of file structure. The issue is that the outputs of these commands are "names" of the objects. Operating on the names does nothing to the objects to which they refer. In a Direct Manipulation situation, we would feel that we had the files in front of us, that the program that "listed" the files actually placed the files before us: Any further operation on the files would take place upon the very objects delivered by the directory listing command. This would provide the feeling of directly manipulating the objects that were returned.

> *When we deal with the conceptually complex objects of computation, we have to remember that each object typically has a number of different attributes. An object can be viewed from different "perspectives," behaving quite differently with each viewpoint.*
>
> *Consider the file example. If the objects are the names of the files (or icons containing the names), this depiction will work well as long as the operations to be performed are things like removing the files, renaming them, or positioning them as sources of input or receivers of output. But it is not a good representation for all possible operations. Changing the contents of the file, for instance, requires a representation of the contents as opposed to its "surface" features: The objects to be manipulated are the words or images within the file itself. This requires a different representation. Or consider what is required to change the properties of the file—its protection status, perhaps. Here we require a representation that makes visible and manipulable these properties of the file. Thus, at least three different representations are required. It is unlikely and probably inappropriate to try to develop a single representation of an object that can serve all purposes.*
>
> *We should recognize that more than one view of an object is often necessary and provide a separate representation for each one. Even a system that gives us the illusion that we are manipulating the objects directly needs different representations when the objects are viewed from different perspectives.*

The point is that when an interface presents a world of action rather than a language of description, manipulating a representation can have the same effects and the same feel as manipulating the thing being represented. The members of the audience of a well-staged play willfully suspend their beliefs that the players are actors and become directly engaged in the content of the drama. In a similar way, the user of a well designed *model world* interface can willfully suspend belief that the objects depicted are artifacts of some program and can thereby directly engage the world of the objects. This is the essence of the "first-personness" feeling of direct engagement. Let us now return to the issue of distance and explore the ways that an interface can be direct or indirect with respect to a particular task.

TWO FORMS OF DISTANCE: SEMANTIC AND ARTICULATORY

Whenever we interact with a device, we are using an interface language. That is, we must use a language to describe to the device the nature of the actions we wish to have performed. This is true regardless of whether we are dealing with an interface based on the conversation metaphor or on the model world metaphor, although the properties of the language in the two cases are different. A description of desired actions is an expression in the interface language.

The notion of an interface language is not confined to the everyday meaning of language. Setting a switch or turning a steering wheel can be expressions in an interface languages if switch-setting or wheel-turning are how one specifies the operations that are to be done. After an action has been performed, evaluation of the outcome requires that the device make available some indication of what has happened: That output is an expression in the output interface language. Output interface languages are often impoverished. Frequently the output interface language does not share vocabulary with the input interface language. Two forms of interface language—two dialects, if you will—must exist to span the gulfs between user and device: the input interface language and the output interface language.

Both the languages people speak and computer programming languages are almost entirely symbolic in the sense that there is an arbitrary relationship between the form of a vocabulary item and its meaning. The reference relationship is established by convention and must be learned. There is no way to infer meaning from form for most vocabulary items. Because of the relative independence of meaning and form we describe separately two properties of the interface language: *Semantic Directness* and *Articulatory Directness.*

Figure 5.5 summarizes the relationship between semantic and articulatory distance. We now examine semantic and articulatory directness separately for the Gulfs of Execution and Evaluation.

Semantic Directness

Semantic directness concerns the relation of the meaning of an expression in the interface language to what the user wants to say. Two important questions about semantic directness are:

- *Is it possible to say what one wants to say in this language?* That is, does the language support the user's conception of the task domain? Does it encode the concepts and distinctions in the domain in the same way that the user thinks about them?

- *Can the things of interest be said concisely?* Can the user say what is wanted in a straightforward fashion, or must the user construct a complicated expression to do what appears in the user's thoughts as a conceptually simple piece of work?

Semantic directness is an issue with all languages. Natural languages generally evolve such that they have rich vocabularies for domains that are of importance to their speakers. When a person learns a new

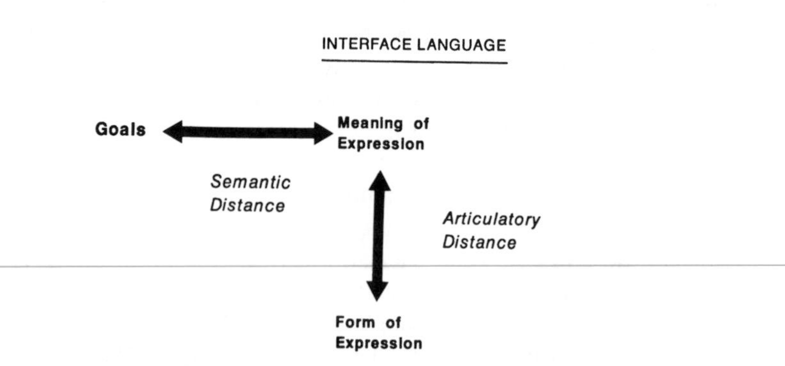

FIGURE 5.5. Every expression in the interface language has a meaning and a form. Semantic distance reflects the relationship between the user intentions and the meaning of expressions in the interface languages both for input and output. Articulatory distance reflects the relationship between the physical form of an expression in the interaction language and its meaning, again, both for input and output. The easier it is to go from the form or appearance of the input or output to meaning, the smaller the articulatory distance.

language—especially when the language is from a novel culture—the new language may seem indirect, requiring complicated constructs to describe things the learner thinks should be easy to say. But the differences in apparent directness reflect differences in what things are thought important in the two cultures. Natural languages can and do change as the need arises. This occurs through the introduction of new vocabulary or by changing the meaning of existing terms. The result is to make the language semantically more direct with respect to the topic of interest.

Semantic Distance in the Gulfs of Execution and Evaluation

> *Beware the Turing tar-pit in which everything is possible but nothing of interest is easy* (Alan Perlis, 1982).

The Gulf of Execution. At the highest level of description, a task may be described by the user's intention: "compose this piece" or "design this building." At the lowest level of description, the performance of the task consists of the shuffling of bits around inside the machine. Between the interface and the low-level operations of the machine is the system-provided task-support structure that implements the expressions in the interface language. The situation that Perlis called the "Turing tar-pit" is one in which the interface language lies near or at the level of bit shuffling of a very simple abstract machine. In this case, the entire burden of spanning the gulf from user intention to bit manipulation is carried by the user. The relationship between the user's intention and the organization of the instructions given the machine is distant, complicated, and hard to follow. Where the machine is of minimal complexity, as is the case with the Turing machine example, the wide gulf between user intention and machine instructions must be filled by the user's extensive planning and translation activities. These activities are difficult and rife with opportunities for error.

Semantic directness requires matching the level of description required by the interface language to the level at which the person thinks of the task. It is always the case that the user must generate some information-processing structure to span the gulf. Semantic distance in the gulf of execution reflects how much of the required structure is provided by the system and how much by the user. The more that the user must provide, the greater the distance to be bridged.

The Gulf of Evaluation. On the evaluation side, semantic distance refers to the amount of processing structure that is required for the user to determine whether the goal has been achieved. If the terms of the output are not those of the user's intention, the user will be required to translate the output into terms that are compatible with the intention in order to make the evaluation. For example, suppose a user's intent is to control how fast the water level in a tank rises. The user does some controlling action and observes the output. But if the output only shows the current value, the user has to observe the value over time and mentally compare the values at different times to see what the rate of change is: See Figure 5.6A. The information needed for the evaluation is in the output, but it is not there in a form that directly fits the terms of the evaluation. The burden is on the user to perform the required transformations and that requires effort. Suppose the rate of change were directly displayed, as in Figure 5.6B. This indication reduces the mental workload, making the semantic distance between intentions and output language much shorter.

Reducing the Semantic Distance That Must Be Spanned

Figure 5.6 provides one illustration of how semantic distance can be changed. In general, there are only two basic ways to reduce the

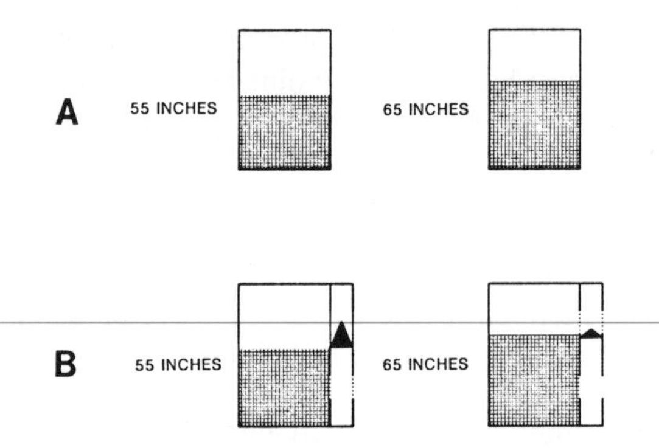

FIGURE 5.6. Matching user's intentions by appropriate output language. The user attempts to control the rate at which the water level in the tank is rising. In **(A)** the only indication is a meter that shows the current level. This requires the user to observe the meter over time and to do a mental computation on the observations. **(B)** shows a display that is more semantically direct: The rate of change is indicated graphically. (These illustrations are from the working Steamer system of Hollan, Hutchins, & Weitzman, 1984.)

distance, one from the system side (requiring effort on the part of the system designer), the other from the user side (requiring effort on the part of the user). Each direction of bridge building has several components. Here let us consider the following possibilities:

- The designer can construct higher-order and specialized languages that move toward the user, making the semantics of the input and output languages match that of the user;
- The user can develop competence by building new mental structures to bridge the gulfs. In particular, this requires the user to automate the response sequence and to learn to think in the same language as that required by the system.

Higher-level languages. One way to bridge the gulf between the intentions of the user and the specifications required by the computer is well-known: Provide the user with a higher-order language, one that directly expresses frequently encountered structures of problem decomposition. Instead of requiring the complete decomposition of the task all the way to low-level operations, let the task be described in the same language used within the task domain itself. Although the computer still requires low-level specification, the job of translating from the domain language to the programming language can be taken over by the machine itself.

This implies that designers of higher-level languages should consider how to develop interface languages for which it will be easy for the user to create the mediating structure between intentions and expressions in the language. One way to facilitate this process is to provide consistency across the interface surface. That is, if the user builds a structure to make contact with some part of the interface surface, a savings in effort can be realized if it is possible to use all or part of that same structure to make contact with other areas.

> *Beware the over-specialized system where operations are easy, but little of interest is possible* (the converse of the Turing tar-pit).

The result of matching a language to the task domain brings both good news and bad news. The good news is that tasks are easier to specify. Even if considerable planning is still required to express a task in a high-level language (Riley & O'Malley 1984; Riley, Chapter 7), the amount of planning and translation that can be avoided by the user and passed off to the machine is enormous. The bad news is that the language has lost generality. Tasks that do not easily decompose into the terms of the language may be difficult or impossible to represent. In the extreme case, what can be done is easy to do, but outside that specialized domain, nothing can be done. The power of a specialized language system derives from carefully specified primitive operations, selected to match the predicted needs of the user, thus capturing frequently occurring structures of problem decomposition.

The trouble is that there is a conflict between generality and matching to any specific problem domain. Some high-level languages and operating systems have striven to close the gap between user intention and the interaction language while preserving freedom and ease of general expression by allowing for extensibility of the language or operating system. Such systems allow the users to move the interface closer to their conception of the task.

> *The Lisp language and the UNIX operating system serve as examples of this phenomenon. Lisp is a general purpose language, but one that has extended itself to match a number of special, high-level domains. As a result, Lisp can be thought of as a system rather than a language, with numerous levels layered on top of the underlying language kernel. There is a cost to this method. As more and more specialized domain levels get added, the language system gets larger and larger, becoming more clumsy to use, more expensive to support, and more difficult to learn. Just look at any of the manuals for the large Lisp systems (Interlisp, Zetalisp) to get a feel for the complexity involved. The same is true for UNIX operating system which started out with a number of low-level, general primitive operations. Users were allowed (and encouraged) to add their own, more specialized operations, or to package together the primitives into higher-level operations. The results in all these cases are massive systems that are hard to learn and that require a large amount of support facilities. The documentation becomes huge, and not even system experts know all that is present. Moreover, the difficulty of maintaining such a large system increases the burden on everyone, and the possibility of having standard interfaces to each specialized function has long been given up.*

The point is that as the interface approaches the user's intention end of the gulf, functions become more complicated and more specialized in purpose. Because of the incredible variety of human intentions, the lexicon of a language that aspires to both generality of coverage and domain specific functions can grow very large. In any of the modern dialects of Lisp one sees a microcosm of the argument about high-level languages in general. The fundamentals of the language are simple, but a great deal of effort is required to do anything useful at the low level of the language itself. Higher-level functions written in terms of lower-level ones make the system easier to use when the functions match intentions, but in doing so they may restrict possibilities, proliferate vocabulary, and require that a user know an increasing amount about the language of interaction rather than the domain of action.

Make the output show semantic concepts directly. An example of reducing semantic distance on the output side is provided by the scenario of controlling the rate of filling a water tank, described in Figure 5.6. In that situation, the output display was modified to show rate of flow directly, something normally not displayed but instead left to the user to compute mentally.

In similar fashion, the change from line-oriented text editors to screen-oriented text editors, where the effects of editing commands could be seen instantly, is another example of matching the display to the user's semantics. In general, the development of WYSIWYG systems ("What you see is what you get") provides other examples. And finally, the spreadsheet program has been so important, in part because its output format continually and elegantly shows the state of the system in the very terms the accountant likes to use, at least so some believe. This format is actually not so good for many users, which explains the rapid proliferation of graphical packages that transform spreadsheet data into a form more readily interpretable by a wider class of user.

The attempt to develop good semantic matches at the system output confronts the same conflict between generality and power faced in the design of input languages. If the system is too specific and specialized, the output displays lack generality. If the system is too rich, the user has trouble learning and selecting among the possibilities. One solution for both the output and input problem is to abandon hope of maintaining general computing and output ability and to develop special purpose systems for particular domains or tasks. In such a world, the location of the interface in semantic space is pushed closer to the domain language description. Here, things of interest are made simple because the lexicon of the interface language maps well into the lexicon of domain description. Considerable planning may still go on in the conception of the domain itself, but little or no planning or translation is required to get from the language of domain description to the language of the interface. The price paid for these advantages is a loss of generality: Many things are unnatural or even impossible.

Automated behavior does not reduce semantic distance. Cognitive effort is required to plan a sequence of actions to satisfy some intent. Generally, the more structure required of the user, the more effort use of the system will entail. However, this gap can be overcome if the users become familiar enough with the system. Structures that are used frequently need not be rebuilt every time they are needed if they have been remembered. Thus, a user may remember how to do something rather than having to rederive how to do it. It is well known that when tasks are practiced sufficiently often, they become automated, requiring little or no conscious attention. As a result, over time, the use of the interface to solve a particular set of problems will feel less difficult and more direct. Experienced users will sometimes argue that the interface they use directly satisfies their intentions, even when less skilled users complain of the complexity of the structures. To the skilled user, the interface feels direct because the invocation of mediating structure has been automated. They have learned how to transform frequently arising intentions into action specifications. The result is a feeling of directness as compelling as that which results from semantic directness. As far as the user is concerned, the intention comes to mind and the action gets executed. There are no conscious intervening stages.

> *The point was made forcibly to us in a class in which these ideas about Direct Manipulation were being described. Erasing a word by using an eraser was defined as direct action. Erasing by appropriate marking with a mouse pointing device in a text editor was also called direct, although slightly less. But using a standard, command-language based text editor was not direct. "But why not?" queried a student. "I am an expert user of vi, and when I wish to delete a word, all I do is think 'delete that word,' my fingers automatically type 'dw,' and the word disappears from the screen. How could anything be more direct?"*

The frequent use of even a poorly designed interface can sometimes result in a feeling of directness like that produced by a semantically direct interface. A user can compensate for the deficiencies of the interface through continual use and practice so that the ability to use it becomes automated, requiring little conscious activity. While automation is one factor which can contribute to a feeling of directness, it is essential for an interface designer to distinguish it from semantic directness. Automatization does not reduce the semantic distance that must be spanned. The gulfs between a user's intentions and the interface must still be bridged by the user. Although practice and the resulting expertise can make the crossing less difficult, it does not reduce the magnitude of the gulfs. Planning activity may be replaced by a single memory retrieval so that instead of figuring out what to do, the user remembers what to do. Automatization may feel like direct control, but it comes about for completely different reasons than semantic directness. Automatization is useful, for it improves the interaction of the user with the system, but the feeling of directness it produces depends only on how much practice the particular user has: It gives the system credit for the work the user has done. Although we need to remember that this happens, that users may adjust themselves to the interface and, with sufficient practice, may view it as directly supporting their intentions, we need to distinguish between the cases in which the feeling of directness originates from a close semantic coupling between intentions and the interface language and that which originates from practice. The resultant feeling might be the same in the two cases, but there are crucial differences between how the feeling is acquired and what one needs to do as an interface designer to generate it.

The user can adapt to the system representation. One way to span the gulf is for the users to change their own conceptualization of the problem so that they come to think of it in the same terms as the system. In some sense, this means that the gulf is bridged by moving the user closer to the system. Because of their experience with the system, the users change both their understanding of the task and the language with which they think of the issues. This is related to the notion of linguistic determinism. If it is true that the way we think about something is shaped by the vocabulary we have for talking about it, then it is important for the designer of a system to provide the user with a good representation of the task domain in question. The interface language should provide a powerful, productive way of thinking about the domain.

> *This is the Whorfian hypothesis of language, transformed to the computer setting. The basic idea is that dedicated users change the very language and manner with which they think about problems, so that the initial goals and intentions are formed in the same language required by the system, thus bridging the gulf from the user side. One has only to listen to devoted system users to be convinced of the validity of the concept: Indeed, the actions available on the computer system have been so thoroughly introduced into the culture of the*

user, that the concepts are used in non-computer applications as well, where they are no longer directly applicable:

"*What movies are playing in town?* Grep *the newspaper, would you?*"

"*Both those ideas are good: Let's* cons *them up.*"

Linguistic determinism takes place at a more fundamental level than the other ways of reducing the semantic distance. While moving the interface closer to the user's intentions may make it difficult to realize some intentions, changing the user's conception of the domain may prevent some intentions from arising at all. So while a well designed special purpose language may give the user a powerful way of thinking about the domain, it may also restrict the user's flexibility to think about the domain in different ways.

That the user may change conceptual structure to match the interface language follows from the notion that every interface language implies a representation of the tasks it is applied to. The representation implied by an interface is not always a coherent one. Some interfaces provide a collection of partially overlapping views of a task domain. If the user is to move toward the model implied by the interface, and thus reduce the semantic distance, that model should be coherent and consistent over some conception of the domain. There is, of course, a tradeoff here between the costs to the user of learning a new way to think about a domain and the potential added power of thinking about it in the new way.

Virtuosity and semantic distance. Sometimes users have a conception of a task and of a system that is broader and more powerful than that provided by an interface. The structures they build to make contact with the interface go beyond it. This is how we characterize virtuoso performances in which the user may "misuse" limited interface tools to satisfy intentions that even the system designer never anticipated. In such cases of virtuosity the notion of semantic distance becomes more complicated and we need to look very carefully at the task that is being accomplished. Semantic directness always involves the relationship between the task one wishes to accomplish and the ways the interfaces provides for accomplishing it. If the task changes, then the semantic directness of the interface may also change.

Consider a musical example: Take the task of producing a middle-C note on two musical instruments, a piano and a violin. For this simple task, the piano provides the more direct interface because all one need do is find the key for middle-C and depress it, whereas on the violin, one must place the bow on the G string, place a choice of fingers in precisely the right location on that string, and draw the bow. A piano's keyboard is more semantically direct than the violin's strings and bow for the task of producing notes. The piano has a single well-defined vocabulary item for each of the notes within its range, while the violin has an infinity of vocabulary items many of which do not produce proper notes at all. However, when the task is playing a musical piece well rather than simply producing notes, the directness of the interfaces can change. In this case, one can complain that a piano has a very indirect interface because it is a machine with which the performer "throws hammers at strings." The performer has no direct contact with the components that actually produce the sound and so, the production of desired nuances in sound is more difficult. Here, as musical virtuosity develops, the task that is to be accomplished also changes from just the production of notes to concern for how to control more subtle characteristics of the sounds like vibrato, the slight changes in pitch used to add expressiveness. For this task the violin provides a semantically more direct interface than the piano. Thus, as we have argued earlier, an analysis of the nature of the task being performed is essential in determining the semantic directness of an interface.

> *We are reminded of a quote from Minsky:*
>
> "*A computer is like a violin. You can imagine a novice trying first a phonograph and then a violin. The latter, he says, sounds terrible. That is the argument heard from our humanists and most of our computer scientists. Computer programs are good they say, for particular purposes, but they aren't flexible. Neither is a violin, or a typewriter, until you learn how to use it.*" (Minsky, 1967)

Articulatory Directness

In addition to its meaning, every vocabulary item in every language has a physical form and that form has an internal structure. Words in natural languages, for example, have phonetic structure when spoken and typographic structure when printed. Similarly, the vocabulary items that constitute an interface language have a physical structure. Where *semantic directness* has to do with the relationships between user's intentions and meanings of expressions, *articulatory directness* has to do with the relationships between the meanings of expressions and their physical form. On the input side, the form may be a sequence of character-selecting key presses for a command language interface, the movement of a mouse and the associated "mouse clicks" in a pointing device interface, or a phonetic string in a speech interface. On the output side, the form might be a string of characters, a change in an iconic shape, an auditory signal, or a graph, diagram, or animation.

There are ways to design languages such that the relationships between the forms of the vocabulary items and their meanings are not arbitrary. One technique is to make the physical form of the vocabulary items structurally similar to their meanings. In spoken language this relationship is called onomatopoeia. Onomatopoetic words in spoken language refer to their meanings by imitating the sound they refer to. Thus we talk about the "boom" of explosions or the "cock-a-doodle-doo" of roosters. There is an economy here in that the user's knowledge of the structure of the surface acoustical form has a non-arbitrary relation to meaning. There is a directness of reference in this imitation. An intervening level of arbitrary symbolic relations is eliminated. Other uses of language exploit this effect partially. Thus, while the word "long" is arbitrarily associated with its meaning, sentences like "She stayed a looooooooooong time" exploit a structural similarity between the surface form of "long" (whether written or spoken) and the intended meaning. The same sorts of things can be done in the design of interface languages.

> *In many ways, the interface languages should have an easier time of exploiting articulatory similarity than do natural languages because of the rich technological base available to them. Thus, if the intent is to draw a diagram, the interface might accept as input the drawing motions. In turn, it could present as output diagrams, graphs, and images. If one is talking about sound patterns to the input interface language, the output could be the sounds themselves. The computer has the potential to exploit articulatory similarities through technological innovation in the varieties of dimensions upon which it can operate. This potential has not been exploited, in part because of economic constraints. The restriction to simple keyboard input limits the form and structure of the input languages and the restriction to simple, alphanumeric terminals with small, low resolution screens, limits the form and structure of the output languages.*

Articulatory Distance in the Gulfs of Execution and Evaluation

The relationships among semantic distance, articulatory distance, and the gulfs of execution and evaluation are shown in Figure 5.7.

Take the simple, commonplace activity of moving a cursor on the screen. If we do this by moving a mouse, pointing with the finger or a light pen at the screen, or otherwise mimicking the desired motion, then at the level of action execution, these are all articulatorally direct interactions. The meaning of the intention is cursor movement and the action is specified by means of a similar movement. One way to achieve articulatory directness at the input side is to provide an interface that permits specification of an action by mimicking it, thus supporting a articulatory similarity between the vocabulary item and its meaning. Any nonarbitrary relationship between the form of an item and its meaning can be a basis for articulatory directness. Where it is not possible to create a nonarbitrary relationship between the form of the item and its meaning, we recognize that not all arbitrary

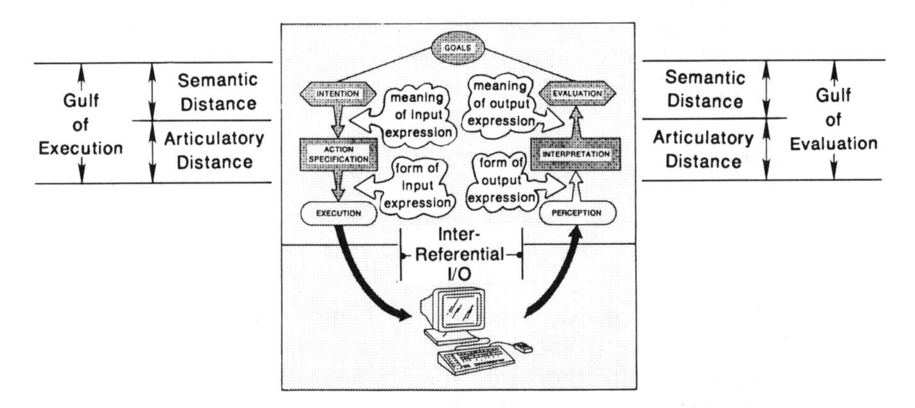

FIGURE 5.7. Forming an intention is the activity that spans semantic distance in the Gulf of Execution. The intention specifies the meaning of the input expression that is to satisfy the user's goal. Forming an action specification is the activity that spans articulatory distance in the Gulf of Execution. The action specification prescribes the form of an input expression having the desired meaning. The form of the input expression is executed by the user on the machine interface and the form of the output expression appears on the machine interface, to be perceived by the user. When some part of the form of a previous output expression is incorporated in the form of a new input expression, the input and output are said to be inter-referential. Interpretation is the activity that spans articulartory distance in the Gulf of Evaluation. Interpretation determines the meaning of the output expression from the form of the output expression. Evaluation is the activity that spans semantic distance in the Gulf of Evaluation. Evaluation assesses the relationship between the meaning of the output expression and the user's goal.

relationships are equally easy to learn. It may be possible to exploit previous user knowledge in creating this relationship. Much of the work on command names in command language interfaces is an instance of trying to develop memorable and discriminable arbitrary relationships between the forms and the meanings of command names.

Articulatory directness at the output side is similar. If the user is following the changes in some variable, a moving graphical display can provide articulatory directness. A table of numbers, although containing the same semantic information, is not articulatorally direct. Thus, the graphical display and the table of numbers might both be equal in semantic directness, but unequal in articulatory directness. Articutlatorally direct interaction can couple the interaction between action and meaning so naturally that relationships between intentions and actions and between actions and output seem straightforward and obvious.

Our students have pointed out that although it is indeed mimetically and articulatorally direct to cause cursor movement by an analogous movement of a pointing device, in many instances the truly direct interface will do away with cursors. That is, cursor movements reflect current technological limitations.

If we expand our imaginations, we can develop interface technologies that are even more direct than pointing movements. Thus, to delete a word, why move a cursor to the spot: Simply look at the word and speak the appropriate command.

In general, articulatory directness is highly dependent upon I/O technology. To make actions and displays articulatorally direct requires a much richer set of technological input/output devices than most systems currently have. We will, of course, need to have keyboards, pointing devices, and high-resolution, bit-mapped screens (which not all machines or terminals have today). The mouse is a *spatio-mimetic* device. That means that it can provide articulatorally direct input for tasks that can be represented spatially. The mouse is useful for a wide variety of tasks not because of any properties inherent in itself, but because we map so many kinds of relationships (even ones that are not intrinsically spatial) on to a spatial metaphor. In addition, we need sound and speech, certainly as outputs, and possibly as inputs. Precise control of timing will be necessary for those applications where the domain of interest is time-sensitive. In general, we will want to open our imaginations to the set of relevant technologies. Perhaps it is suggested and carried out a set of experiments on doing arithmetic by sense of smell (Galton, 1894). Less fancifully conceived, input might be sensitive not only to touch, place, and timing, but also to pressure or to torque. (See Chapter 15 by Buxton.)

Articulatory distance can also be manipulated by the mental model adopted by the user. An example of this comes from the following story, told us by Yutaka Sayeki. His motorcycle had a switch on the left handlebar for controlling the turn-signals: Moving the switch forward signaled a right turn, backward a left turn. The switch control was semantically direct in that it had a single unambiguous item for each relevant intention regarding the turn signals, but Sayeki's understandings of the forms of the actions encouraged him to map the action "push switch forward" onto the intention "turn left," which is wrong. As a result, it was difficult to remember which switch direction was associated with which direction of turn. A mimetically direct switch would have moved to the right for the right turn, to the left for the left turn. Sayeki solved the problem through the creation of an appropriate mental model, in part by realizing that the motorcycle switch motion was analogous to the required movement of the turn signal lever in an automobile: The required direction is parallel to the direction of movement of the handlebar (or steering wheel).

Consider the way in which the handlebars of the motorcycle turn. When making a left turn, the left handlebar moves backwards. For a right turn, the left handlebar moves forwards. The turn switch was located on the left handlebar, and the required switch movements exactly paralleled the handlebar movements. By reconceptualizing the task as signaling the direction of motion of the handlebars, the switch suddenly became mimetically, articulatorally direct: The switch movement mimics the desired movement. With one mental model, the motion of the switch seems arbitrary, indirect, and difficult to remember. With a different mental model, the switch motion is articulatorally direct and, therefore, easy to learn and to use.

Iconographic languages. Pictographic and iconic languages are examples of articulatory representation in which the form of the expression is related to its meaning. By definition, an icon is a representation that stands for its object by virtue of a resemblance to it. However, even when the form of the icon is very like its intended meaning, the mapping is not complete. Instead, certain features of the referent are abstracted and preserved in the form of the icon, while others are discarded. And even those features that are preserved may be established by convention. For example, while the international iconic symbols for male and female (see illustrations) may seem imitative of a fundamental distinction, they are scarcely interpretable to societies in which trousers and skirts are not worn. Even the interpretation of widely recognized icons requires background knowledge of conventions. Most icons of necessity are abstractions of the thing they depict. Thus, the articulatory directness of even iconic or pictographic representations is not complete: Their interpretation requires knowledge or explanation.

DIRECT ENGAGEMENT

Direct Engagement occurs when a user experiences direct interaction with the objects in a domain. Here, there is a feeling of involvement directly with a world of objects rather than of communicating with an intermediary. The interactions are much like interacting with objects in the physical world. Actions apply to the objects, observations are made directly upon those objects, and the interface and the computer become invisible. Although we believe this feeling of direct engagement to be of critical importance, in fact, we know little about the actual requirements for producing it. Laurel (Chapter 4), discusses some of the requirements. At a minimum, to produce a feeling of direct engagement the system needs:

- Execution and evaluation to be direct in the senses discussed in this chapter.

- Input and output languages of the interface to be interreferential, allowing an input expression to incorporate or make use of a previous output expression. This is crucial for creating the illusion that one is directly manipulating the objects of concern.

- The system to be responsive, with no delays between execution and the results, except where those delays are appropriate for the knowledge domain itself.

- The interface to be unobtrusive, not interfering or intruding. If the interface itself is noticed, then it stands in a third-person relationship to the objects of interest, and detracts from the directness of the engagement.

In order to have a feeling of direct engagement, the interface must provide the user with a world in which to interact. The objects of that world must feel like they are the objects of interest, that one is doing things with them and watching how they react. In order for this to be the case, the output language must present representations of objects in forms that behave in the way that the user thinks of the objects behaving. Whatever changes are caused in the objects by the set of operations must be depicted in the representation of the objects. This use of the same object as both an input and output entity is essential to providing objects that behave as if they are the real thing. It is because an input expression can contain a previous output expression that the user feels the output expression is the thing itself and that the operation is applied directly to the thing itself. This is exactly the concept of "interreferential I/O" discussed by Draper (Chapter 16).

In addition, all of the discussion of semantic and articulatory directness apply here too, because the designer of the interface must be concerned with what is to be done and how one articulates that in the languages of interaction. But the designer must also be concerned with creating and supporting an illusion. The specification of what needs to be done and evidence that it has been done must not violate the illusion, else the feeling of direct engagement will be lost.

One factor that seems especially relevant to maintaining this illusion is the form and speed of feedback. Rapid feedback in terms of changes in the behavior of objects not only allows for the modification of actions even as they are being executed, but also supports the feeling of acting directly on the objects themselves. It removes the perception of the computer as an intermediary by providing continual representation of system state. In addition, rapidity of feedback and continuous representation of state allows one to make use of perceptual faculties in evaluating the the outcome of actions. We can watch the actions take place, monitoring them much like we monitor our interactions with the physical world. The reduction in the cognitive load of mentally maintaining relevant information and the form of the interaction contribute to the feeling of engagement.

A SPACE OF INTERFACES

Distance and engagement are depicted in Figure 5.8 as two major dimensions in a space of interface designs. The dimension of engagement has two landmark values: One is the metaphor of interface as conversation; the other the metaphor of interface as model world. The dimension of distance actually contains two distances to be spanned: semantic and articulatory distances, the two kinds of gulfs that lie between the user's conception of the task and the interface language.

The least direct interface is often one that provides a low-level language interface, for this is apt to provide the weakest semantic match between intentions and the language of the interface. In this case, the interface is an intermediary between the user and the task. Worse, it is an intermediary that does not understand actions at the level of description in which the user likes to think of them. Here the user must translate intentions into complex or lengthy expressions in the language that the interface intermediary can understand.

A more direct situation arises when the central metaphor of the interface is a world. Then the user can be directly engaged with the objects in a world, but still, if the actions in that world do not match those that the user wishes to perform within the task domain, getting the task done may be a difficult process. The user may believe that things are getting done and may even experience a sense of engagement with the world, yet still be doing things at too low a level. This is the state of some of the recently introduced direct manipulation systems: They produce an immediate sense of engagement, but as the user develops experience with the system, the interface appears clumsy, to interfere too much, and to demand too many actions and decisions at the wrong level of specification. These interfaces appear on the

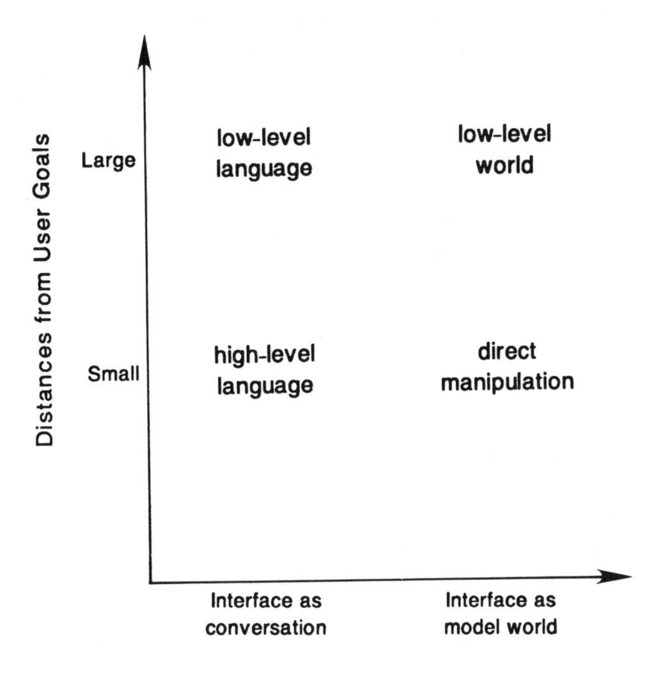

FIGURE 5.8. A space of interfaces. The dimensions of distance from user goals and degree engagement form a space of interfaces within which we can locate some familiar types of interfaces. Direct manipulation interfaces are those that minimize the distances and maximize engagement. As always, the distance between user intentions and the interface language depends on the nature of the task the user is performing.

surface to be Direct Manipulation interfaces, but they fail to produce the proper feelings of Direct Engagement with the task world. Beginners tend to like them: Opinions among experts vary.

Closing the distance between the user's intentions and the level of specification of the interface language allows the user to make efficient specifications of intentions. Where this is done with a high level language, quite efficient interfaces can be designed. This is the situation in most modern integrated programming environments. For some classes of tasks, such interfaces may be superior to direct manipulation interfaces.

Finally, the most direct of the interfaces will lie where engagement is maximized, where just the correct semantic and articulatory matches are provided, and where all distances are minimized.

PROBLEMS WITH DIRECT MANIPULATION

Direct Manipulation systems have both virtues and vices. The immediacy of feedback and the natural translation of intentions to actions make some tasks easy. The matching of levels of thought to the interface language—semantic directness—increases the ease and power of performing some activities at a potential cost of generality and flexibility. Not all things should be done directly. Thus, a repetitive operation is probably best done via a script, that is, through a symbolic description of the tasks that are to be accomplished. Direct Manipulation interfaces have difficulty handling variables, or distinguishing the depiction of an individual element from a representation of a set or class of elements. Direct Manipulation interfaces have problems with accuracy, for the notion of mimetic action puts the responsibility on the user to control actions with precision, a responsibility that is often best handled through the intelligence of the system, and sometimes best communicated symbolically.

A more fundamental problem with direct manipulation interfaces arises from the fact that much of the appeal and power of this form of interface comes from their ability to directly support the way we normally think about a domain. They amplify our knowledge of the domain and allow us to think in the familiar terms of the application domain rather than those of the medium of computation. But if we restrict ourselves to only building interfaces that allow us to do things we can already do and to think in ways we already think, we will miss the most exciting potential of new technology: to provide new ways to think of and to interact with a domain. Providing these new ways and creating conditions that will make them feel direct and natural is an important challenge to the interface designer.

Direct manipulation interfaces are not a panacea. Although with sufficient practice many interfaces can come to feel direct, a properly designed interface, one which exploits semantic and articulatory directness, should decrease the amount of learning required and provide a natural mapping to the task. But interface design is subject to many tradeoffs. There are surely instances when one might wisely trade off directness for generality, or for more facile ways of saying abstract things. The articulatory directness involved in pointing at objects might need to be traded off against the difficulties of moving the hands between input devices or of problems in pointing with great precision.

It is important not to equate directness with ease of use. Indeed, if the interface is really invisible, then the difficulties within the task domain get transferred directly into difficulties for the user. Suppose the user struggles to formulate an intention because of lack of knowledge of the task domain. The user may complain that the system is difficult to use. But the difficulty is in the task domain, not in the interface language. Direct Manipulation interfaces do not pretend to assist in overcoming problems that result from poor understanding of the task domain.

Reassessing the Claims for Direct Manipulation

What about the claims for direct manipulation? Do these systems really meet the hopes? Alas, it is too early to tell. We believe that direct manipulation systems carry gains in ease of learning and ease of use. If the mapping is done correctly, then both the form and the meaning of commands should be easier to acquire and retain. Interpretation of the output should be immediate and straightforward. If the interface is a model of the task domain, then one could have the feeling of directly engaging the problem of interest itself. It is sometimes said that in such situations the interface disappears. It is probably more revealing to say that the interface is no longer recognized as an interface. Instead, it is taken to be the task domain itself.

But are these desirable features? Are the tradeoffs too costly? As always, we are sure that the answer will depend on the tasks to be accomplished. Certain kinds of abstraction that are easy to deal with in language seem difficult in a concrete model of a task domain. When we give up the conversation metaphor, we also give up dealing in descriptions, and in some contexts, there is great power in descriptions. As an interface to a programming task, direct manipulation interfaces are problematic. We know of no really useful direct manipulation programming environments. Issues such as controlling the scope of variable bindings promise to be quite tricky in the direct manipulation environments. Will Direct Manipulation systems live up to their promise? Yes and no. Basically, the systems will be good and powerful for some purposes, poor and weak for others. In the end, many things done today will be replaced by Direct Manipulation systems. But we will still have conventional programming languages.

On the surface, the fundamental idea of a Direct Manipulation interface to a task flies in the face of two thousand years of development of abstract formalisms as a means of understanding and controlling the world. Until very recently, the use of computers has been an activity squarely in that tradition. So the exterior of Direct Manipulation, providing as it does for the direct control of a specific task world, seems somehow atavistic, a return to concrete thinking. On the inside, of course, the implementation of direct manipulation systems is yet another step in that long formal tradition. The illusion of the absolutely manipulable concrete world is made possible by the technology of abstraction.

Earlier in this chapter we reprinted two sets of descriptions of the virtues of direct manipulation systems, claims made in the early enthusiasm of their first discovery. Now that we have examined the many aspects of directness, it is time to go back and re-evaluate those claims. Let us see what we can say about them now. First, Shneiderman (1982) pointed at three "essential features" of Direct Manipulation. Let us examine each of the three features: Shneiderman's comments in *italics,* our assessment in regular font.

1. *Continuous representation of the object of interest.*

   This relates to the directness of evaluation, and as our analyses show, seems an essential aspect of a direct engagement system.

2. *Physical actions or labeled button presses instead of complex syntax.*

   The feature of "physical actions" we interpret to refer to articulatory directness, and more particularly, to what one might call "mimetic directness," actions that mimic the desired changes on the objects of interest, especially movement or size changes. "Labeled button presses" must refer to semantic directness. The former clearly is important in establishing direct engagement. The latter is not. The latter operation really seems irrelevant to arguments about this form of directness, whatever the virtues for ease of learning or ease of use.

3. *Rapid incremental reversible operations whose impact on the object of interest is immediately visible.*

   This is perhaps the essence of direct engagement: It reflects what we and Draper (Chapter 16) call the importance of "inter-referential I/O." The important point to us, implicitly stated in this assumption, is that the objects upon which the actions are taken are exactly the same as those upon which evaluation is made. The reversibility of the operations is desirable, but not necessary. Not all operations can have this feature in a natural way. So too with immediacy of the result: Where immediacy is a natural part of the domain, then this is essential. Otherwise, it is desirable, but may not always be necessary—as we have discussed at length within this chapter.

If we reinterpret Shneiderman's claims to be about *Direct Engagement*, then his three features fare well: They do seem to help define what is necessary to develop a direct engagement system. Now let us look at Shneiderman's six claims for the results of such a system:

1. *Novices can learn basic functionality quickly, usually through a demonstration by a more experienced user.*

   This may or may not be true. We think it really derives from the fact that a good Direct Manipulation interface is invisible—the user feels as if operations are actually done directly on the task domain. And if the computer novice is already knowledgeable in the task domain, then much of what is needed to use the interface is already known.

   Why might training be possible through demonstration? Well, two reasons. Lewis (Chapter 8) argues that demonstration is, in general, a superior method of instruction. Think of a demonstration as a dynamic example. In this case, then, the superiority of demonstration has nothing to do with direct manipulation. But there is a second reason. Owen (Chapter 17) points out that demonstrations with normal, command language systems, are often puzzling because the actions are not visible. That is, the expert waves hands over the keyboard and mysterious wonderful results appear on the screen. The learner often has little notion of what operation was performed. But with a typical direct manipulation (read direct engagement) system, the actions themselves are visible, and their results are both visible and also direct reflections of the operations done upon them. Whenever these cases apply, we believe the claim will be valid.

2. *Experts can work extremely rapidly to carry out a wide range of tasks, even defining new functions and features.*

   We are suspicious of this claim. In fact, we would not be surprised if experts are *slower* with Direct Manipulation systems than with command language systems. We suspect that the virtues of Direct Manipulation lie elsewhere: Speed at execution is not likely to be a relevant factor. Real experts can probably type a few lines of obscure code much faster than they could move objects around a screen, position them properly, and do the necessary pointing operations.

3. *Knowledgeable intermittent users can retain operational concepts.*

   This could be true, but if so it probably reflects two things: The expertise at usage really reflects expertise in the subject matter, which is probably well established and apt to fade slowly from memory, if at all; and a good semantic mapping of actions leads to slower forgetting and also easier rederivation of the operations that are forgotten. We suspect that these claims are true, but derive from aspects that would be true for many well-designed systems, not just Direct Manipulation systems.

4. *Error messages are rarely needed.*

   Yes and No. Error messages are often not needed because the results are immediately visible, and because some classes of errors may not even be possible. But Direct Manipulation systems have their own problems. It is possible to make new classes of errors, some of them potentially serious. Worse, because these are apt to be errors in the task domain (but legal operations as far as the interface is concerned) they are hard to detect. Finally, Direct Manipulation systems sometimes simply don't bother with error messages, assuming that the ease of evaluation will make it obvious to the user that the desired operation was not done. It's not that the message wasn't needed, it's just that the system didn't bother to present one. This what Lewis and Norman call the "do nothing" strategy. This strategy is not always desirable (see Lewis & Norman, Chapter 20).

5. *Users can see immediately if their actions are furthering their goals, and if not, they can simply change the direction of their activity.*

   The first part of the claim results from the properties of Direct Evaluation. The second part has already been discussed in our response to feature 3 of the first list. But basically, the ability to "change the direction of their activity" results from the natural reversibility of many actions. For those actions that are not so naturally reversible, the systems do not fare any differently than more conventional systems. Immediate feedback as to the outcome of an operation and "undo" commands are valuable for any system, not just Direct Manipulation systems.

6. *Users have reduced anxiety because the system is comprehensible and because actions are so easily reversible.*

   This is outside the domain of our analyses and difficult to assess. A fair comparison would require two systems, both equally well-matched in capability and in semantic directness.

In conclusion, some of the claims seem too strong, some represent features of many well-designed systems, not just direct manipulation systems, and some seem to be correct. All in all, the early claims seem to fare reasonably well, although today we can bring more sophistication and depth of analysis to bear upon them then was possible at the time they were made.

CONCLUDING REMARKS

Direct Manipulation implies a directness of action, a directness of the translation of intentions to actions, and a directness in the feedback and knowledge of the system. The feeling of directness also implies a feeling of control over the objects in the task domain—a Direct Engagement which results, in part, from the feeling that one is operating directly upon the objects, where the objects upon which actions are performed are the very same as those from which output is received.

In this chapter we have concentrated upon the various aspects of directness with respect to the performance of a task, with little mention of the problems of learning. Nonetheless, a system that is properly designed should also be one that is easy to learn. Our discussions of the nature of the Gulfs separating a person's intentions and evaluations from the system's inputs and outputs provide a foundation for making the learning task easier. If the Gulfs of Execution and Evaluation are properly bridged from the system side, then the system should be relatively easy to learn: Minimum processing is required to get from intention to action and from outcome to interpretation. The learning required should be concentrated in the task domain. The hope is that people will be able to spend their time learning the task domain, not learning the computer system. Indeed, a properly developed Direct Manipulation interface should appear to the user as if it is the task that is being executed directly: The computer system and its interface will be more or less invisible. Much empirical work remains to be done to substantiate these claims.

The understanding of Direct Manipulation interfaces is complex. There are a host of virtues and vices. Do not be dismayed by the complexity of the arguments. To our mind, direct manipulation interfaces provide exciting new perspectives on possible modes of interactions with computers. It is our intention to explore these modes further, through the building of working systems, expanding the theoretical analysis, and performing empirical evaluations. We believe that this direction of work has the potential to deliver important new conceptualizations and a new philosophy of interaction.

ACKNOWLEDGMENTS

We thank Ben Shneiderman for his helpful comments on an earlier draft of the chapter, Eileen Conway for her aid with the illustrations, and Julie Norman and Sondra Buffett for extensive editorial comments.

CHAPTER 6

Notes on the Future of Programming: Breaking the Utility Barrier

ANDREA A. diSESSA

In preparing this chapter I felt obliged to look at a number of other discussions of the future of programming. Few of them mention anything resembling the issues discussed here. Instead, probably the most prominent theme is the "complexity barrier"—the tremendous difficulty and cost involved in creating and maintaining huge programs. This is, no doubt, a serious and enduring problem. But it is the most serious one for professional programmers, not nonprofessionals who, at least in terms of numbers, will dominate future use of computers. More than numbers, I believe the ultimate social and cultural impact of computation will be determined to a great extent by what we can cause to happen when technologically unsophisticated users sit down at a machine. The hope I share with many others is that computation can significantly enhance intellectual development and productivity for most, if not all, people. (Two exemplary references are Winograd, 1984a, 1984b.)

What role will programming play in this? Some of my favorite antagonists in this regard feel that computers will totally disappear into the woodwork as far as ordinary people are concerned, the way electrical relays and motors and control micro-processors have already disappeared. Even if they don't disappear, certainly, it is said, the ordinary person will need to know as little about programming as about repairing an automobile—that's a job for specialists. Then the future of programming would indeed fall back into the province of specialists and strike the complexity barrier head-on.

If one were to judge from present programming languages, I might well agree. They are clumsy, inelegant, hard to understand, harder to learn, and just don't do very much very easily. So I am saying the future of programming in the large-scale scheme of things is not evident in what we currently have, but will depend largely on future forms and contexts for programming that we invent for nonspecialists.

What might computers do for nonspecialists? What should programming be like for them? If I were to pick a single image, it would be the computer as an interactive, constructible and reconstructible medium. "Interactive" deserves a great deal of discussion, but it does not need it. The degree of interactivity is the well-recognized key difference between computation and all previous media. "Constructible and reconstructible," however, are just as important, but much less discussed. The written word would not be half as powerful as it is if most people couldn't write. As a tool for accomplishing information related tasks, for thinking, for inventing, for developing oneself intellectually, text would be next to useless, except as a way to convey to others what some "expert," who could afford to hire a scribe (read "programmer") wished to convey. We need in interactive media the equivalent of notes to yourself in margins, crossings out, personal summaries, and your own essay or even book—little things in bits and pieces that can be put together into a masterpiece, but don't need that scale of effort to warrant doing. Even great masters of interactive media, the equivalent of a great author or poet, would be limited unless they could play, tinker and create in pieces, rather than waiting for the long-loop to the scribe and back. Surely there will always be the equivalent of technical editors and publishing houses, but just as surely, there needs to be the equivalent of pencils and paper for everyone.

So programming will mean being able to construct and reconstruct in interactive media. This is why the automobile engine metaphor is an inappropriate way of thinking about all computers. If a machine is to serve one very specialized role, such as providing mechanical power, it ought to be hidden and of no concern. But if the purpose of the machine is flexibility and personal adaptability, we had better figure out how to give users maximum control. Expressive power and nuance are incompatible with invisibility and inaccessibility.

Programming languages viewed in this way will certainly be, in many ways, much different than those of today. In the first instance they had better be far superior to present languages at interaction. Long, silent, invisible algorithms are much less the point than being able simply to arrange to observe and control an ongoing process. Languages had better make modification of existing structures and processes very easy. They had better accept bits and pieces of whatever the medium supports, certainly text, programs, and pictures, to be recombined easily into a new product. They had better respond to the complex, interleaved activity structures of humans much better than any set of isolated programs can do. See the chapters in the section on User Activities, Section IV of this book.

There will, however, be many similarities between present and future languages as well, at least for some time to come. There are certain things one simply needs in order to describe a complex process. Representing state, controlling availability and flow of information, deciding what happens when: These seem very likely to remain salient, in one form or another. The meaning of programming will change, but I am certain that the means of construction and reconstruction will be powerful and complex enough that no one will be embarrassed to call them programming languages.

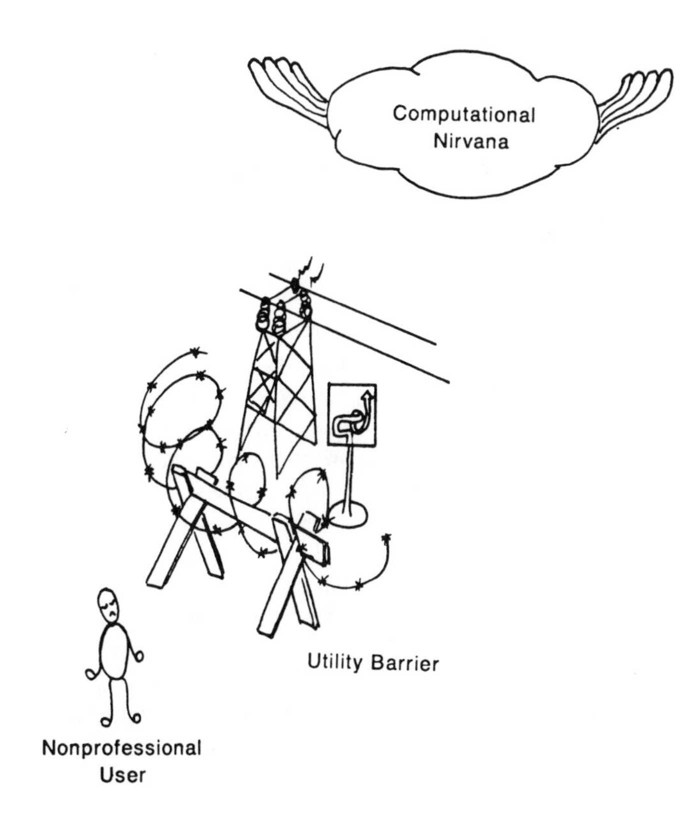

The complexity barrier has surreptitiously changed into the utility barrier. The challenge for the future is to make programming into something that is simple enough and useful enough that everybody will want and be able to learn how to grasp and stroke with this new pencil. Complexity is still an issue. But it is complexity with an emphasis on a different range of the spectrum. In contrast to battling huge programs, if we can allow a broad range of simple but useful things to be done transparently, we will have won more than half the battle. And complexity is by no means the full story. Please note that utility is the ratio of value to effort expended. Utility can be increased by decreasing the denominator, with which complexity is intimately involved, or by increasing the numerator. We need to worry as much about value, about the uses of programming, as about making it easier to do.

I do not wish to play futurist here and wax poetic about possibilities we can barely imagine. Instead, I wish to look conservatively to the near future—in some cases, the recent past seems not yet to have been noticed—and make the simple remark: From an engineering point of view, it is clear that, far from reaching an equilibrium, we have at best crossed a threshold in terms of making computation accessible and useful to everyone. If we wish to, we will be able to make rapid changes in what programming looks like, and in the context for learning construction and reconstruction in an interactive medium.

> *Putting programming in terms of utility rather than as good in itself is a position that might surprise some who know the history of our work. "We" are the Logo Group at MIT, and its descendants. The language Logo, our best known product to date, was developed in the aura of general intellectual skills that can be developed in computational environments, and of deep mathematical and scientific ideas that can become part of everyday experience through programming. Thus, many might expect me to write this note about designing programming languages on the basis of what powerful ideas one could build into them. But I have come to believe several things that move utility to first place.*
>
> *First, as a matter of fact, we have had much more success building microworlds, computational environments for open exploration of particular clusters of ideas, on top of a programming environment rather than into it. Interest and the many things that we want to teach people are mostly at a different level than the generalities of programming per se. Second, the need for a lingua franca of interaction is great.*

Multiple microworlds, each strongly tuned to its own particulars, may be too disparate to allow any economy of learning the interactive medium. In fact, the pressures of commonality go beyond the individual. Designing a new interactive medium will certainly be a long term and gradual social process in which what eventually solidifies must serve the everyday purposes of many, yet be tailorable gracefully into technical domains. Natural language, for example, is useful for everybody, but easily accepts technical terms and conventions of exposition for specialists. Third, no matter how valuable it might be to have learned programming, it will be harder, and in other ways less enticing a job, without perceived utility along the learning route. This is a major theme of this chapter. Finally, it seems to me that the good things intrinsic to programming are very robust and will not go away if we put utility in a more prominent position as a design goal.

I consider three dimensions of change in programming. Each can bring us noticeably closer to breaking the utility barrier.

Since complexity is a problem (bad) and utility is a goal (good), perhaps a more parallel wording would be the "uselessness" barrier. But complexity is bad like sex in a Puritanical society (interesting, nonetheless), and uselessness is bad like yesterday's garbage. So I'll stick to the utility barrier.

PRESENTATION

The first type of change is *presentational.* These are not modifications to the underlying structure of programming, but only to how it is visually (and, potentially, through other senses) presented and manipulated. Presentational changes will not, in general, affect the ultimate perceived power of a language, but most certainly they can dramatically reduce the effort involved in learning and understanding it. My points will be in the form of examples.

Beyond Logo

Logo's roots are in the teletype interfaces that were available when it got its start. The communication format between user and machine is linear and "conversational." The user says (types) something, and the machine says something back. One small problem with this format is simply that you cannot even point to something you "said" a few lines ago in order to "say" it again. The trace of what was done is, in computational terms, a useless artifact and cannot be, for example, turned into a procedure. Initially, things even got worse with the demise of hardcopy terminals; in addition to having to type over tried out commands, users had to remember what they just tried if it scrolled off-screen. Of course, versions of *concrete programming* (making a program essentially by doing, step by step, what you want to have happen in the program) can be built in Logo, albeit somewhat clumsily. Various popular versions of what I started calling "instant" (single key activation) programs a number of years ago incorporate this feature. But this is not the ordinary way one programs in Logo.

A second disadvantage of conversational interaction is that large-scale structures are difficult to notate and manipulate as a whole. The Logo END command is really no command at all, but a syntactic marker of the boundary of an object (program). Unfortunately, END can easily be confused with parts of the program because it must look just like another piece of the conversation and cannot be connected visually with the previous part of the conversation, the start of the procedure definition, with which it should really be connected.

A more profound but subtler problem is that you cannot directly see and manipulate the state of the system on the screen. Instead, you must send a request to see some state (e.g., PRINTOUT), and if you want to change it, you must send another request (e.g., MAKE or TO). Recent microcomputer implementations of Logo incorporate a screen editor to ameliorate these problems to some extent, but this is only a patch. One still must make a major mode switch, entering the editor, if you want at all to pretend the screen shows the state of the system. And the details of the relation of the editor buffer, the definition process and the defined state of the workspace are both invisible and subtle.

> *Here's a hack to show the subtlety. Suppose you want to delete all but one procedure from your workspace. It is painful to type ERASE APPLE, ERASE BEAR, ERASE CAT, ... So what is frequently done is to load the editor with the whole workspace, EDIT ALL. Then use a few simple edit commands to delete everything but the wanted procedure. Then exit the editor, clear the whole workspace, reenter the editor (the buffer is not lost) and reexit again, which redefines the one procedure you wish. This is the very opposite of a direct manipulation system.*

The bitmap display, a pointing device, and enough memory can solve all these problems. In Boxer (it's all boxes!), the language we are designing as a successor to Logo, we believe we have done this. Computational objects such as programs are visual units, boxes, and are trivially manipulable as a whole, somewhat like a large character in a text editor. The program that appears as a box in Figure 6.1 can be deleted by pointing to it and pressing the rubout key. Similarly, it can be picked up and moved around as a unit.

In Boxer we have changed the conversational interaction paradigm to "looking at and directly altering the state of the system." Boxer is "editor top level": You are always able to change directly or use anything you (or the computer) have put on the screen. This automatically gives you a simple form of concrete programming since you can simply select a set of lines you have typed and executed, and box them to make a program out of them. More fundamentally than that, once one learns the few basic editing actions to point, pickup, move and delete, one can inspect, construct, or modify anything in the system without learning any more commands. The programs in Figure 6.1 can be edited just by moving the cursor into them, deleting a few characters and adding some new ones. The editor constitutes your feet and finger tips for moving around and building your computational world.

We call this profitable illusion that one sees and directly manipulates the system on the screen *naive realism.* It is, in a sense, an old principle, but it is rarely applied to computational structure. The principle has many implications. It makes the system easier to use in that you can always see what you are doing in changing or adding to the system. A small set of editing commands can replace all the separate structure creating and modifying commands. If you can remember or imagine

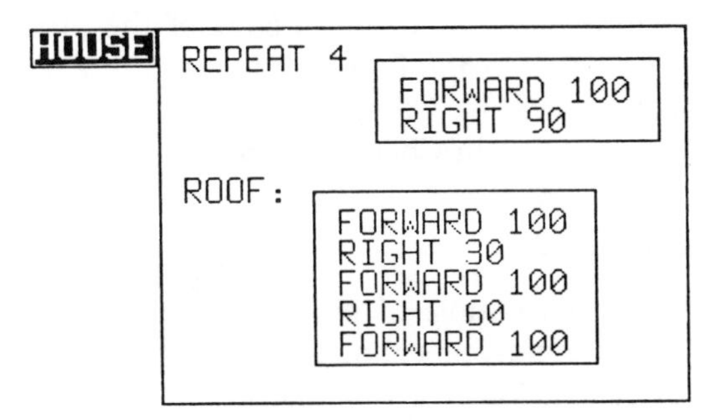

FIGURE 6.1. A simple program definition in Boxer. The tab is the procedure's name. The box labeled ROOF is a subprocedure written in place.

what some structure should look like, you can create it. The visibility that comes with naive realism also makes many aspects of the system easier to understand, particularly its large scale organization. For example, procedure/subprocedure relations may be made absolutely explicitly (Figure 6.1).

Over long time scales, seeing the structure of the environment rather than only a trail of a recent "conversation" should pay handsome dividends, especially for those who may have difficulty keeping system organization in their heads, children for example. In order to find something in the system, you may simply wander around looking for it. (See also Figure 6.4 and surrounding discussion.) Note that the ability to make an audit trail is not lost from Boxer. All it takes is the discipline to type rather than point. More elaborate and automatic audit trail mechanisms, of course, may be programmed.

Boxer makes another contribution toward presentational change that goes hand-in-hand with the naive realism just described. The geometric configurations one sees on the screen express fundamental semantics of the language. I call this the *spaital metaphor*. For example, containment implies inheritance. Procedures and variables that are defined in a box are accessible only in that box, and in recursively contained boxes. Thus, the important modularity constructs of environments and namespaces are plainly visible. This use of regions to represent namespace environments is a concretization of the "contour model," which is frequently used to explicate these ideas.

We have also taken pains to make sure that dynamic as well as static structure of the language is visible. If one chooses, the execution of a procedure can be watched. The essential features of this visibility are very simple. Commands making up a procedure are highlighted successively as they are executed; returned values appear in place of the procedure that returned them; and one has the rule that to watch the execution of a procedure called by name, a copy of the procedure definition will replace the name as the first step in executing it. (Details can be found in diSessa, 1985). Compare Figure 6.2 to the little man model shown in Figure 10.1 in Chapter 10 in this volume.

An Iconic Presentation

Let me give one other example of potential presentational changes in programming. Quite some years ago, Radia Perlman and Danny Hillis constructed a device known as a slot machine. Children programmed by inserting cards representing commands into rows of slots, which represented programs. Not only could one see and directly manipulate the programs and their pieces, but also the sequential activation of

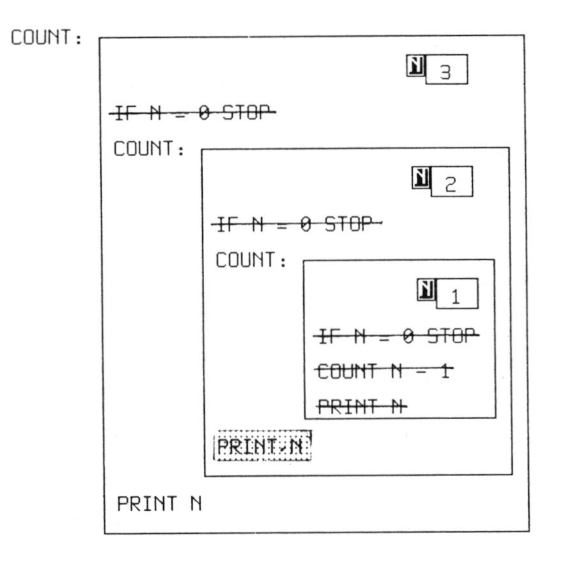

FIGURE 6.2. The display of a procedure stopped in midstream.

program steps, subprocedure calls and returns, could be directly observed as lights lit up under each card when that step was executed. Apart from avoiding typing, and adding concreteness to computational objects, the slot machine provides substantial help in making a mental model of the operation of a computational system like Logo.

With the advent of high resolution bitmaps and graphical objects like sprites, one can easily implement a slot machine on the video screen, moving icons around like cards with a joystick or mouse. The screen slot machine can be freed of many of the limitations of the physical one. On a screen, it is easy to invent spatial/graphical representations for general input parameters and conditionals which did not exist on the slot machine. Adding symbols, e.g., by typing a new word, is trivial in the screen version whereas making a new, physical card is not so easy. Most important, the process of abstraction can be represented easily on the screen—e.g., a spatial sequence of card-icons forming a program can slide together over one another like a spread out deck of cards being pushed back together and be given a top "cover" icon, becoming a single unit like all the supplied primitive card-icons. There does not seem to be any reason that nearly the whole of a computer language like Logo could not be presented in this graphical, concrete way. Though it has limitations that will be discussed in the section on direct manipulation, it is a very attractive introductory presentation to programming.

In passing I note an interesting theoretical issue. This iconic form of programming represents an attempt to use spatial and object knowledge to replace the linguistic means Logo uses to promote comprehensibility (see Chapter 10). Looking at the set of bugs and difficulties students have with such a system as compared to present Logos should provide an insightful comparative study.

STRUCTURE

The second type of change in programming is change in *structure*. This goes beyond the presentational changes mentioned above and significantly alters the underlying computational mechanisms and structures. While it might not be apparent to the user because of syntax or other presentational issues, the structure of a system provides some bedrock characteristics.

Traditional motivations for changes in structure have been increased modularity, preventing and catching bugs, precision of expression, mathematization of programming, and the like. In view of mounting an assault on the utility barrier, priorities become reordered. Looking back at the ratio of value to effort, one comes immediately (in the numerator) to expressive power—how quickly and simply does the language describe situations users can immediately perceive in terms of existing goals and needs. And one comes (in the denominator) to understandability. To be sure, ease of use plays a role in defining the work needed to accomplish something, but I have been constantly impressed with how much effort humans will expend if they value the result and understand what they are doing. In this section, I largely ignore ease of use. As long as it is not confused with understandability and perceived value, ease of use may decide which product succeeds in the marketplace, but not which paradigm of computation will succeed. (Compare Norman's remarks on first-and second-order issues in Chapter 3.)

When it comes to expressive power, one must talk about the tasks a user wishes to accomplish. Structure can be important. While all programming systems may be formally equal in power (Turing equivalent), some may be radically better adapted to some purposes than other systems. However, I defer discussion of this to the section on context, except where it is unavoidable.

When it comes to understandability, the structure of a language does not stand in isolation, but relies in general on presentation as a major channel of communication to the user. Thus the shift from print- to heavily graphics-based interfaces can precipitate a shift toward visual–spatial programming languages, as exemplified by the icon programming system. But there, no change in structure was implied. With Boxer, however, we have found many small, and some not so small, ways that the structure of programming can be profitably changed to mesh with new presentational means, to which we turn briefly.

Structural Innovation in Boxer

Variables are different in Boxer than in any other languages for reasons of understandability and usefulness. The meaning of variables in Boxer includes the familiar set and fetch protocol, but variables are turned into genuine places in which different things can be stored. The place metaphor cannot be supported very well by the simple association of a name with a data object. For example, a Boxer variable can be shared not only by having another procedure use the same name, but also by an actual reference to the place of the variable. This reference is what we call a *port* in Boxer. Even if you change the name of the variable, the program will still share its value through having a port to it. Ports as part of data objects give a hyper-text functionality. This is as valuable concretely, as it is in programs. For example, it allows one to create cross referenced and non-hierarchically organized documents. One can think of ports as analogs of traditional pointers, except objects shared with pointers do not have a unique place of existence like the target of a set of ports. Instead, objects shared with pointers belong equally to all owners of a pointer to them. To get a grip on what this slight change in structure means to comprehensibility, imagine what a bizarre world it would be if possession were not largely synonymous with physical location, if several people could hold, look at, and even change the same object at the same time. (To be fair, Boxer does not eliminate this kind of thing. But it makes it an advanced topic rather than an entry level one. One needn't program with ports. At the same time, the issue is better presented visibly.) Figure 6.3 shows the display of a port. Though in this case the target of the port is on screen, that need not be true.

This is a variable: Data Bananas Peaches

and this is a port to it: Port Bananas Peaches

FIGURE 6.3. A port and its corresponding target box.

We could continue the list of subtle, but not insignificant differences between Boxer structures and those of other contemporary languages. For example:

1. Boxer does not have the same QUOTING mechanism as Lisp, Logo or Smalltalk, but instead has one with a simpler interpretation. The equivalent of quote is a type marker, *data*, and data objects are simply unevaluated by the interpreter. The quote is not stripped in evaluation.

2. Boxer does not distinguish intrinsically between atomic and compound data objects in order to eliminate a class of type bugs beginners have.

3. Actor-oriented programming with class and subclass hierarchy is not primitive in Boxer as it is in Smalltalk, but can be built with box structure.

Each of these structural changes was motivated by utility, making the language simpler and more powerful. But they are not dramatic. I don't believe Boxer will succeed or fail on the basis of its innovative structures—presentation and context (to come) are its strong points.

Beyond Procedural Languages

So far only procedural programming languages (Basic, Logo, and soon, we hope, Boxer) have made significant inroads with nonprofessional programming. To be sure, spread sheets border on offering general computational power in a new form. See Kay's article (1984). But at this stage it is more tease than substance. Visicalc is more than usually flexible for an applications program, but not yet a really programmable medium. Actor-oriented programming, as represented by Smalltalk, has some significantly different structures than run-of-the-mill procedural languages. But, Smalltalk's strengths are really at the scale of systems—it is the first genuine example of a fully integrated medium taking advantage of high resolution graphics. Its aim as a personal dynamic medium puts it squarely in the line of development of interest here. But, by and large, the designers of Smalltalk have attacked the complexity barrier rather than the utility barrier. The basic computational mechanism is more complicated in Smalltalk than in Logo. Actors, classes, and instances are undisputably useful, but they are levels of organization on top of functions and variables, not simplified replacements. The things one can do quickly and easily with the basic system and unprofessional skills are not advanced enough to be a popular medium in the sense of this chapter. Thus, Smalltalk is more a meta-medium than a medium.

Prolog is also not a breakthrough in understandability. Its logical semantics seem quite orthogonal to advances in visual presentation that motivated Boxer and, in a different way, Smalltalk. But Prolog really does offer a potentially dramatic shift in the kind of thing done with programming, so it will warrant a second look later.

Let me turn to a "new," as yet to be fully defined, example to illustrate another class of near future possibilities for changes in structure.

Device Programming

Device programming is motivated by the image of an electronic or mechanical device consisting of a number of components of a few classes, like resistors, transistors and capacitors; or pipes, pumps and reservoirs. These components each have relatively simple behavior and achieve the functionality of the device by being hooked together at their terminals into a network. Computationally, we want to have graphical components that can be manually assembled into devices by connecting their inputs and outputs with "wires" (lines drawn on the screen). The wires communicate messages of an arbitrary symbolic kind, which could, for example, represent flow of substance or electricity by passing numbers representing amounts. Each component knows when it gets a message at an input and can compute and send output messages as it sees fit. Device programming is structurally a significantly different form than contemporary procedural languages because of its explicitly parallel nature of computation, and its explicit representation of data flow rather than sequence. On the other hand, device programming can simulate a function as a component with one input and one output. Activation of the function amounts to simply giving it an input. Furthermore, a component can be built out of very little more than a procedural programming language in which to express the actions to be taken to compute outputs from inputs.

Naturally, it would be important to have a general abstraction mechanism so that a network of devices could be made into a component. Some set of free inputs and outputs in a device could "extend beyond the boundary" of the device to act as terminals of the abstracted component. Visually, the parts of the device-become-component can shrink and/or acquire a new surface form to hide detail. Likely one would like the surface form to show some small part of the internal state of the device.

Device programming is attractive because it has such a simple and graphic method of combining elements to make compound things. There is reason to believe it can have some intuitive accessibility that the hidden data flow and complex sequencing of pure procedural programming does not. Lastly, it opens doors to more easily simulating an important class of physical computations, thus aiding in our understanding of the world. One can even engage in the delightfully recursive task of constructing a computer out of computationally implemented components, emulating every level of abstraction of a physical computer. All of these characteristics of device programming—intuitive accessibility, important and interesting application, and even ease of use—are illustrated in the contemporary computer games *Rocky's Boots* and *Robot Odyssey*.

Device programming, like most new ideas, is not entirely new by any means, but a crystallization in a new context of a collection of old ideas. In fact, the first system I know of that substantially followed this outline was made 20 years ago by Sutherland (1966). The reason the idea is timely again is that powerful processing and high resolution graphics, along with the particular concern for creating an easily accessible popular medium, define a niche into which device programming may well fit nicely. It is interesting to remark on how this niche redefines past conceptions similar to device programming. UNIX pipes, by their very name, suggest the right topological metaphor. They are also a parallel processing system. However, pipes are structurally one dimensional and presentationally nongraphic. They provide no easy opportunity for output, let alone input, at intermediate stages of the pipe. Pipes are not intended to be a means of implementing programs on a small scale, but rather, they are a way of combining existing rather large chunks. Most other parallel processing constructs, *Simula* coroutines for example, are at best intended to control graphical objects, not to be graphical, nor do they make use of device topology to define the communications network in a program.

CONTEXT

Continuous incremental advantage. The third dimension of change in programming is context, what you do with your programming system. Turtle graphics, whereby drawings may be created by issuing commands to a mobile graphics cursor, is a crucial part of the advance of Logo over previous languages. It is motivating; it allows children to set goals immediately that they understand (drawing pictures), yet it can evolve naturally and slowly into a medium of contact with profound mathematics (Abelson & diSessa, 1981; Papert, 1980). Again the old story of natural language tells the tale. It is an incredibly complex and large learning task which, nonetheless, nearly every child masters because it can be mastered one tiny bit at a time, and is useful to the child at every step along the way. The steps are small, but the range is large: A child gets a cookie by learning how to ask; a university professor gets recognition not only for his ideas, but also for presenting them well. The principle really deserves a name—*continuous incremental advantage*. If we want anyone to master any complex but powerful tool, it had better offer continuous advantage to the learner for learning more, and the bits of learning had better be in manageable (incremental) chunks. The importance of the principle is not only having small steps in learning, and motivation for those steps, but so that at each stage the learner gets structured feedback on competent performance, feedback in terms of achieving understood goals.

In learning Logo we have found some blocks, or at least apparent plateaus in continuous incremental advantage (Papert, diSessa, Watt, & Weir, 1980). For very young children, procedures seem to constitute a small barrier. They are more difficult than necessary to master, and don't do all that much beyond packaging the picture the child is drawing. Many children prefer to type out a stereotyped set of commands over and over rather than to make them into a procedure. This is one case where more work is done to avoid a not firmly understood but more efficient process. For older children, procedures are more productive, but variables seem to stall them for a while. Effective use and understanding of list processing has similar problems.

Table 6.1 shows how Boxer compares to Logo with respect to transparency and incrementality in the important early activity of making a definition.

Data objects can be built in a similar way and turned into a variable simply by adding a name. The value of the variable can be changed at any time under program control, or directly with the editor.

Boxer Data Worlds

With regard to continuous incremental advantage, Boxer's concreteness not only provides smaller, more understandable increments, but it also enlarges the scope of programming's context to include text production and manipulation, and the organization and manipulation of many other sorts of data. If a programming system is literally also a child's book and pencil (text editor in modern parlance), and if he can, bit by bit, modify, extend, and personalize not only what comes in his book, but

TABLE 6.1
COMMANDS

Logo	Boxer
1. Type a set of commands to try them out.	1. Type a set of commands, or select and collect from previously typed text.
2. Type TO <name>. Screen goes blank. You're in the editor.	2. Mark the commands (push a button and draw pointer across). Then press a key to make a box.
3. Recall or type over from pencil and paper the list of commands tried out.	3. You may now point to the box to execute it.
4. Exit editor.	4. If you like, add a name to the procedure.
5. Type the name to execute.	5. Type the name to execute.
	6. If you like, shrink the box to hide its contents. Or you may move it entirely off the screen which is the currently visible part of the Boxer world.

also the form of the presentational medium, then programming becomes a learnable-in-tiny-increments and constantly useful extension to written communication, something with which children are in constant contact. A simple example of such modification is to add a new editing command, or to use a variable as a means of keeping around a template for electronic mail messages or other "forms." Hierarchical structure, boxes in boxes, which Boxer makes so prominent and easy to generate, can be used by children to organize and reorganize their personal computational world (Figure 6.4). Note that there need not be any such things as files and filing commands. Such use may seem trivial in view of the the power and subtlety of variables and hierarchical structure in the hands of expert programmers, but simple instantiations provide for continuous incremental advantage. More advanced use of Boxer's degree of integration are also possible. Because everything in the system is a computational structure, computing on any text (say, computing a reorganized format for a notebook) is simple to arrange.

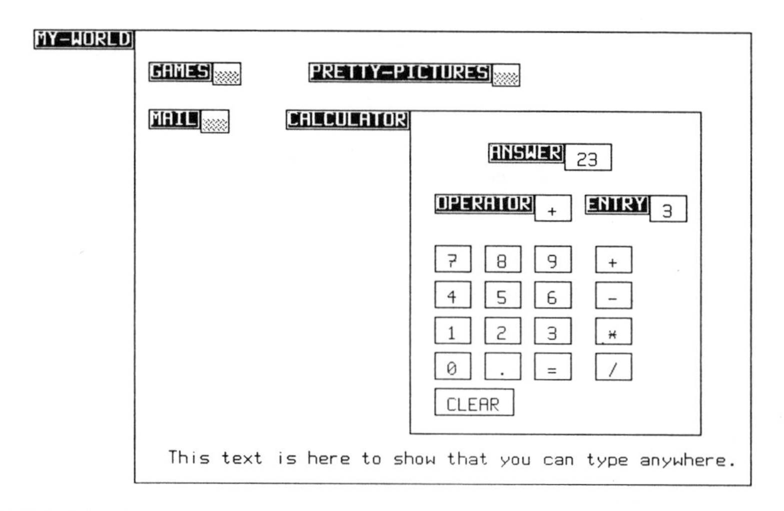

FIGURE 6.4. The top level of a Boxer world. Small gray boxes have been shrunk to hide detail. A subbox may be "entered" by expanding it (with a keypress) to full screen size.

In comparison to iconic programming or programming by direct manipulation, Boxer's text orientation may seem conservative, if not reactionary. But the main reason we chose to move in that direction was to promote a synergy between the written word, including all the incremental advantage it offers, and the advantage that programming and programming structure offers to written communication. Our judgment is that text will not fade away as a dominant medium, but will be transformed and improved by computation and programming.

To solidify this notion, consider a thought experiment. Imagine Boxer as a future publishing medium for educational materials. A student might buy a digital optical disc containing a Boxer book. Figure 6.5 shows a "page" (box) from that book. Box structure allows hierarchical presentation with details suppressed at each level by shrunken boxes. The table of contents of the book may *be* the book, viewed from the top level. Within a "section" (box), graphics boxes provide illustrations. Graphics boxes are like most boxes in that they can be made with a keystroke, expanded and shrunk, moved around or deleted like a large character. They are computational objects. They can be named, ported to, and referenced in programs. But graphics boxes also come equipped with a set of primitives for making and modifying pictures. As an example, generalized turtles—movable, touch sensitive (to the mouse and to each other) graphical objects—may be created in graphics boxes. (A paint program in Boxer is not

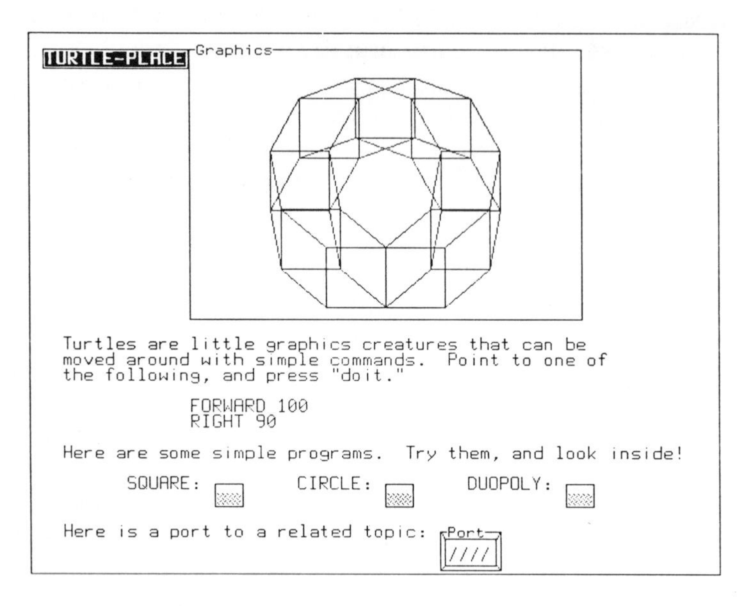

FIGURE 6.5. A page from a Boxer book.

viewed as defining graphics capability, but as an optional program written to simplify construction of a certain kind of data object, a graphics box, the way one might write a program to simplify construction of programs or more familiar data objects such as records.) So the illustrations may easily be dynamic, controlled by programs written into the text. The simulation programs, of course, are inspectable, and changeable, so students may experiment freely, learn from the way the simulation is written, and they may even clip pieces of the program out of the book for their own purposes, such as game program writing.

Citations to other sections of the book may be made through ports. Students may write annotations into the book. They may provide their own summary view of the book by collecting ports to selected sections into a box, together with their own notes. Any tables written into the text are also automatically data bases, available to be used for further analysis or computation.

Note how much work it would take to produce such a book. Basically, the author would just type in the text (including box organization), assemble pictures and type the programs that control the graphics or that provide particular facilities not built into Boxer. Instead, or in addition to programming, the author might hire a programmer or take programs from public domain Boxer books.

It is easy to imagine teachers making fragments of Boxer books, dynamic worksheets, in the course of a day's work. Undoubtedly, professionally produced books would be published containing fragments of programs and ideas the same way worksheets or lab kits that constitute a limited, but potentially very significant, part of today's learning culture for teachers.

To the level that I have described this hypothetical book, the standard Boxer interface is the means of construction, and, as well, the means of consumption and reconstruction. That this is true is an important characteristic of the kind of medium we want to have. One can step easily from the role of consumer to that of producer. The fact that Boxer has this property is in large part due to its visibility and principle of naive realism. Constructing an interface to an application is often trivial because, for example, output from a program can be produced simply by putting a program variable or port to one on the screen. Input can be equally trivial, since typing into that variable changes it.

The constructable interface. Naturally, for special purposes, one may want input and output farther from the default paradigm. We are in the process of designing a slice of Boxer that I call the constructable interface through which more specialized interactions can be made with incremental programming. Indeed, this is a new context of continuous incremental advantage for programming. In the past, programming environments like Logo, Basic, and Pascal have been conceived of as relatively or absolutely fixed, not seriously user adjustable. In Smalltalk, the user interface is totally redefinable, but the cost of this flexibility is that fiddling with the interface often can involve fiddling with complex, highly refined system code, or it might be a huge project of supplying a complete replacement for that code. Our hope is to steer a middle course that avoids these extremes and provides more realistic incremental advantage.

The constructable interface in Boxer at present has two parts. The first consists of "bells and whistles" and adjustable parameters on boxes with user interface functionality. For example, keystrokes and mouse clicks can be redefined to run arbitrary Boxer functions when performed within a given box, boxes can have demons associated with them to take definable actions on entry or exit of the cursor, and boxes can be frozen at any size (overridable by explicit "expand" or "shrink" commands) to allow selectable display of the interior of a box, not just all or none.

With definable mouse clicks alone, it takes about 10 minutes to make a screen-based calculator from scratch, where you point and click to press the buttons (boxes). An arbitrary amount of the internal structure of the calculator, registers and so on, can be shown. The surface form of the calculator shown in Figure 6.5 involves no graphics at all; it was simply typed in using the Boxer editor. More generally, we imagine users building or modifying little Visicalc-like interfaces for doing jobs like data presentation (graphing) and analysis (easy entry and sorting of data fragments).

The second part of the constructable interface implements more extreme changes in the interface. For this we use the interior of graphics boxes, which structurally do not allow precisely the same "text editor" interface as the rest of the system in any case. Thus, the interior of a graphics box will not have the very well-elaborated and universal interaction of the rest of Boxer, but default behaviors and a few appropriate high-level primitives to build interactions. Examples of the former are: (a) Graphics objects are automatically highlighted (selected) as the cursor passes over it; (b) clicking "expand" on an object expands it to show the data that define its state (position coordinates, special procedures to define behaviors like animation, or other change of visible presentation); (c) any Boxer code can be activated on selecting an object and pressing a mouse button or key. Examples of the latter include the process of following the mouse cursor, changing default behavior, etc.

I would like to close this section with two small, less parochial notes on the important topic of context. The first is about Prolog, or, more generally, about logic programming. Much has been made of the elegance and conciseness of such languages in expressing certain facts and relationships in a form that is executable. Just as much has been made about this as a proper way of using the high performance, multiple-processor hardware that is soon to be available. In relation to the complexity barrier, this is appropriate. In relation to the utility barrier, it is less so. I have already stated that Prolog does not obviously accept present technological advances that can improve understandability, such as integration of visible properties with semantics (e.g., Boxer's spatial metaphor) or presenting images of the computational mechanism in process. Arguments have been presented, in fact, to the contrary: that the invisibility of mechanism and relative uncontrollability of Prolog makes it difficult to program and, especially, to debug. But Prolog offers the glimmerings of a substantially different and exciting context for programming.

In Prolog, one makes relational assertions in a data base such as "John loves Mary," or "John is-a man." Rules of general value may be written, such as "If X is-a man, X is-a person." Then the data base may be queried by making variabilized statements like "X is-a person," which returns all values of X that make the statement true. Thus logic programming is in substantial ways about expertise, and the hope can be that Prolog or future versions will allow incrementally learnable advantage in "expertise about something" over common-sense reasoning or noncomputable "writing down facts and generalities." Ennals and Kowalski are pursuing this possibility where children may, in the end, play with and even write their own tiny expert systems (Ennals, 1983; Kowalski, 1979).

Finally, to close an open loop, anyone who develops a general device programming system may well provide an exciting new set of contexts in which to learn and appreciate programming. (This is not to slight present versions, like *Robot Odyssey*, but their aims are quite limited with respect to being a general medium.)

DIRECT MANIPULATION—WHERE'S THE PROGRAMMING?

The concept of direct manipulation as a future kind of programming has acquired a good deal of support in recent years, particularly because of its more attractive features in simplifying the programming process. (Chapter 5 by Hutchins, Hollan, & Norman discusses these hopes in its first sections. Readers would be advised to look over those sections before continuing here.) I would like to comment on the status of these hopes.

The plan for this section is to understand direct manipulation by situating it along the three dimensions of change in the meaning of programming: presentation, structure, and context. The overarching question is what exactly does direct manipulation have to do with programming as defined here? By and large, I shall assert that direct manipulation does not define or even suggest any major change in computational structure or context beyond present programming. So my major discussion will be on how direct manipulation can re-present programming. We shall find that direct manipulation has difficulty with particular aspects of functionality intrinsic to programming. There are limits that may not be evident at first sight on what we can expect direct manipulation to do for programming. The section concludes by rediscovering direct manipulation in what, given the perspective on the future of programming in this chapter, is a proper relation to programming.

Direct Manipulation of Standard Programming Objects

To begin, it is important to distinguish between tasks accomplished by direct manipulation. The task might be to operate on computational objects, which might be more or less conventional objects in a programming system. If this is the case, then what we are talking about is not a change in structure, but a change in presentation. Boxer is a good example of a system that permits direct manipulation on more or less conventional computational objects. Boxer makes all computational objects visible and manipulable in a uniform way. Adherence to naive realism assures that the visible representation and the underlying reality are minimally disparate. And the uniform text representation of all objects assures a rich manipulation language (the Boxer editor). Visible reality and a rich manipulation language bode well for a successful direct manipulation system in the straightforward sense of those terms. Naturally, I see an important future to such systems. But Boxer is not what direct manipulation advocates have in mind. It is structurally more or less a garden-variety, general purpose, computational system. It contains a lot of "code," and has such things as syntax errors.

One may, of course, try to turn procedural structure more or less directly into nontextual forms, like iconic programming, slot machine style. But in consideration of a simple-minded "conservation of complexity," this tack will have limited success compared to the prospect of eliminating error and providing instant accessibility to first-time users. Arguments below detail other limitations of this general line.

Manipulation as a Presentation of Programs

On the other hand, the task accomplished by manipulation might be not to construct computational objects, but to demonstrate or represent them. This is a considerably more optimistic image of what direct manipulation might bode compared to replacing words on a one for one basis with icons. But, again, this does not necessarily offer substantial changes in structure, but only presentation. If one considers paradigms such as procedural programming, logic, and constraint-based programming, then it appears that manipulation as a programming language belongs firmly in the procedural camp: The basic programming metaphor is demonstrating a sequence of actions.

How much of the procedural paradigm can direct manipulation carry? Here, the programming language community is in the midst of substantial research. One can point to such examples as Gould and Finzer's programming by rehearsal (Gould & Finzer, 1984), where routines are taught (programmed) by manually running through the sequence of actions that one wishes to teach. The actions are essentially all selecting from among preprogrammed actions of a troupe of computational actors. The system gets much of its character and power from the set of initially supplied actors. In its present state of development, the system doesn't feel like a general programming system; not even its implementers chose to build the supplied troupes with the system itself. My suspicion is that this will not give those who learn only the rehearsal level of the system enough flexibility for it to be widely adopted as a style of programming, that is, a reconstructible medium.

Halbert's (1984) programming by example system and Lieberman's Tinker (1984) attack more directly some of the fundamental problems of direct manipulation as a representation of programs. To understand what these are, we need to look at some of the fundamental functionalities that define procedural programming. This will put us in a better position to make general comments than examining particular systems. It also seems appropriate in a paper on the future of programming to say something in detail about what the essence of programming is, and what are the invariant structures we will find, in one disguised way or another, in any near-future programming system.

I will use a description of a general computational system provided by Newell (1980, recharacterized and slightly reorganized). Table 6.2 shows a taxonomy of functions. The four major areas, abstraction, representation, computation and I.O. appear in column 1. Column 2 contains the operators Newell uses to exemplify key subfunctionalities, and column 3 contains my characterizations of the subfunctionalities from column 2. The bracketed enumerations mark functions that may be problematic for direct manipulation, and they are discussed below.

Abstraction. Abstraction is the first functionality. In its simplest interpretation, abstraction means elevating from instances to classes. Variables and the literal/computed distinction are very simple mechanisms that provides much of this functionality to contemporary programming languages. But these pose a fundamental problem for direct manipulation. How do we express in the manipulative paradigm, which selected items are to be regarded as instances, which as classes, and along what dimension of generalization are we to define the class, if variabilization is intended? Much of Halbert's work is to solve this problem. Even this work assumes that the objects manipulated are the items to be abstracted, rather than the potentially much more difficult problem of abstracting on procedures. How does one say, "Look, do what I just did, but such and such a part of that was just an example of what might be done."

TABLE 6.2
TAXONOMY OF FUNCTIONS

Abstraction	*Quote*	controlling reference; variable/ literal distinction **[1]**
	Assign	naming **[2]**
	Copy	supporting isomorphism as well as identity
Representation [3]	*Read* symbol at position *R* from an expression	fetch
	Write to position *R* in an expression	mutate
Compute	*Do*	sequence
	Exit if ...	Control **[4]**
	Continue if ...	
I.O.	*Behave*	output
	Input	input

Naming. Abstraction also means having complex entities as units, substantially functioning objects in the system. Naming is at the core of this functionality. Here, direct manipulation is on a slippery slope. Pointing can be quite effective in some circumstances, but in order to have broad access to many entities, which is often needed in present computational systems, one would like to have some easily reproducible symbol. Once one allows this, however, text becomes such a powerful competitor, that it is unlikely any gestural or other method of specification could effectively compete. After all, text evolved over eons to serve precisely this role. Once one allows textual names, and has any sort of written representation for programs, then it seems to me a text-based programming is not far behind.

Even a seemingly strong point of manipulation, demonstrating sequence, begins to appear problematic without suitable abstraction mechanisms like mnemonic naming. For debugging purposes, reproduction of the gestures made to demonstrate a program is a very weak replacement for all the intention and context that existed in the head of the direct manipulation programmer. Not being able to use the expressiveness of language within a program seems a dubious improvement. Not being able to annotate beside the "code" seems unnecessarily restrictive. Not having a visible notation seems fatal. Debugging was a weak point of an early and ambitious piece of work that mixed manipulation with static program structures to represent programs (Smith, 1975). For reasons of debugging alone, I have much more hope for a hybrid text and direct manipulation system than for any attempts at pure manipulation systems.

Symbolic representation. If one thinks of representation (in the sense of building complex objects that model noncomputational systems through attributes or assertions) as a separate functionality, then again one seems to slide directly into symbolic presentations for the same reasons as for naming. It is plausible to construct these representations and write programs that manipulate them by direct manipulation, but easily reproducible symbols and variabilization are still issues. A strongly hybrid system is again the likely outcome.

Control. Control is a serious problem for manipulation. I know of no workable gestural indication of conditional branching. Compared to a simple IF ... THEN ..., it seems unnecessary to try. So once more, it seems some symbolic representation enters into the system with this functionality.

Direct Manipulation as a New Context

As far as contexts are concerned, direct manipulation does not by itself suggest anything new. It, of course, motivates a particular class of domains, in particular those where substantial interaction or processing can be assembled by organizing visible units. But that, by itself, is a weak heuristic.

Where have the high hopes for direct manipulation gone? Could it be that direct manipulation has nothing much to do with programming, but is instead a very general, thus weak, heuristic for constructing interfaces, or a trend in applications programs to draw icons in high resolution, to allow pointing, poking and moving those icons around in order to do a few things, but certainly not to program?

Let me sketch a class of ways to achieve most of what I believe realizable from direct manipulation while maintaining a true programming medium. To start, we would like a system which can support an "object metaphor" in the sense that graphical objects with a standardized set of manipulations on them such as *copy*, *move*, *connect* and *recognize a pointing operation*. This is meant to be the surface level of most "programs" written in the system. However, beneath this level must exist a full computational system to define the semantics of the objects and specialized meaning of their object manipulations. For example, the objects could open into data structures, programs or complete environments that contain multiple programs and data. So far so good, but the key is to establish a rich set of connections between the computational semantics and the object manipulations so that object manipulations are, in fact, reconstructing the computation. One must be able to bind interface operations to computational ones: touching to activation, connection to data flow, copy to copy, movement to context switching.

If all this sounds familiar, it should. This is precisely the intent of the constructible interface of Boxer, particularly its graphics box component. So, what I have in mind is opening up graphical objects, the generalized turtles described earlier, into Boxer structures that define them. Some of the connections between interface operations and computational ones are already in place in our current implementation. So the plausibility of these arguments on the relation of direct manipulation to programming will soon have a strong test. If we can do our design responsibly, Boxer should become, in part, a medium for building, using and modifying what everyone will recognize as direct manipulation systems.

Summing up. Direct manipulation does not appear to offer any substantial new structure or even, of its own, any new context for programming. Even as a different presentation of procedural or actor-based forms, there are several areas of functionality where text or other symbolic presentations are so strong, that at best we should expect a general purpose language to be strongly hybrid in ways akin to the relation of Boxer to its constructible interface. Indeed, I believe that direct manipulation of symbolic computational structures will become standard, especially for nonprofessionals. But except for a few very special cases, I do not believe direct manipulation will soon become even an acceptable substitute for symbolically presenting or representing computational structures.

I cannot claim any finality to this brief critique of direct manipulation. The line of argumentation presented here is subject to at least three criticisms, which I will very briefly counter. First, the functionalities abstracted may be abstracted too directly from present day programming so that the conclusion that form is hard to change is essentially built in. Independent of details, however, I believe these functionalities are indicative of a class that could be the basis for refined argument. Second, one may ask what difference it makes if direct manipulation is not programming. Flexibility and other virtues that we need in a medium are not synonymous with programming. Here, the more experience with particular systems, the better. When we have examined the empirical constraints on flexibility, etc., of many systems,

I think it may well be easier to convince the direct manipulation enthusiast that those problems fall into categories like those above.

Finally, invention has a way of revamping our assessments of possibility. Two examples of invention that might have impact here are the advent of quite substantial machine intelligence and radically enhanced input devices. Intelligence might, for example, figure out a programmer's intended level of generalization in pointing to an object, and represent it, even for the programmer's use, in other ways. Gestural input at a level significantly beyond pointing and poking (in concert with voice, intelligence, etc.) might give new meaning to manipulation. In comparison to the rich manipulation language of the physical world (consider a watch-maker's or a machine tool-maker's craft), the manipulation available with present machines is a travesty. See Buxton's contribution, Chapter 15. But these innovations are quite beyond the conservative, near future stance I took in this chapter.

SUMMARY

This chapter centers around the notion that a cluster, including ease of use, comprehensibility and perceived value—in short, utility—will be a major factor determining the extent to which programming will enter the lives of most people in the future. Although most chapters in this book deal, appropriately, with questions of "how" with respect to interfaces, I have attempted to bring to center stage the prior question of "interfaces to what?" The image of computation as an interactive medium of unprecedented breadth and power, with programming as the means of construction and reconstruction in this new medium, establishes a context in which we may set our goals and judge the results of our work.

The concept of utility that must be employed in this context is not simple. To succeed in enticing individuals and society at large to learn a complex and subtle device, that device must offer "continuous incremental advantage," motivation to take each step forward at each stage and at each level: from day one, to expertise; from immediate gratification to the noble goals of intellectual advance of civilization. In this regard, no individual or group of designers can pretend to substitute for the experiences and judgment of a hugely diverse society. But, while computer professionals are rapidly crystallizing computational environments out of their needs and aesthetics, few are attempting to give nonprofessionals a chance to experience a medium of minimally constrained possibilities. We must do better.

In changing programming from its present to future forms, we have three major dimensions of change at our disposal. We can alter the way we present computation to the user; we can alter the structures of computation themselves; or we can alter the contexts of application of these structures to do different things. While they are conceptually distinct, these are not independent dimensions. Each can influence the other. Means of presentation may select different structures as optimal; different structures may make new applications possible; and important applications may carry with them suggestions for presentation.

This chapter has looked at a span of near future changes to the meaning of programming. This yields some expected conclusions, and some more surprising. Programming is a relatively complex task that is not likely to become trivial through any near future change in any of these dimensions. On the other hand, there are several promising moves we can make that offer improved utility with changes along all three dimensions. I believe we can look forward to an exciting future for programming as a popular medium that can extend the reach and grasp of us all.

ACKNOWLEDGMENTS

This paper contains many ideas that have been developed and honed by a group of people associated with Logo and Boxer, a group too large to enumerate. Discussions with Brian Silverman about device programming have been provocative. The constructible interface in Boxer is being developed jointly with Jeremy Roschelle. The discussion of direct manipulation was stimulated by suggestions by Don Norman, Ed Hutchins, and Jim Hollan. The presentation of this chapter has been improved by suggestions from the Asilomar group, particularly Bill Mark, Don Norman, Steve Draper, Clayton Lewis, and Mike Eisenberg.

SECTION III

USERS' UNDERSTANDINGS

This section concerns the understanding users can have of a system. It starts with three chapters that attempt to characterize the *nature* of understanding relevant to a user of an interface: Chapters 7, 8, and 9 by Riley, Lewis, and Owen. Then we turn to diSessa and his focus upon the (noncomputational) sources of knowledge brought to bear by users. He discusses three types of understanding: structural, functional, and distributed. These three types of understanding differ in their inherent nature and in the applications to which they mainly relate. We conclude the section with a chapter that applies ideas on the nature of understanding to interface design: the chapter by Mark. This is only one dimension underlying the topic however. Others include the origin of users' understanding, its nature, and how to support it in system design.

Riley (Chapter 7) argues that knowledge relevant to the understanding of a topic has three major components: internal coherence, validity, and integration. Internal coherence reflects how well the components of the representation that is to be learned are related in an integrated structure, validity concerns how well the representation reflects the actual behavior of the system, and integration reflects to what extent the representation is tied to other components of a user's knowledge. Riley is concerned with the nature of the understanding that people form of the systems they are learning. A major issue is how to present the information so that it becomes understood—which is where the concepts of internal coherence, validity, and integration become important. Her other concern is with the robustness of the knowledge so formed, robust in the sense that it will not collapse in the face of unexpected outcomes.

The discussion by Lewis (Chapter 8) draws attention to the role of explanations and their construction in understanding and learning about a system. Lewis emphasizes false explanations, explanations that satisfy the learners that they truly understand and follow what is happening, but that in fact sometimes bear suprisingly little relationship to the truth. In some sense, Lewis shows what happens when the knowledge structures are *too* robust—so robust that they survive and prosper despite repeated evidence that they are erroneous.

Owen (Chapter 9) proposes that we should consider the idea of a "Naive Theory of Computation," analogous to the "naive physics" that underlies our understanding of physical events. This is another way of characterizing a kind of informal knowledge underlying a user's practical understanding. He also stresses that the problem of how to help users acquire it may be by making things visible: by "Answers First then Questions" (a suggestion he follows up in detail in Chapter 17) and by Direct Manipulation or Visible Computation design techniques that make a system self-revealing, rather than relying on users taking the initiative and asking for explanations.

DiSessa (Chapter 10) offers a typology of the kinds of understanding users may have—"structural," "functional," and "distributed." He stresses that understanding is, in practice, often much less systematic than may be presupposed by researchers, and the examples he gives of each model type illustrate this. DiSessa also shows that the common notion in Human–Computer Interaction of a "mental model" is much too simple: There are different kinds of models, each kind having quite different functional and computational properties.

Several chapters in the book talk about the need for a good, consistent design model and the need for it to be reflected in a consistent, coherent System Image (see especially Chapters 3 by Norman and 7 by Riley). Mark (Chapter 11) accepts this challenge, but points out that this consistency is difficult to achieve unaided, especially when the system to be designed is large and complex, involving a team of programmers working over a period measured in months or years. As a result, he proposes a design tool, one based on modern knowledge representation techniques from Artificial Intelligence (the knowledge representation language KL-ONE by Brachman, 1978).

Mark starts from the assumption that a consistent system image is a good thing. He describes a system that helps programmers develop an explicit and consistent "system image model"—a formal model of the behavior of a system as observable from the user interface. The subsequent implementation is then constrained by requiring programmers to link each concept in the model to a specific software data object: The relationship between model concepts must hold between the software objects. The knowledge representation technique is used to define similarity of concepts as a basis for consistency in the interface, and its implementation gives a support system to allow designers to explore consequences of choices of concepts. It can furthermore support the interpretation of user inputs.

Bootstrapping

Many of the chapters in this section touch on the concept of *bootstrapping* (see glossary) when dealing with how new users can get started at acquiring understanding. The phrase "pulling yourself up by your own bootstraps" applies with some force to the problem facing new users. It is of course physically impossible to lift oneself by one's own bootstraps, and there is a corresponding philosophical problem of how one can come to know anything without first already knowing something. However, many of the problems that new users face hinge upon the fact that in order to understand the new concepts they have been exposed to, they need already to understand part of those concepts, or at least already to have a framework for acquiring and anchoring the new concept. All the chapters in this section are concerned with the kinds of understanding that users must and do acquire, but it is the consideration of new users that forces attention to the issue of "bootstrapping." For instance, Owen and Lewis, in different ways, are both concerned with the way users interpret their experience of events (and "errors") at the interface. On the one hand, experience can lead to new understanding and a change in concepts. On the other hand, the interpretation of events, and hence the possibility of new understanding, depends in part on the user's existing concepts. Owen suggests that the first seed of a new piece of understanding must be delivered "Answers first, then questions": The initial information must be delivered as an unsolicited display, because the user doesn't yet know enough to formulate and pursue precise questions.

CHAPTER 7

User Understanding

MARY S. RILEY

My goal in this chapter is to explore how much understanding a user needs to perform skillfully. My interest in the relationship between understanding and skilled performance is currently focused on two areas: basic electronics and human-computer interfaces.

In the study of how people learn basic electronics, I am analyzing how theoretical understanding of physics might influence skilled performance in analyzing and troubleshooting electronic circuits. A traditional assumption is that formal physics is necessary for skilled task performance. Yet, many skilled electronics technicians probably do not use formal physics to perform these tasks, and some may have had no formal training at all. However, the issue is complicated. Electronic technicians spend years of apprenticeship and practice. Perhaps knowledge of formal physics provides a more direct way to develop understanding early in learning. Just because formal physics may not show up explicitly in skilled performance does not mean it would not facilitate learning.

My interest in human-computer interactions focuses on how much understanding is required to become a skilled user. Clearly, understanding must be defined relative to the task or set of tasks the user wants to perform. Someone who is using a text editor to write a paper needs to understand commands at a very different level than someone who intends to modify the editor. To the person who uses the editor only in order to write a paper, a detailed understanding of the principles of systems design would probably have little beneficial effect; it may even be harmful if the ideas distracted the user.

We must also consider tasks other than the immediate ones the user will perform: users also do experiments; they make mistakes; at times the system fails. For a user to be able to interpret the effects of an experiment or the nature of a mistake requires additional understanding. Even more understanding is required to distinguish between the behavior of a functioning system and a nonfunctioning system. Furthermore, with the increasing proliferation of computers, most users need to learn several systems. We must be concerned with identifying the kinds of understanding a user needs to transfer skills between systems.

In the next two sections I focus on a framework for characterizing user understanding and discuss the role of understanding in performance and learning. I propose that (a) the level at which a user interacts with a device is determined by the tasks being performed; (b) the functions and structures that are understood differ from level to level (see Miyake, in press); (c) a uniform set of criteria are appropriate for evaluating understanding at any level.

My discussion of understanding is informal, focusing primarily on examples from human-computer interfaces to illustrate the various points. Cognitive theories are being developed that more accurately specify the nature of understanding. The framework for characterizing understanding used here is based on current theories of problem solving and language understanding. Ideally, analyses of user understanding will reach a similar level of description, maximizing the extent to which work on human-computer interfaces can benefit from, and contribute to, other developing theories.

A FRAMEWORK FOR CHARACTERIZING UNDERSTANDING

I view the user's task in interacting with a complex device as a problem-solving episode. The user constantly sets goals and must plan how to achieve these goals with the available commands. This view is generally consistent with several other current analyses of human-computer interactions (e.g., Card, Moran, & Newell, 1983; Kieras & Polson, 1983; Moran, 1983; Norman, Chapter 3).

Figure 7.1 presents a typical planning episode in the form of a hierarchical goal structure, or *planning net* (cf. Sacerdoti, 1977;

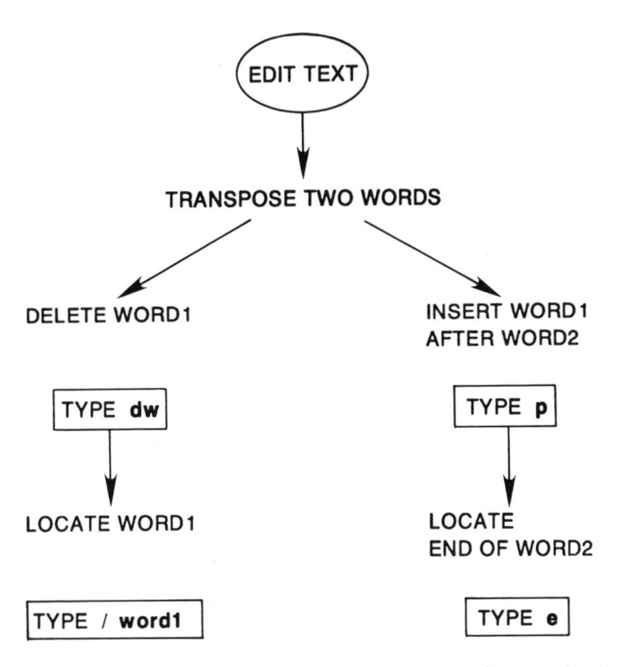

FIGURE 7.1. Planning net for the task "*transpose two words*" (from Riley & O'Malley, 1985).

VanLehn & Brown, 1980). At the higher levels of the planning net are global goals. Here the person is using a text editor: The current goal is to edit the paper. This, in turn, generates the additional goal to "*transpose two words.*" Since this goal does not correspond to an executable action, further goal specification and planning is required. "*Transpose two words*" is broken down into the subgoals "*delete word1*" and "*insert word1 after word2*" which correspond to the actions of typing "*dw*" (delete word) and "*p*" (put), respectively.

Planning does not necessarily stop with the selection of the primary actions. Associated with actions are requisite conditions that must be taken into account in the planning process:

- *Prerequisites* are conditions that must be satisfied before an action can be performed. The prerequisite of "dw" and "p" is that the cursor be at the appropriate location. Additional goals must be generated to ensure that these prerequisites are satisfied.

- *Consequences* are the changes that result from performing an action. In this example, the consequence of "dw" is that the word is deleted from the text. The consequence of "p" is to put the most recently deleted word at the location of the cursor. These consequences define the *order* in which "dw" and "p" must be executed and, furthermore, place restrictions on interleaving plans.

- *Postrequisites*, relevant to some commands, are conditions that must be satisfied after performing an action. For example, the action of inserting text must be followed by pressing the ESCAPE key, to return to command mode.

It is one thing to solve a problem, but quite another to solve it with "understanding." As Greeno (1977, 1978) suggests,[1] understanding depends on three important criteria for evaluating the representation generated during problem solving:

Internal Coherence.
Are the components of the user's representation mutually coherent?

Validity.
Does the representation accurately reflect the behavior of the system?

Integration.
To what extent is the user's representation integrated with the user's knowledge of other areas?

Of course, devices can be described at different levels. Devices can be hierarchically decomposed into structures with each structure serving one or more function at a level of the hierarchy (cf. Brown, Burton, & de Kleer, 1983; Miyake, in press). For example, at one level of description, we can talk about editors, mail systems, and directories; at another level we can talk about the underlying programs; at yet another level we can talk about shuffling bits. The objects differ at different levels, as do the commands and procedures available for operating on those objects. Thus, depending on the particular task a user wishes to accomplish, the objects in the user's problem representation will differ. Nevertheless we can use the same criteria of understanding to evaluate a user's problem representation.

[1] Greeno used the terms "coherence," "correspondence," and "correctedness."

Furthermore, understanding continues to be important regardless of the amount of experience. True, skilled performance of routine tasks probably involves having a large store of automated procedures for accomplishing frequent goals, but not all of a user's interactions with a computer are routine. Among other things, users experiment, make mistakes, and may also have to regenerate procedures that were once automated but forgotten through disuse. For these reasons, a user needs to have more knowledge about a system than simply a list of procedures for accomplishing specific tasks.

WHAT KIND OF UNDERSTANDING DOES A USER NEED?

In this section I discuss in more detail the three criteria for understanding and their roles in performance and learning.

Internal Coherence

Internal Coherence is the extent to which components of knowledge are related in an integrated structure. I argue that a coherent knowledge base facilitates learning and increases the likelihood that commands will be remembered or can be regenerated.

There are many kinds of knowledge that contribute to internal coherence:

- Knowledge about the action structure of a command;
- Knowledge about the syntactic structure of a command;
- Knowledge about how a command works;
- Knowledge about objects (e.g., files, programs, buffers);

The knowledge of the action structure of commands is *coherent* in that goals are associated with a command's requisite conditions and component actions. Furthermore, during planning, this knowledge can be used to generate a coherent representation of the hierarchical relations between related sequences of commands. In Figure 7.1, there is a central high-level goal, "*transpose two words,*" and commands are related to that goal as either primary or enabling goals.

Improved performance (i.e., increased flexibility and efficiency) could be achieved by memorizing additional goal-command pairs. But still, important components of understanding would be missing—components that could have an important influence on performance and learning. Improved understanding could be achieved with the addition of knowledge about the *syntactic structure* of commands, as shown in Figure 7.2.

In the particular editor used in this example (Berkeley UNIX *vi*), command sequences are defined as the cross-product of an action and a text object: "dw" is the cross-product of the action "d" (for delete) and the object "w" (for word). Furthermore, the text object can be preceded by a number, e.g., "d4w" deletes four words. Other rules involve systematic changes in the scope or direction of a command. Whatever the form of these rules, they add *coherence* to the knowledge, facilitating learning and increasing the likelihood that commands will be remembered or can be regenerated.

> *Note that the coherence of a user's representation depends largely on the system itself. If there is no high-level rule to describe the syntactic structure of a command language, the user is restricted in forming a coherent representation of the relation between command sequences. Indeed, several studies (Payne & Green, 1983; Reisner, 1981) have linked the degree of syntactic coherence to users' ability to learn, remember, and regenerate commands.*

Knowledge about how a command works also adds internal coherence

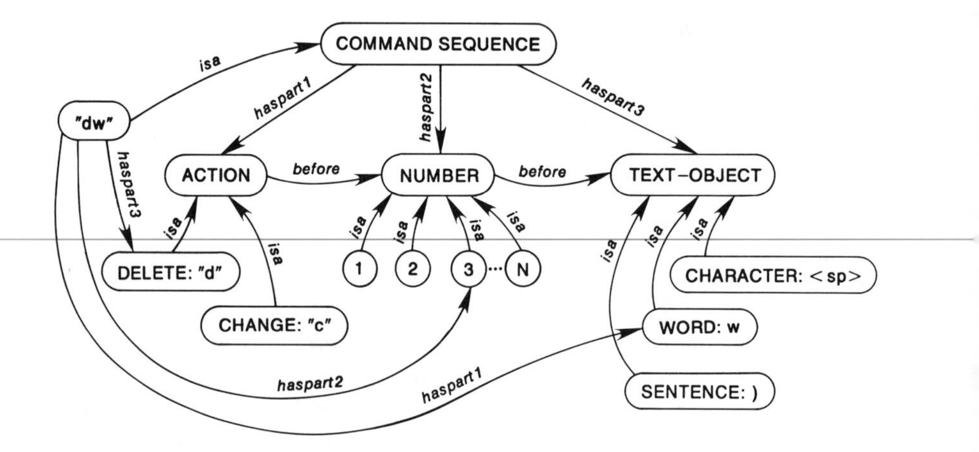

FIGURE 7.2. Knowledge about the syntactic structure of commands.

to the user's representation of the editor command structure. For example, in Figure 7.1, the consequence of deleting a word is not related directly to the action of putting the word in a new location. That is, there is no explicit representation of the relation between the consequence of "dw" and the prerequisite of "p." In fact, the consequence of "dw" is to place the word in a temporary storage place, or buffer. The prerequisite of "p" is that there be something in the buffer and its consequence is to put the buffer contents at the location of the cursor. Figure 7.3 shows how this additional knowledge adds coherence to the user's representation by explicitly identifying the relation between the consequence of "dw" and the prerequisite of "p."

Other kinds of knowledge also contribute to coherence. Most editors keep the modified version of the text in active memory and do not update the permanent copy on the user's disk until the user leaves the editor. In order to understand this fact, the user has to know that the editor is a program and be familiar with concepts of memory and disk storage (see Kieras & Polson, 1983, for further discussion). Users also need to know about the properties of the objects that are operated on by these programs and commands. A coherent representation of why the command to edit a file is sometimes preceeded by a command to change directories and/or a command to change the file protection depends on the user knowing that files are organized in directories

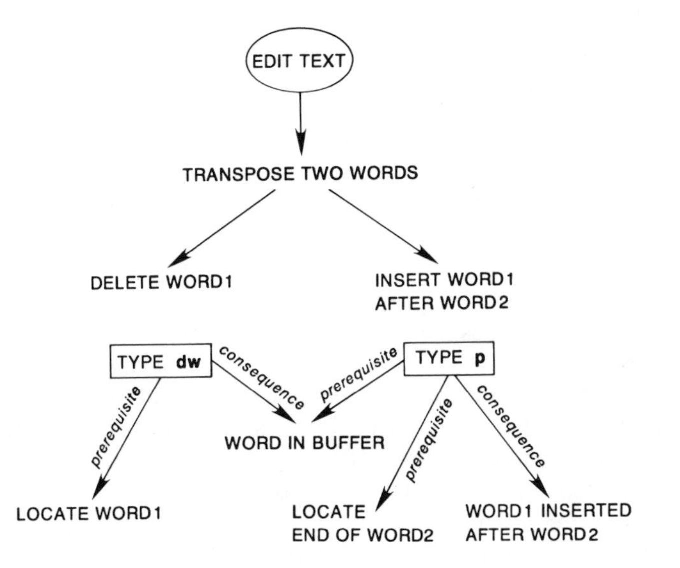

FIGURE 7.3. Knowledge about buffers.

(which are also files) and that each file has a protection status that is checked by the editor program. A coherent representation of the effects of editor commands requires knowledge of how text objects are specified, e.g., the different notions of a "line" discussed by Owen (Chapter 9).

An interesting issue concerns whether there is a tradeoff between internal coherence and learnability. Coming up with a single coherent view of a system may involve structural models and rules that are quite difficult to learn. Multiple "distributed" models may be less coherent overall but may be easier to learn: See diSessa's discussion of this point in Chapter 10.

Another issue concerns whether knowledge at any level of description increases coherence. Would an explanation of the editor in elaborate programming terms or design terms increase coherence? Or would it simply provide an alternative perspective (see Miyake, in press), without directly affecting the connections between the objects of concern? That is, this additional knowledge about underlying mechanisms and design constraints may add more nodes and links to the knowledge representation, but it is not clear that the density (i.e., coherence) of the representation would increase.

Several studies (e.g., Halasz & Moran, 1983; Kieras & Bovair, 1984) show that having some model of how a device works facilitates learning, retention and/or invention of procedures for operating a device. However, the benefits of a device model depend on whether it allows the user to infer the exact steps required to operate the device. Specifically, inferring procedures requires information about the system topology (what is connected to what) and the principle of flow of control. Thus the critical "how-it-works" information is the specific description of the controls and their path relations to the internal components. Neither details about the nature of the components, nor general principles about how a system works enable users to infer procedures. How-it-works information must be selected so that it is actually relevant to the user's task—it must explain how or why a goal must be accomplished.

Notice that this does not constitute a general argument against the role of formal principles in learning complex skills. In basic electronics, physics principles can be used to constrain the quantities in the problem representation, adding coherence. Physics principles also provide a way of relating different procedures, thereby embedding them in a higher-order organized structure that could be beneficial for retention and application. (See Riley, 1984, for a detailed discussion of the role of formal principles in learning basic electronics.)

Validity

Validity is the extent to which the user's components of knowledge are consistent with the behavior of the system. To the extent that the user's knowledge corresponds to the system's behavior, the user will be able to *explain* and *predict* the effects of commands and *generate* new sequences of commands to achieve desired effects. I identify different kinds of validity that are roughly equivalent to diSessa's distinction between functional models, distributed models, and structural models (see Chapter 10).

Distributed and structural models correspond to how-it-works knowledge at varying levels of description. How-it-works knowledge enables users to predict the behavior of any sequence of commands, justify rules when they are correct, explain their limitations, and go beyond them. However, as diSessa (Chapter 10) and Young (1981, 1983) point out, structural models are not always sufficient to enable fluent interaction with the system. A significant amount of problem solving may be required to invent a way to achieve a goal. It seems highly unlikely that users would continually rederive procedures through explicit reference to models and rules. For example, when transposing two letters for the 100th time it is unlikely that the user explicitly represents the letter going into the buffer and then being moved from the buffer to a new location in the text. Rather, the keystrokes corresponding to this procedure are probably done automatically, as a unit (cf. Robertson & Black, 1983).

> *Note that validity does not depend on whether the user's models are identical to (contain same objects and relations as) the Design Model, as long as they lead to the same predictions. At the same time, if correspondence is maintained as a user's knowledge encompasses more and more of the system, the space of functionally equivalent models probably decreases.*

Users often generate context-dependent validity to achieve coherence at the expense of accuracy. Lewis (Chapter 8) shows how users generate coherent explanations of a command's function in a specific context but these explanations often do not predict the command's function in another context (see de Kleer & Brown, 1983). Robust models contain no implicit, context-specific assumptions about a command's function. Such models should predict the command's effects and functions, regardless of the particular command string it is embedded in. Robust models correspond to the behavior of the system for sequences of commands that were unanticipated and therefore not precompiled.

Consider robustness in the context of the buffer model. Knowledge about which commands change the contents of the buffer and which commands access the contents of the buffer suffices to predict and explain the effects of many command sequences. For example, the command for changing a word, "cw," also has the consequence of replacing the contents of the buffer with the word that was changed (and therefore deleted). Following a "cw" event with the command "p" has the effect of putting the changed word at the current location of the cursor. If "dw" were executed between "cw" and "p," the model correctly predicts that consequence of "p" would be different.

An interesting issue concerns the degree of robustness acquired with experience. I use diSessa's terms "structural" and "distributed" to characterize two different perspectives. de Kleer and Brown (1983) suggest that with experience, users acquire increasingly robust device knowledge—approaching what diSessa refers to as "structural models." Initial component models invariably include many implicit assumptions about the overall functioning of the device, which may or may not be correct. With experience, component models become more robust by making implicit assumptions explicit. This transition is primarily motivated by discovering violations of consistency, validity, and/or robustness constraints. That is, the causal model may contain conflicting models for a single component type, the causal model may not correspond to the observed behavior of the device, or the causal model may only correspond when the device is functioning correctly. Discoveries like these encourage the learner to identify an underlying implicit assumption in a component model, gradually making the component models more robust. These assumptions (and future ones) need not be discarded—they lead to very efficient reasoning about the correctly functioning device from which they originated. Thus with experience, the learner probably stores more and more device-specific assumptions along with the component models. These new component models are robust in that, if assumptions are violated, the learner can automatically distinguish these assumptions from the actual model and proceed to envision a correct causal model of the device.

I extrapolate from diSessa's notion of distributed models to propose an alternative perspective. It is possible that users continue to have many models to account for different aspects of the system and no single model is robust (in the strong sense used by de Kleer and Brown). Does an expert electrician ever acquire robust models of the circuits being tested? Does an auto mechanic ever acquire a robust model of a car engine? What about operators of steam plants and nuclear power plants? Often there may be no single model that is perfect for describing the behavior of the system (cf. Bott, 1979; Rumelhart & Norman, 1981). The distributed perspective says that improvements in skill result from learning to apply the right model at the right time and perhaps refining models with context-specific knowledge.

It is likely that the distributed view of learning describes most users. For one thing, inferring a robust causal model is difficult, if not impossible, even for relatively simple devices (cf. Miyake, in press; Young, 1981). Young's (1981) account of how he derived a robust model of the stack calculator shows the importance of systematically generating, testing, and revising hypotheses. In contrast, many users probably form hypotheses on the fly on the basis of isolated examples, fail to experiment systematically, or may be afraid to experiment (see Lewis' Chapter 8 for further discussion). Especially when a detailed model is not required to operate a device, users probably do not take the time to generate one. One of Miyake's subjects had used a sewing machine for years without generating a detailed model of how the machine worked—although when forced, she generated and refined a very robust model.

There are several reasons why understanding more about the kinds of models users generate is interesting from a design perspective. Knowing more about the kinds of models people develop with experience would tell us something about the limits on the complexity of the models with which humans reason—or at least the kinds of models with which they prefer to reason. It would also provide us with information about how multiple mental models are coordinated in learning and performance. Furthermore, information about how those models are acquired, modified, or abandoned, may be useful in guiding users through progressive layers of competency.

Integration

Integration is the extent to which the components of knowledge in one domain are tied to other components of a person's knowledge. Ease of learning depends on the extent that objects and relations in a new domain can be connected to familiar components of knowledge.

One kind of integration that is especially important for new users is integration with general knowledge. For example, many editor goals are *connected* with goals familiar to the user—changing and deleting characters and words, inserting new text, correspond to general goals in editing text, regardless of whether the text is handwritten, typewritten, or written using a computer text editor.

> *However, the semantics of the actions are not necessarily the same. For example, on a typewriter, the actions associated with Change-word involve erasing the old word, making space for the new word, and then typing in the new word. The goals "Transpose-Two-Letters" and "Repeat-Last-Command" have no direct counterpart in typewriting or handwriting (but transpose-two-words is used in proof-reading; also, these actions are frequently carried out in several steps, e.g., with white-out). See Owen's discussion of the same issue (in Chapter 9).*

A second kind of integration is with other knowledge of the other systems. Buffers and cross-product rules are not specific to this particular text editor or to text editors in general. At the right level of description, these concepts can be *connected* to other systems, leading to efficient transfer of knowledge. For example, most editors use buffers to store text that has been recently deleted or inserted. Even though editors may differ in the specific ways they use buffers, simply knowing that buffers are used constrains hypotheses, and explanations.

Integration is not necessarily beneficial. As Lewis points out (Chapter 8) learners unfamiliar with a domain often make connections that are not valid, resulting in inefficient performance and errors. [2] Users learning to use a text editor for the first time connect commands for inserting text to their knowledge of inserting text using a typewriter (or to their knowledge of inserting text using paper and pencil). As a result, users often think that space has to be made before text can be inserted. Similarly, users think that any text visible on the screen is in the file and, vice versa, any text not visible on the screen is not in the file. This leads to predictable confusion in editors where inserting text has the consequence of typing over existing text until a special key is pressed to terminate the input mode. Subjects are likely to think that the over-typed text is no longer in the file. Predictable confusion also results in editors that leave deleted text on the screen during input mode, even though the user has backspaced over the text to delete it. In this case the user is likely to think that the text has not been deleted, when in fact it has been.

A major problem with naive models is that they are surprisingly persistent (cf. Clement, 1983; diSessa, 1983). Among the frequently suggested reasons for the persistence of these naive models are that (a)

---

[2] (cf. Bott, 1979; Douglas & Moran, 1983; Halasz & Moran, 1982; Lewis & Mack, 1982; Riley & O'Malley, 1985; Rumelhart & Norman, 1981).

students may misinterpret or distort information to fit their naive views; (b) students may have several models for different instances of the same phenomenon and shift between models to interpret the various situations; (c) students may focus only on the salient aspects of an event and ignore less salient (or invisible) factors. Clearly, further empirical and theoretical analyses are required to identify the cognitive processes that could lead to necessary changes in user's models.

SUMMARY

Understanding facilitates learning, provides predictive and explanatory power, increases the likelihood that procedures will be remembered or can be regenerated, and enables the transfer of skills. In unfamiliar situations, understanding improves the efficiency, flexibility, and reliability of performance, permits and constrains generation of new procedures, and facilitates checking answers.

A direct outcome of the analyses presented in this chapter is a view of understanding as a multidimensional quality rather than as something one has or one does not have. Understanding is related to three characteristics of the user's knowledge: internal coherence, validity, and integration. Coherence concerns the degree to which the user's components of knowledge are related in an integrated structure. Validity concerns the extent to which the user's components of knowledge accurately reflect the behavior of the system. Integration concerns the degree to which the components of knowledge are related to other components of user's knowledge.

The degree of internal coherence, validity, and integration does not depend on a single aspect of knowledge, but upon several. This emphasizes that a user should not be considered as either performing with or without understanding, since it clearly is possible for the user to have acquired some components of knowledge and not others.

CHAPTER 8

Understanding What's Happening in System Interactions

CLAYTON LEWIS

Explanation-Building as Part of Learning

In a series of studies of how people learn word-processing (Lewis & Mack, 1982; Mack, Lewis, & Carroll, 1983) we noticed that learners were able to devise explanations which made the effects of even disastrous errors seem reasonable. As a result, they would sometimes continue working without any attempt to correct errors, since they had explained them away.

One learner misread the instructions for a typing excercise and instead of entering text, simply did a series of cursor movements. The result was to move the cursor around an empty screen, a result quite different from that described in the instructions. The learner was puzzled by the fact that there was no text, but concluded that the cursor movements were part of a process of creating a template into which text would later be entered. As a result, she concluded that she was doing the operation correctly.

Other episodes reveal other confusions. A particularly interesting case is illustrated in Figure 8.1. Here, the learner got out of synchronization with the system. In this example, the learner was attempting to generate form letters. The letter had variables inside it (which allowed the form letter to be "personalized," addressing the recipient by name

Visible Actions by Learner (L) & System (S)	The System's View	The Learner's View
	LETTER 1 BEGINS	LETTER 1 BEGINS
L: Issue PRINT command	Learner wants to start letter 1	letter 1 is printed
		LETTER 1 ENDS
		LETTER 2 BEGINS
S: Ask user about variables	Request user to supply values for variables for letter 1	System wishes to start letter 2
L: Provide Requested Information	System now prints letter 1	
	LETTER 1 ENDS	
	LETTER 2 BEGINS	
L: Issue PRINT command	Learner wants to start letter 2	letter 2 will now be printed
		LETTER 2 ENDS
		LETTER 3 BEGINS
S: Ask user about variables	Request user to supply values for variables for letter 2	System wishes to start letter 3
L: Provide Requested Information	System now prints letter 2	
	LETTER 2 ENDS	
	LETTER 3 BEGINS	
L: Issue PRINT command	Learner wants to start letter 3	letter 3 will now be printed
		LETTER 3 ENDS
		LETTER 4 BEGINS
S: Ask user about variables	Request user to supply values for variables for letter 3	System wishes to start letter 4
• • •	• • •	• • •

FIGURE 8.1. Two views of the same interaction. The left column shows the visible actions. The middle column shows the system's point of view, and the right column shows the learner's interpretation. Note that the actual printing of the letter is done on a remote printer and is not visible to the user.

and putting the recipient's name and address in the salutation). When the learner issued the PRINT command, the system began a dialog culminating in the printing of the letter, but first eliciting the recipient's name and address. However, the learner thought that the actual printing of the letter occurred as soon as the PRINT command was issued, so that the request for a name and address made no sense in this context. The learner therefore decided that the request must be referring to a new copy of the form letter. Therefore, after responding to the system's requests, the learner issued another PRINT command. This caused another round of prompting, which the learner capped with another PRINT, and so forth. The learner's explanation was that (for some reason) the system was spontaneously continuing the exercise by issuing more prompts. The learner worried aloud about how the sequence would end. Figure 8.1 presents a summary of this sequence of events and the learner's interpretation of it.

Such episodes suggest that explanations may be an important mediating structure in learning to use a system. Rather than translating the events of an exercise directly into a prescription for how to perform a task, it appears that learners may build an explanation of the events. The explanation indicates how their actions contribute to the goal of the exercise, and indicates what has caused the system to respond in the way it does. During subsequent tasks the learner could refer to this analysis to find what actions might lead to the accomplishment of a given goal.

Who Cares?

Understanding how these explanations are built is important for these reasons:

- The explanation process appears to play a role in failure to detect errors: Learners are able to construct a satisfactory (to them) explanation of events that actually are reflections of serious problems (Norman, 1984b, 1986). We might be able to reduce the incidence of undetected errors if we could control or anticipate the workings of the explanation process.
- Some sequences of events are easier to explain, or require less knowledge to explain, than others. If we understood this we could design systems that are easier to understand.

- Explanation-building probably involves some top-down processing, in which users ideas of how goals might be achieved influences the explanations they build. Better knowledge of this process could enable us to anticipate and avoid problems due to violated expectations.

- Since more is happening than could ever be explained, explanation involves selection of what to explain. If we knew how selection is done we could do a better job of designing the information that systems provide, so that explanation-building is made easier and more accurate.

What Is an Explanation?

An explanation tells why something happened. In the context of explaining a computer interaction, an explanation is a description of a sequence of events, including what the user did and how the system responded, that indicates what the user actions accomplished and what caused the system to do what it did.

In a meaningful interaction, the user does things in order to accomplish some goal. Therefore, to explain a user action is to say how it contributes to the goal of the sequence of activities of which it is part.

The system's responses occur either as a result of earlier events or as spontaneous happenings: The system just does something. In this latter case, the whole sequence makes sense only if one can see how the spontaneous system action contributes to the goal, just as is necessary for user actions.

Relating actions to goals. In considering a given goal the user is likely to have some idea of what must be done to accomplish it. An experienced user will have detailed knowledge of the actions that are needed. A novice may have no specific notion of what must be done, but may still be able to guess the outline of the required procedure using very general ideas about how the interaction might work.

Consider a novice using a graphics editor. When told that the editor can be used to move an object from one place on the screen to another, the novice can guess the following, using the basic idea that the computer cannot know what the user wants without the user telling it: It will be necessary to designate somehow the object to be moved, where it is to be moved, and that moving is the desired operation. I call the elements of this kind of analysis, or the more complete analysis of a more experienced user, *goal requirements.* A user action, or a spontaneous system action contributes to the goal requirements for the goal of the overall intention.

Note that the user's idea of the goal requirements may well be inaccurate. The analysis of the move operation just given, for example, does not apply to all systems. In the sequence called "cut and paste," there is no event corresponding to telling the system that "move" is the desired action. Instead, the system is told separately to "cut" the object, that is, to move it to a holding area off the screen, and then to "paste" it, that is, to move it from the holding area to a new location on the screen. A user trying to understand a "cut and paste" sequence using the basic analysis of "move" would encounter a mismatch: The designation of either "cut" or "paste" would seem superfluous, and hence not explained.

I can identify at least three ways in which actions can contribute to goal requirements:

- An action can *cause* a goal requirement to be satisfied, as when the action of pressing the POWER button causes a machine to be turned on.

- An action can *contribute* to a goal requirement, as when designating an object enables the system to do something to that object as distinguished from other objects.

- An action can *form part of* a goal requirement, as when a mouse click is part of a sequence of actions that is interpreted by the system as designating an object.

In establishing these kinds of connections between actions and goal requirements the user must rely on knowledge about when such relationships are likely to hold. For example, realizing that pressing the POWER button turns on the system relies on experience, something users do not always have.

Detecting which actions *contribute* to a goal often seems to be based on noticing identities. If an action features some entity, and the same entity figures in a later event, it is plausible that the first action supplied the identity of the entity to the event. Thus pointing at an object could contribute the identity of the object to a later event in which the system highlights the object, for example. Temporal proximity seems also to be important: If the pointing and highlighting were separated by much other activity it might seem less likely that they were connected.

Seeing when an action is part of a larger event also requires knowledge. For example, issuing a command in some systems is done with a pop-up menu. One presses and holds down a mouse button (which causes a menu of items to appear on the screen). To make the selection, one moves the mouse so that the word for the desired command is highlighted and then releases the mouse button. While it might be possible for a novice to interpret the elements of that sequence by identifying causal and information-providing connections among them, the experienced user can identify each simple action as part of a stereotyped sequence, and explain all of them in terms of what the sequence does, thus greatly simplifying the process of building the explanation.

An action need not be connected directly to a goal requirement to be satisfactorily explained, but may be connected via one or more intervening events. If action A contributed information to event B, which in turn causes goal requirement G to be satisfied, the point of A is clearly established.

Assigning causes to system actions. In addition to justifying user actions, a complete explanation must establish the origin of things the system does. Knowledge of possible causal connections, as used to explain user actions, can be used to assert that the system did something as a result of some earlier event. Knowledge of information flow, of when one event contributes information to another, also plays a part in explaining how a system action comes about. When the system highlights an object, it is likely, but not certain, that the identity of the object was supplied by some earlier user action.

When a cause for a system action cannot be found, or when information is used in a system action that has not been provided by an earlier event, the action, or the choice of information, must be regarded as spontaneous. As with user actions the explanation must motivate such spontaneous events. Why did the system do what it did, given the goal? All of the knowledge that is available to connect user actions to goals is applicable here.

Building Explanations

This analysis suggests the following picture of the overall process of building an explanation of an interaction. Assume that a learner is viewing a demonstration, or is working an exercise from a training manual and the overall goal of the interaction is stated. The learner develops a set of goal requirements, and attempts to identify connections between the actions and the goal requirements, using the kinds of knowledge just described. The learner also tries to establish the causes of system actions, or, failing that, tries to determine how spontaneous system actions connect to the goal requirements.

An explanation is complete if all events are appropriately connected to the goal requirements, or if the events have causes assigned from among earlier events that are themselves appropriately explained.

Building a complete explanation may require the learner to construct events that were not directly observed. Consider, for example, the learner who explained moving the cursor around the empty screen as "creating a template." This imaginary event served to connect the observed actions of moving the cursor with the assumed goal of typing a letter. Without it the sequence of actions made no sense (as indeed it should not have).

Just as knowledge of likely causal connections, information flows, and stereotyped sequences are used in building explanations, so it seems likely that building explanations can help to create or extend this knowledge. If the learner is viewing a demonstration which can be explained except for some missing connections, the missing connections can be hypothesized in such a way as to complete the explanation. In this way novel actions could be understood if seen in a meaningful context.

How Could Explanations Be Used in Task Performance?

While there is evidence that explanation-building does occur during learning, as just illustrated, there is no clear evidence that explanations are referred to during later performance. There are two arguments that can be made that such a role for explanations is probable. First, since memory for arbitrary sequences is poor, it is reasonable that learners would attempt to encode event sequences in a meaningful way, such as by embedding them in explanations. There is limited evidence that explanation-building can dominate simple retrieval of event sequences: Learners sometimes "remembered" false sequences of events that were suggested by explanations they were currently devising.

Jonathan Grudin (personal communication) describes a related phenomenon. A learner was dealing with two related commands, say A and B, that had different conditions of appropriate use. Early on the learner successfully used both commands and made correct decisions about which to use. Later, the learner encountered a situation that looked like a case for A, but for which A failed and B worked. Apparently concluding from this that B should be used even when A looked right, the learner stopped using A altogether and substituted inferior methods involving B followed by fixup commands. The earlier successful experience with A was apparently not accessible.

Among various possible meaningful encodings, explanations have the special virtue that they help to indicate how a sequence of actions must be changed to accomplish a variation on a goal. If exercises were encoded simply as a series of uninterpreted events, there would be no way to construct the actions necessary to revise a document different from that referred to in an exercise, say, or to carry out a different set of revisions. Explanations show how each action contributes to a goal, and to causing system responses. These contributions can be analyzed to discover how to accomplish related goals.

Faced with a goal to achieve, the learner can construct the requirements for this goal, and use the connections represented in one or more explanations to find actions that would contribute to them. As long as no new goal requirements are involved even novel goals can be accomplished in this way; there is no necessity to find a stored event sequence for that particular goal or even a similar goal.

For example, a learner who has successfully explained a moving exercise in a graphics editor could attempt a copying exercise as follows: The goal requirements for copy might be "designate object," "designate destination," and "tell system that *copy* is the desired operation." Except for the last, these are the same goal requirements as for move, so the learner can construct a reasonable plan of action based on how the object and destination were designated according to the explanations built up for move. How to specify the desired operation would remain to be worked out, and of course it may be that objects and destinations are not designated in the same way for the two operations, but at least the learner has a plausible start on the problem.

As this example shows, there is a kind of duality between explanations and plans: If I can explain how given actions contribute to a goal, I can then build a plan for the goal based on tracing the contribution connections backward. This duality expresses the value of explanations as encodings of experience.

Norman (1984b, 1986) has proposed three hypotheses to explain why people often do not detect their false explanations: relevance bias, partial explanation, and the overlap of the explanation with the events in the world. These three hypotheses are closely related to the analyses presented here.

Relevance bias. *Psychologists have noted an apparent bias toward seeking confirmatory evidence when looking for disconfirming evidence may be more useful. Norman suggests that this bias may be forced by the need to select evidence* *from a large pool, too large to permit complete examination. In selecting evidence one would have to make judgments of relevance, since obviously one does not wish to process irrelevant information, and these relevance judgments must be made on the basis of one's current conception of the state of affairs. If disconfirming evidence is less likely to seem relevant that confirming evidence, the bias can be explained not as a bias toward confirming evidence but as a bias toward relevant-appearing evidence.*

Partial explanation. *The second of Norman's conjectures is that errors are not detected because people accept crude agreement between analyses and evidence. The example just discussed shows this mechanism at work as well, in that the explanation builder accepts an analysis that accounts for only part of the data. Whether stronger mismatches are accepted, requiring a further liberalization of the interpretation rules, is not clear. It may be that all mismatches could be analyzed as partial matches, in which case the present machinery might be adequate.*

Overlap of model and world. *Norman's final suggestion is that error detection is impeded by the fact that the subject's model of a situation will tend to be similar in many respects to the actual state of affairs in the world, since it explains real observations. Therefore many actions the subject might take will produce similar outcomes in the model as they do in actuality. Exploring this idea would require some extension to the present analysis since the explanations treated so far do not constitute full models of the world.*

My work might be extended to address this issue without great change by using the present machinery to build explanations for hypothetical sequences of events with respect to given goals. To predict the consequences of a sequence of user actions one would search for a goal and a sequence of system actions so that the combined sequence of user and system actions would have a complete explanation with respect to the goal. One could then examine how often such predictions would be violated, or fail to be violated, when based on erroneous explanations of earlier events.

Implications for Interface Design

Hard and easy sequences. Explanation building will be more difficult, and hence learning will be harder, if learners cannot find interpretations of actions that make clear their contribution to the goal requirements. For example, if objects were designated by pointing not at the object, but at a point determined by reflecting the object through the center of the screen, the connection between the action and its effect would be obscured.

Less drastically, compare an interface in which objects are designated by pointing at them with one in which they are designated by indicating the corners of an enclosing rectangle. Explaining the more complex sequence requires some insight into the relationship between two points on a plane, the rectangle whose diagonal they define, and objects lying in that rectangle. It seems plausible that systems requiring complex interpretation rules like these would be more difficult to learn. In the terminology of Hutchins, Hollan, and Norman (Chapter 5), pointing at the object is more referentially direct than pointing at the corners of an enclosing rectangle. In my framework, referential directness influences the likelihood of discovering the correct connection of the action to its effect.

Note that directness must be defined with respect to the learner's ideas about the goal requirements. If the learner thinks of copying as acting not on an object but on a rectangular area, then the more complex corner procedure becomes more intelligible than pointing at the object. Thus the learners' expectations, at the level of imagined goal requirements, will influence how easily understood an interaction will be.

Effects of learner expectations. While the role of user expectations in determining success of a design has been often discussed, little concrete work has been done. Mack (in press) found it difficult to find robust regularities in expectations or to base design decisions on them. In the absence of a process model that assigns a particular role to expectations, Mack collected low-level expectations about the sequence of actions needed to accomplish a goal on a given system. The framework in this chapter suggests that, at least for novices, there may exist robust expectations at another level: system-independent goal requirements. There is a possibility that these expectations may be consistent across users, because they are based more closely on a logical analysis of an operation than on the steps required by any given system. It may also be more straightforward to use them in design, since the model indicates just what level of correspondence between actions required by a given system and goal requirements is required to facilitate explanation.

Figure 8.2 shows two explanations for the same sequence of events, given two different sets of goal requirements. The goal requirements reflect two different ideas on the part of the learner of what has to happen to attain the goal, and these different ideas produce different analyses of the events, even though the same knowledge is used to assign connections between events. In the first explanation, moving is thought of as a cut and paste sequence, as described earlier. In the second, the simpler goal requirements of designating object and destination and indicating the desired operation are used.

Note that the simpler goal requirements do not lead to a complete explanation, since the explanation builder cannot figure out what the PASTE designation is for.

The example shows that the learner's expectations determine the explainability of the sequence of events, so that either the system should be designed to be analyzable with respect to those expectations, or the expectations should be changed. In the example, the clipboard idea of an intermediate destination for copies (as used in the Apple Macintosh) would allow the overall copy operation to be analyzed as two copy operations each fitting the simpler expectation, if the learner knew that was what was happening.

The framework suggests that correspondence between the learner's conception of goal requirements and actual operations is more important than length of operation sequence as an influence on learning. The actual number of operations required for either a simple-move style or a cut-and-paste style interface could be freely varied by changing the details of how objects and locations are designated. But difficulty of building explanations would be influenced not by the variation in length (assuming equally intelligible details) but by whether the learner's expectations about the goal requirements were correct.

Consistency. According to my framework, in planning a method for a task users consult their stored explanations looking for examples of goal requirements related to those of the present task. Therefore, if learners are confronted with a system in which different methods are used for the same goal requirements, new pairings of task and method would be expected during later performance.

For example, if in one task objects are designated by pointing and clicking, while in another they are designated by indicating the corners of a surrounding rectangle, I would expect learners to retrieve either one of these methods when examining their stored explanations, and hence to make errors if in fact the choice of method is constrained.

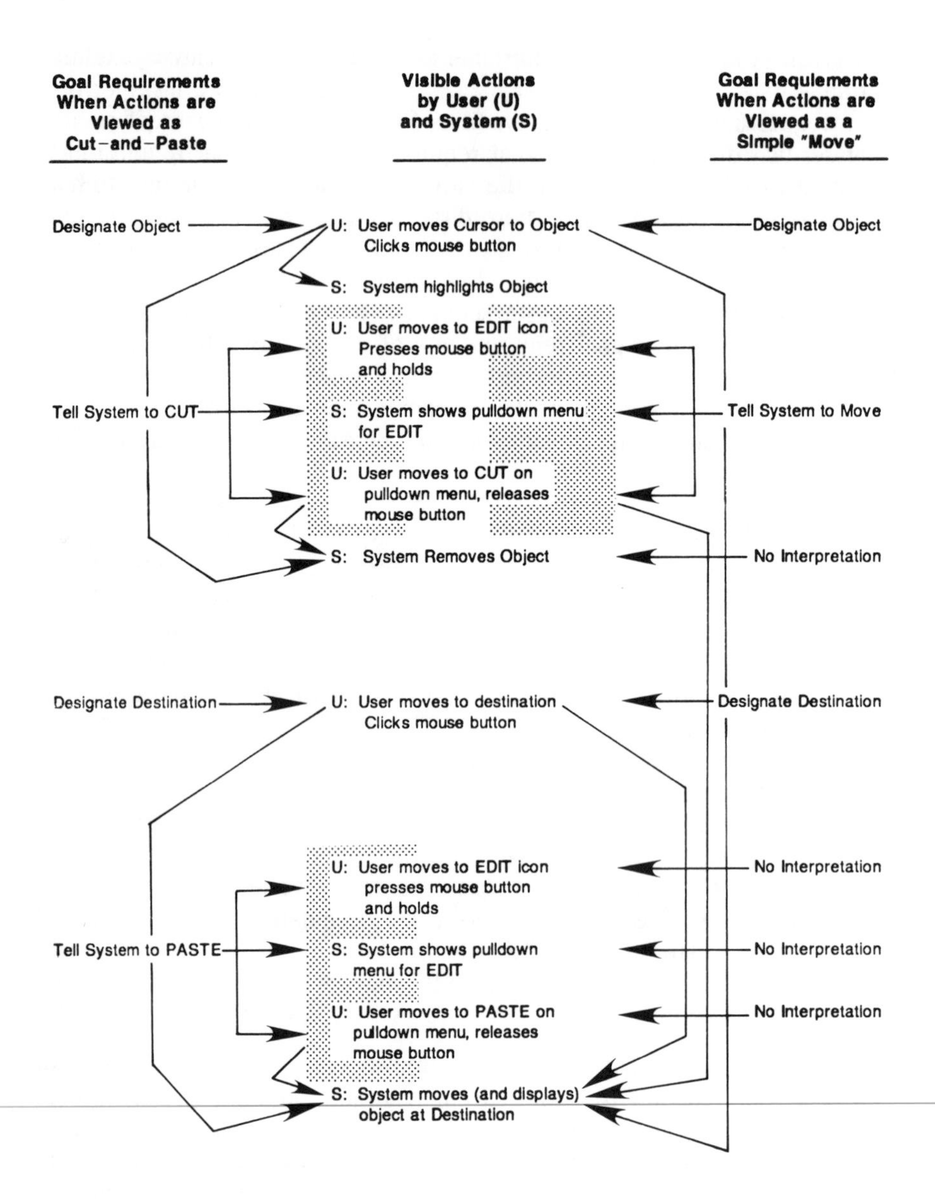

FIGURE 8.2. Two explanations of the same events, beginning from different goal requirements. Shading indicates actions that are grouped together to accomplish the goal. Arcs show how earlier events contribute to or cause later ones.

Consistency may also be expected to affect the explanation-building process itself, not just the later use of explanations during task performance. Knowledge of stereotyped sequences allows the explanation builder to chunk events for purposes of analysis. Knowing the right chunking rules is obviously important, and a system that requires few distinct rules would seem to be good. Once the rule that designating the corners of a rectangle can be seen as a way of designating an object is known, that sequence can be recognized elsewhere in the system with low cost.

Selection of Information

Since a great deal of information is presented in a typical system interaction learners face a serious problem of selection. They often make errors by assuming that actually irrelevant information is related to their concern of the moment. Thus, I once observed a learner who wanted a page number assume that a field on the screen which actually contained a line number was the page number. After all, if you are looking for a number, and find a number, it's the right number.

This example illustrates three features common to many similar situations. First, the analysis works top-down. The desire for a page number field drove the examination of the screen. Second, identity, in this case category identity, is a powerful cue. The number looks relevant because it's a number. Third, the evaluation of relevance is very lax. The label on the supposed "page number" made no sense on the assumption that it was a page number, but the learner accepted the supposition anyway. Apparently learners do not expect the world to be completely interpretable; if part of it makes sense, that's enough to go on.

These aspects of information selection can lead to design directions when coupled with the idea that learners are building explanations. Words and labels should be used to allow learners to form the right connections between (say) menu selections and actions, but that much is obvious anyway. What is less obvious is the importance of avoiding spurious connections that can be built into spurious explanations. Probably the best way to do this is to limit strictly the information that is displayed. One could go further and examine the content of what is displayed for coincidental identities to important notions for the learner, but these will be difficult to predict.

Implications for Training

As I have already illustrated, standard approaches to unassisted learning of computer systems have serious problems. On the one hand, "cookbook" style instructions are not robust. Learners have trouble getting back on track after making errors. Moreover, there is poor transfer to other tasks. Learners complain, "I did it but I don't know what I did." One learner successfully completed an exercise in which an underscoring operation was used several times. But when the operation appeared on a later test the learner did not remember ever having seen it. On the other hand, material intended to explain what is going on often goes unread. One learner paged through a manual saying, "This is just information." Yet the goal was for instructions that told what to do.

One response to these problems is to encourage active exploration of a system (see work by Carroll, 1985, and others) so that the learner takes responsibility for setting goals and finding system facilities to accomplish them. But even when a system is designed to permit this mode of learning (so that mistaken actions can be undone, for example) such learning can be inefficient and frustrating. Learners may waste a great deal of time looking in the wrong parts of a system, or looking for things that really won't help.

Another approach to unassisted learning that offers some promise in overcoming these difficulties is demonstration. If you show how to perform tasks, they do not have to discover the method themselves. In a sense, cookbook instructions are demonstrations that learners are supposed to produce themselves, but not only can these self-produced demonstrations have errors, but learners have to divide attention between following the instructions and observing the resulting demonstration. These problems are avoided when canned demonstrations are used. Demonstrations may also be cheaper to implement as a training approach than other methods that require more voluminous materials.

To be effective, the events in a demonstration have to be encoded by the learner in an effective way; this means in a meaningful way. This raises a series of questions: What determines whether a demonstration can be encoded meaningfully? How could a system be designed so that demonstrations of it could be readily understood? What information could be added to a demonstration to enhance its effectiveness?

If explanations are an effective and natural encoding for sequences of events, these questions can be reformulated as follows. What determines whether a sequence of events can be easily explained? How could a system be designed so that the sequences of events associated with tasks can be easily explained? What information could be added to a sequence of events to make it easy to explain?

If the explanation-building framework is correct, the critical aspect of a demonstration of a system is the ability of the learner to assign roles to the observed actions. This will depend in turn on the correspondence of the actions observed to the goal requirements *as conceived by the learner,* and on the ability of the learner to identify the role of each action. Both of these factors could be influenced by auxiliary information. For example, the sound track of the demonstration, or on-screen commentary, could identify both the actual goal requirements for the given system and the contribution made by each action.

Further Implications

If learners often encode their experiences with a system in the form of explanations, this has implications for other work on procedural learning. Analyses that center on the construction of production rules during learning, such as that of Kieras and Polson (1985), would have to be elaborated to accommodate explanations and the factors that affect their construction and use. Work on learning procedural information from text may proceed differently if explanations are a key encoding structure; rather than extracting conditions and actions for production system encoding of procedural knowledge, for example, readers may be building explanatory structures during comprehension.

CHAPTER 9

Naive Theories of Computation

DAVID OWEN

In my house I have a well-developed sequence for making tea. I go to the tall cupboard and get out the yellow kettle. From the right hand tap, I put enough water in it for one pot, put the kettle on the front lefthand ring, turn the bottom left-hand knob to "HI," etc., etc. In my neighbors house, I can still manage even though all the details differ. My neighbor's kettle is green and is already on the stove. Although that means I might just turn on the heat, I still pick it up to check if it is empty. There appears to be only one strange looking tap, but I know everyone has both hot and cold water so...etc., etc. The point is that in addition to my ability to formulate and remember a plan for making tea, I am relying on a body of knowledge concerning the properties of various physical objects and of concepts like "pouring" and what it means for something to be "full." I know that the color of the kettle does not influence its capacity to hold water and that by shaking the kettle I can get an idea of how full it is.

In the late seventies, the term "naive physics" (Hayes, 1978) was used to describe the rather unsophisticated understanding of the concepts and properties of everyday objects that guide us in our day-to-day activities. The motivation for studying it was that if people can manage without an explicit representation of Newton's Laws, then an AI program might also be able to. Physics, the science, provides us with a formal detailed account of some of the rules underlying the way the world works. "Naive physics" can be regarded as an informal, often phenomenological, precis of the real laws of physics, which people develop from encounters with the world. This chapter argues that a similar theoretical construct can provide useful insights into the problems people face when learning how to use a computer. To perform tasks in the computational domain equivalent in sophistication to making tea, people need a body of knowledge equivalent to naive physics: *a naive theory of computation*. There is the opportunity, at least in principle, to design interfaces which foster the development of these theories.

The first half of this chapter examines how we acquire a naive physics of the world and its role in allowing us to make sense of physical systems. It provides a new perspective on the problems confronting people as they develop similarly naive theories of computation. The study of errors in people's naive physics suggests ways of minimizing the chance of similar errors in the computational domain. The last half of the chapter discusses some aspects of what a naive theory of computation must cover, including a detailed analysis of a simple task.

HOW IMPORTANT IS IT FOR THINGS TO MAKE SENSE?

When people claim to "understand" something, they often mean that they can follow a chain of cause and effect down to some particular level. At that level, which depends on the individual, an article of faith precludes the need for them to pursue the chain further.[1] For example, I understand that when I turn the power/volume knob on my radio, it completes a circuit allowing electricity to flow from the battery to some electronics. I neither know nor care what the electronics does. But just on this level of understanding, it would not make sense to me if one day the radio started to emit sounds without my having touched it, or without there being a battery in it. It depends on a considerable amount of naive physics. I have implicitly asked myself the questions: "How is a radio possible? What do I have to accept about the world for the phenomenon of a radio to make sense." Electricity and its storage in

[1] In Riley's terminology (Chapter 7) their understanding is "internally coherent."

a battery are examples of a concept and an object attribute that I have to include in my naive physics as part of the answer to that question.

I also ask myself what the point of it is. There is some evidence that people run into trouble if they can not find an answer for this question that they find adequate. In an experiment designed by Wason (Wason & Johnson-Laird, 1972) and reported by Johnson-Laird (1983), four cards are laid out in front of a subject.

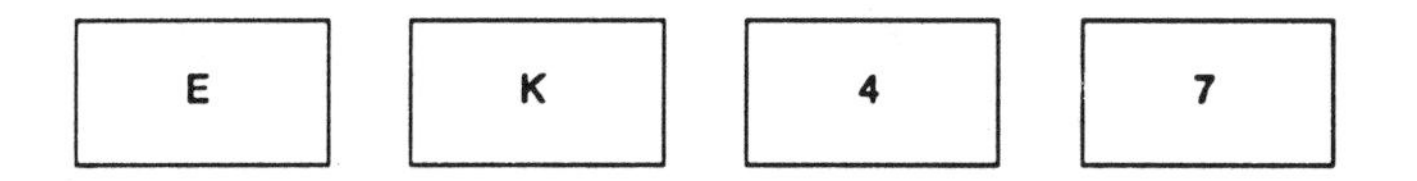

Each card has a number on one side and a letter on the other. The rule is that if a card has a vowel on one side, it has an even number on the other. Question: What is the minimum number of cards that must be turned over to determine whether the rule is obeyed? The problem is easy to understand, but hard to solve. However, subjects had considerably less difficulty solving the following, formally identical problem. In this problem, the letters and numbers of the previous experiment were replaced with destinations and modes of transport respectively:

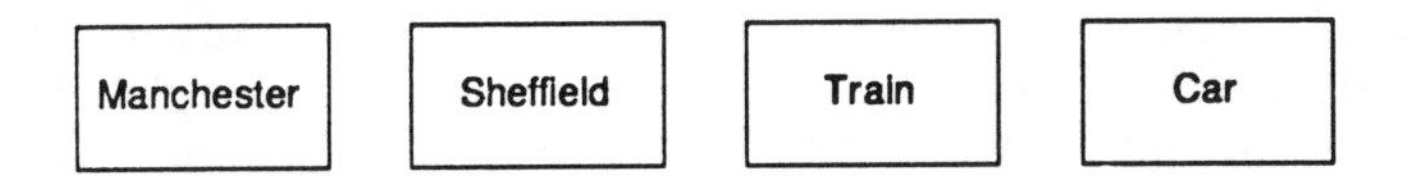

and the rule was changed to: Whenever I go to Manchester I travel by train. The question is the same: What is the minimum number of cards that must be turned over to determine if the rule is obeyed?

Exactly what gives rise to the performance difference is the subject of considerable debate. In both cases people understood the problem, but in trying to solve it: "They showed insight with the 'sensible' principle but they lacked insight with an arbitrary principle" (Johnson-Laird). This is a crucial observation to which I shall make reference again later. The ability to determine that it might be 'sensible' to relate a destination with a mode of transport relies to a considerable extent on naive physics. For instance it relies on a knowledge of the concepts of time and distance and the relationship between them in the context of movement. (In the first example, it is only necessary to turn the E and the 7 cards over. In the second, it is only necessary to turn over "Manchester" and "car.")

We are confronted with physical (and nonphysical) phenomena of the world at least as soon as we are born. By splashing around in water, playing in a sandbox, and scribbling on paper, we discover and come to trust some basic primitives: consistency, texture, flow, even the conservation of matter.

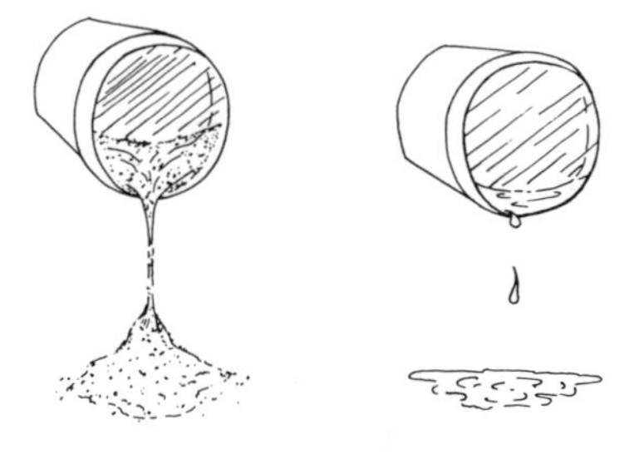

sand versus water

Each new exposure contributes something to our developing naive physics, resulting in a core of knowledge that helps us to determine relevance, to generalize our knowledge to new but similar situations, and even to recognize situations as similar. More complicated systems can be understood in terms of primitives with which we are already familiar. The channeling of water through a hose is an example. We start out knowing something about water, and that makes it easier to understand which of its properties are being exploited. That in turn enables us sort out mistakes ("debug") simple systems which we know from past experience should work. Most people will be able to formulate some theories as to why water isn't coming out of the end of a hose. We know that a hose is just a channeling device and that water will only come out at one end if water is being put in at the other and that the channel must not be blocked.

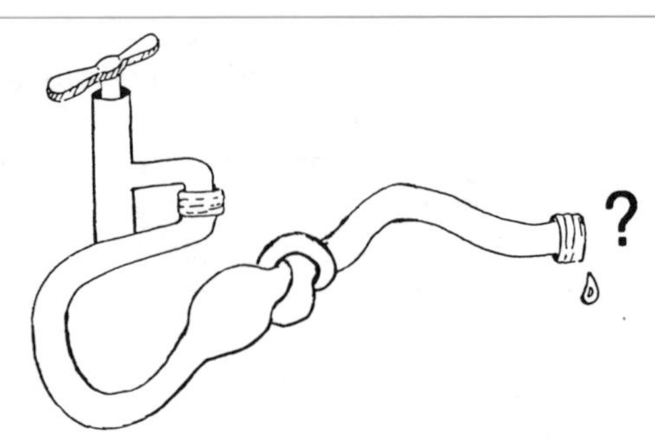

Our naive physics also supports innovation. We may notice that a screwdriver happens to have the right properties for opening a tin of paint, or at least quickly understand the idea if it is pointed out by someone else.

innovation

A Bootstrapping Problem

The way we are introduced to the world makes it, in general, a naturally supportive learning environment. We are allowed to inch our way over a long period of time to a workable understanding of essential concepts and relations. Our experience of its properties tends to start with relatively primitive phenomena like water or pencil and paper. And they really are "simple"; they do not conceal complexity that will later be relevant. Only later are we exposed to more complex systems like hoses and typewriters which involve several of the more primitive phenomena with which we are already familiar. In contrast, many people are necessarily introduced to the computing domain via interfaces which disguise an enormous amount of complexity. Users have to answer for themselves many implicit questions of the form: "What do I have to accept about the nature of the objects and concepts involved here for this to make sense?" This has to be done at the same time as understanding how they are exploited in this particular system. They are confronted with a "bootstrapping" problem. To understand the system, they need to know something about the primitives on which it is based, and their only access to those primitives is through that system.

In the final section of this chapter, an analysis of a simple editing task gives more substance to what this means for the user.

But clearly, people have no trouble using some extremely complex devices on the basis of the most superficial degree of understanding. Radios, washing machines, and televisions sets are examples. However, they will be restricted to a few operations with very limited flexibility. The complexity underlying the device is glossed in an article of faith which never has to be unpacked. In the computing domain too, there are interfaces which require an understanding of the domain which they address and only a minimal grasp of how that domain is represented in the machine. Often, they succeed by restricting the range of interaction to the point where familiar artifacts of the domain can be mimicked with the introduction of very few new concepts. Systems that provide enormous flexibility are bound to present most people with a much more severe challenge.

WE DON'T ALWAYS GET IT RIGHT

The argument is that people will necessarily develop a naive theory of computation for much the same reason that they develop a naive theory of physics, but that the context is much less supportive and the available learning time much shorter. If an adequate, coherent and comprehensible naive theory of computation could be articulated it would be an invaluable pedagogical tool. But even if a complete *formal* theory were available, its distillation into a naive precis which matched the cognitive mechanisms of people would be, to say the least, non-trivial. However, understanding more about conditions which foster the development of robust naive theories does not necessarily demand such a complete articulation.

Existing studies of naive physics, particularly examples of misconceptions, offer some insights. McCloskey, Washburn, & Felch (1983) analyze the commonly held misconception that an object carried by another moving object (a person running with a ball) will, if dropped, fall to the ground in a vertical straight line. They hypothesize that the frame of reference in which people are naturally inclined to interpret similar events they encounter in the world, leads to a misperception. One may debate the form of the resulting error in their naive physics, but of equal interest is the reason the misconception survives. People can straighten out or at least accommodate (there being no commitment to any particular degree of understanding here) similar common misconceptions, without formal physics training, if there is sufficient motivation. Spear fishermen for centuries have been able to cope with position distortions caused by the different refractive indices of air and water. A significant difference between these two cases is that the straight line misconception rarely interferes with achieving commonly held goals. The limitations of the adopted reference frame are never exposed. Whereas for the fisherman intent on eating, the persistent and conspicuous absence of a fish from the end of the spear is an unambiguous indication of the failure of the subgoal of catching the fish. Moreover, his understanding of the causal relations in the entire plan enable him to identify precisely where the plan failed.

The point is simply that the reference frame suggested by a situation may conceal potentially variable parameters or implicit subgoals of an action sequence, and thereby have an adverse affect on the robustness of the naive theories that situation invites. The concepts and properties which have to be inferred, accepted, and subsequently built on for this situation to make sense may collapse painfully when used piecemeal in another context. It is not an unfamiliar problem to schoolteachers, but the computer is particularly unrevealing of the many novel phenomena to be grasped. Clearly not all the subgoals will be of significance, but at least three kinds of subgoals are important: those which the user may have to achieve separately in a different context; those whose failure results in the failure of the whole goal (cf. the failure to eat a fish); and those which expose some crucial concept or property of an object.

In the UNIX operating system for example, if the text editor is invoked on a nonexistent file, the file will usually be created. Moreover, it will be assigned some protection status. However, the editor gives no evidence that a complicated protection structure is being automatically developed until one attempts to transgress it. The intent here is not to reiterate the "more meaningful error messages" chestnut, but to suggest that the user be allowed to absorb the notion of (in this case) protection by making its presence as a subtask apparent. Users will not always know enough to either formulate or express appropriately revealing question and so the development of their naive theories will often depend on what the interface presents to them.

> *The role of the "answers first, then questions" paradigm for information delivery is discussed in Chapter 17. In particular the DYK program described there provides one possible alternative to overloading the interface with this information or relying on conventional tutorial material.*

Understanding at least some of the crucial aspects of what a naive theory of computation must include will in any case contribute to the optimal resolution of the tradeoff between confusing users by swamping them with information and confusing them by concealing it.

DiSessa (1983) found that although subjects could "understand" that tennis balls bounced, they had great difficulty accepting that a steel ball bearing could bounce in the same way.

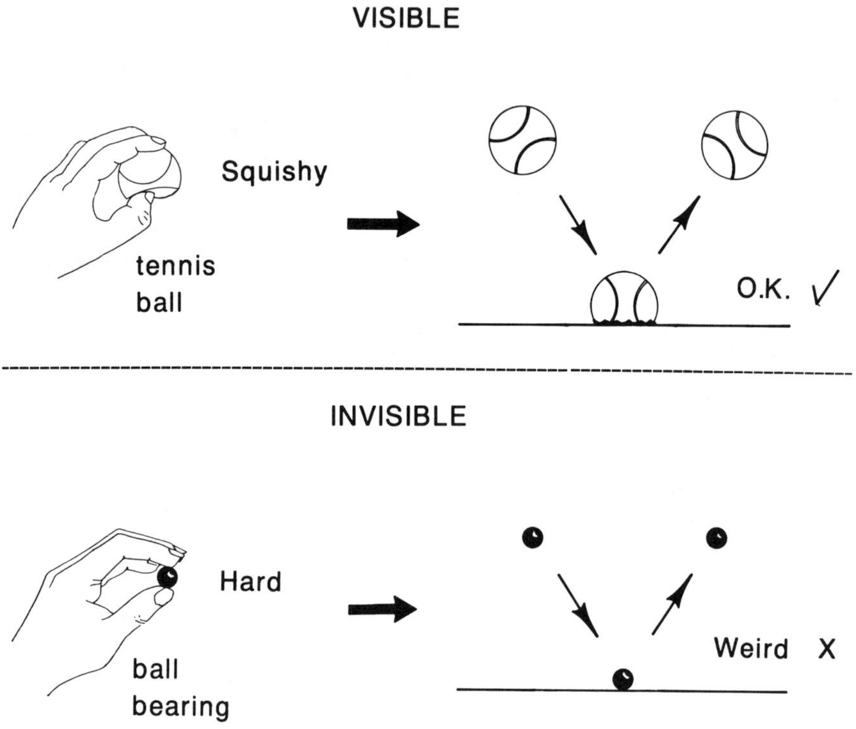

One problem appeared to be that the elastic properties of a tennis ball are detectable, whilst those of the ball bearing are not. The argument is that to exploit people's capacity to infer robust naive theories, both the nature of a procedure and the properties of the objects involved have to be apparent. This may just equate to slowing something down or making it bigger.[2] In fact, the investigator pointed out that with strobe photography the equivalence between the squishiness of a tennis

[2] At Brown University for example they are experimenting with what is effectively a slow motion graphical representation of the steps in a computer program.

ball and of a ball bearing could be seen. A second point which is suggested by both misconception examples is that diversity in the situations in which phenomena are confronted also encourages robustness in the theories that are developed. The subjects' stubborn disbelief that ball bearings could bounce may be based in a long held, and never challenged confidence in a theory about elasticity which was formulated in a limited context. It may be seen as an example of a limitation on the pedagogical value of some artificially simple systems.

I am not making an argument that computing systems should mimic their real world physical counterparts. On the contrary, the use of icons that look like wastepaper baskets is of questionable value if associated concepts are not supported. (Can it fill up? Who empties it? Can things be retrieved?) Metaphor crucially depends on knowing enough about the attributes of the objects in both the central domain and the domain to which it is being likened, to determine what it could possibly make sense to import. (A similar point is made by Greeno, 1983.)

WHAT ARE THE THEORIES ABOUT?

Postulating that people necessarily formulate and base their understanding of computing systems on a naive theory of computation raises questions about the form and content of that theory. Determining the form is probably more intractable than speculating about its content, although of course both are open questions. Theories have been developed in other domains; diSessa (1983) for example has argued that much of our naive physics is encoded as phenomenological primitives. These are a kind of cliché which can embody the causality and object attributes inherent in some small aspect of the physical world, like the bouncing of a ball. One of the reasons this question is important is concerned with the accessibility the form offers, a point which again is stressed by Greeno (1983). Roughly, the argument is that if a characteristic of some concept or a parameter of some situation is not discriminated in the form it takes in our representation, then it cannot be subsequently used in a situation in which it may prove crucial. So for example the concept of a "bag of groceries" would be of limited value if we could not also unpack the elements of that concept. Similarly, the general concept of a buffer is inaccessible if its sole representation is in the form of an editor command which restores the last thing deleted.

Two aspects of the content of a naive theory of computation are immediately suggested by the fact that computation is concerned with the manipulation of *symbols* by *procedures*.

Symbols

For documents or airplane wings, the difference between working with the object and working with a representation of it will inevitably be apparent in the interaction. In general people are used to this. Only in cartoons do people knock nails in with a photograph of a hammer. Similarly the difference between using written digits and using beads to represent a quantity is clear, and in natural language we use symbols in a sophisticated way. However the nature of the symbols used by a computer is new, and so is the degree of detail to which an object may be represented using them.

Procedures

The nature of the interaction is also affected by any new device which may be used to manipulate the representation. For example, an abacus is a manipulative device which may be "applied" to beads used to represent quantities. As such the abacus fosters the use of particular procedural strategies. The procedures with which we are most familiar are compatible with the time and energy constraints of physical manipulative devices, shuffling beads or chess pieces directly for example. In contrast, the computer is a manipulative device for which the constraints that apply to physical representations are relatively unimportant, thus making a completely new unfamiliar class of procedures practicable. The implications of this for a user are far from obvious.

One may argue that understanding the semantics of procedures like recursion is inescapably very hard and, in fact, unnecessary for most purposes. In examining this argument, Sheil (1981) describes a system which only provides the simple linear structuring device of allowing the output of one predefined operation as the input to another. Although he finds this leads to a desirable simplicity in some contexts, he also points out that the lack of flexibility leads to different kinds of complexity, incomprehensibly long reasoning chains in his example, and does not exploit the skills people do have. DiSessa makes the same point in Chapter 10. This suggests that there may be two aspects of the procedural novelty introduced by a computer that are worth distinguishing. Some procedural formalisms might always be comprehensible only by programming specialists. For others, the problems people experience may not be with the semantics of the procedures per se, but with their use in the computational context.

Now reconsider the four-card experiment. People seemed to have trouble with procedural reasoning when they could not "make sense" of the situation. By virtue of its time and energy constraints, the computer not only allows new procedural formalisms, but also introduces new criteria of how it is "sensible" to use more familiar ones. This, coupled with the unfamiliarity of the objects to which the procedures are applied, provides fertile soil for confusions affecting both the users' abilities to recognize the procedural strategies used by the machine and also to generate appropriate ones for themselves. Furthermore, the intermediate states of a procedure being used by a computer are in general far less apparent than is commonly the case in the physical world. They are either not visible at all, as in the case of temporary editor storage buffers, or flash by too quickly to be registered, as in the manipulation of characters on the screen by an editor, or even the rapid typing of key sequences by practiced fingers (note the resemblance to the imperceptible squishiness of the ball bearing). There is another novelty introduced by the need to exploit the lack of time/energy constraints of the computer to manipulate the enormous number of symbols. Whether writing assembly code or giving instructions to an editor, "programming" requires the intent of a procedure to be specified separately from its execution, a point made by Sheil (1981). The procedure has to be translated into an abstract form and the ramifications of each step in the procedure anticipated. Although the need for this is minimized in many existing interfaces, (Chapter 5 on direct manipulation gives some examples), it still occurs, for example, in making global edits to a file, and in the use of a mouse to indicate first the action and then the object of that action.

A simple example shows the effect of these novel aspects of the computer and also shows more clearly the form of the bootstrapping problem discussed earlier. When people confront a screen editor for the first time, on the face of it they are just learning specific commands to drive a particular system. But a closer examination reveals many examples of new concepts, object attributes and procedural strategies that they must also grasp over and above the specifics of the command language. To begin with, the separation of specification and action takes the form of a novel conversational convention with an intermediary. Then there are the concepts of a line, a character, and a display surface, which differ in their phenomenology from the forms we have been accustomed to since we first started to scribble on paper:

- Paper limits in a very obvious way where we can make marks.

- It provides a fairly unambiguous definition of what a line of text is and how long it can be.

- We learn that characters are fixed relative to each other and to the limits of the paper.

A screen editor enforces different constraints:

- The boundaries of the writing surface are not at all clear. In many editors the the cursor can be moved from the rightmost character of a line towards the rightmost edge of the screen, *only* by typing spaces ("constructing paper") and not by cursor control keys or by the mouse.

- There are at least two definitions of a line of text. One is fairly obviously established by the limits of the screen. The other is the essentially infinite internal representation.

- The characters are not fixed in place but can be shuffled around.

Grasping all this reveals something of what it is *possible* to do. It provides a partial answer to the question: "What does one have to accept about the (computational) world for this editor to make 'sense'?" But one also has to understand the different constraints of the computer as a manipulative device to determine what it is *sensible* to do. For example, based on the constraints typical of physical manipulative devices a sensible strategy for inserting a missing word in a sentence is shown in Figure 9.1A.

First make enough space for the whole of the new word by moving some text to the left. Then type in the missing word.

However, the strategy typically adopted in a screen editor is to work a character at a time.

Make space for a single character, insert it, make space for the next one and so on.

In the long run this is more convenient for users. They don't have to work out and declare how long the new word is. But this is sometimes difficult for people to learn (see Riley, Chapter 7; Riley & O'Malley,

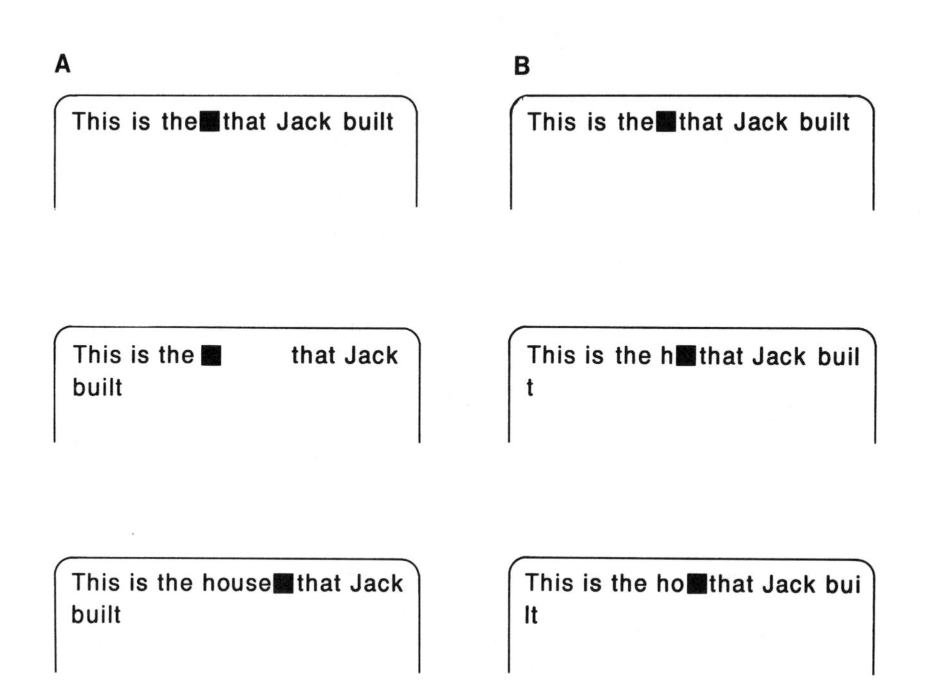

FIGURE 9.1. **(A)** A Sensible Strategy: *First make enough space for the whole of the new word by moving some text to the left. Then type in the missing word.* **(B)** The Intermediary's Way: *Make space for a single character, insert it, make space for the next one and so on.*

1985; Roberts & Moran, 1983 for examples). The absence of appropriate naive theories of computational objects and procedures invites criteria for what is both possible and practical to be drawn from naive theories developed in other domains. Thus, the analysis of what users have to grasp in terms of naive theory in this editor example suggests that discriminating between editors on the basis of command complexity or how well they match what users want to do, may only be part of the story. What may also be significant, at least to begin with, is the extent to which the commands reveal to users the characteristics of the domain on which those commands rely. It may be that this correlates to some extent with command complexity, but it remains a distinct criterion.

The computer is capable of using many arbitrarily complex mappings between symbols, including those displayed on the screen, and the objects, situation, or events they signify. The relationship between what appears on the screen and "the file" in the editing example above is already nontrivial; it is a copy which is being manipulated (although this is not apparent) and failing to "save" it, a common mistake, may result in the loss of many hours work. Such relationships become more sophisticated with the use of icons and the kind of customized direct manipulation work benches that are discussed in Chapters 5 and 16. There is the potential for confusion between actions performed on the object itself and its manifestation in one of several displays. For example, if I remove the icon for a file from this display, is the file lost or is it only finally removed with the removal of its last iconic manifestation? Users familiar with more conventional interactive systems are in a reasonable position to anticipate the possibilities or at least understand the convention for a particular system (cf. the idea of using a screwdriver to open a tin of paint). A novice on the other hand must also accept the potential arbitrariness of the relationship between what happens to an object and what happens to its screen manifestation.

CONCLUSION

What I have argued here is that people necessarily develop a naive theory of computation in much the same way and for much the same reasons that they develop a naive physics of the world. It is a perspective which seeks to expose more clearly the way in which complexity and novelty may confound users, and thereby contribute to the design of systems which foster the development of robust naive theories. The two theories are similar in that neither is a formal theory with explicit representations of universal "laws" but is nonetheless concerned with the nature of the objects, concepts, and the recurring relations between them in the respective domains. It is an essential core of knowledge in terms of which people can accept a more complex device or system as "making sense" and make reasoned interpretations of its behavior. A significant difference between the two domains is the contexts in which people develop their naive theories. In the computing domain, essential elements of the theory are combined and opaquely embedded in a working system. Users seeking to understand that system are thereby confronted with a bootstrapping problem.

ACKNOWLEDGMENTS

My thanks to the other contributors to this volume for many helpful comments on earlier drafts.

CHAPTER 10

Models of Computation

ANDREA A. diSESSA

Our theories of how people understand computational systems are in a very primitive state. I am convinced that this is more than ignorance, that many researchers are unnecessarily constraining their image of how understanding develops, thus missing important issues and phenomenology. In particular, I believe there is a general overestimate of how systematic people's understanding is, and even how systematic it should be. Part of my intention is to begin to take steps beyond preconceptions we might have about understanding.

I will approach understanding by reifying three kinds of models, three very different ways of understanding a computational system. Actually, programming per se is my target. These kinds of models proceed from very systematic, toward more and more fragmented ways of understanding. Each has different advantages and disadvantages, and a different learning curve. I believe all of these models can and should be cultivated simultaneously in a learner.

In developing examples of these models, I will draw heavily on (largely informal) experience in teaching children about Logo, a language designed from the beginning to be understandable. Historically, reflection on this experience was motivated by the attempt to build systematically on Logo, designing a new and more understandable system. The relation between the framework built here and that, as yet unfinished design task is reported in diSessa, 1985, which also contains an abbreviated form of the arguments made here. Some features of the new system are described in my chapter on the future of programming, Chapter 6.

Structure and Function

I begin with a fundamental distinction that is crucial to understanding any complex device, that between structure and function. The distinction focuses on either (a) characteristics of an object or action which are defining and independent of specific use (structure); or (b) characteristics which have to do with specific use, consequences or intent (function). The point is to separate descriptions according to whether the implied descriptive frame universally applies (structure) or not (function).

For example, the structural aspects of a variable in a computer language are given primarily by the rules for setting their values and for getting access to their values. These rules apply in all contexts. In contrast, a variable's functions might vary. Sometimes they might be described as "a flag" or more generally, as "a communications device." At other times a variable might function as "a counter," "data," or "input."

My intent is to use structure and function to seed the conception of two paradigmatic kinds of models that capture some substantial intuitions about how people ought to understand, and some of the realities about how they really do.

STRUCTURAL MODELS

The words "mental model" often conjure up the image of a sort of replacement machine located in the mind on which one can run experiments and envision results without touching the actual machine. Such coherent, runnable conceptions have been called *surrogate models* by Young (1983). For example, one typically models a "push-down list" as a physical stack on which objects may be piled and removed. A push-down list, of course, does not behave identically to a pile of objects (e.g., "overflow" versus gravitational instability), but the image of a pile manipulated with *put* and *remove* operations does allow one to simulate its important behavior.

For present purposes, I prefer the nearly identical concept of *structural model* to that of surrogate model. A structural model is one that is intended to capture the computational mechanism in such a way as to offer explanation and correct predictions in uniform terms. It establishes the bottom line for the user for saying what will happen, given the present state, and for saying what state must have existed, given a particular behavior. Structural models are what one usually sees when a model is explicitly taught. A tutor will typically invoke one when a novice's expectations have gone awry. Two of the most elaborate structural models that are attempts to encompass a large part of the operation of a computer language are the actor model of Smalltalk (Tessler, 1981) and the versions of Papert's little man model of Logo done by the Edinburgh Logo Project (du Boulay, O'Shea, & Monk, 1981) (See Figure 10.1).

One possible structural model of a language would be a specification of its implementation, though this is hopelessly inadequate pedagogically for technically unsophisticated users. "Meta-circular evaluators,"

FIGURE 10.1. A picture of a little man model unfolding procedure calls. On calling a subprocedure, the currently active little man (procedure activation) calls a new little man to action, then sleeps. He is awakened to finish his job when the new little man reports his job done.

in which details of a language can be shown through an implementation of the interpreter of the language, written in that same language, is a trick available to some languages to bootstrap into a structural model on the basis of a selected understanding of the language.

Having a structural model would seem to be the ideal way to understand a system. What could be better than to have a universally valid and comprehensive view of a system? Yet this notion is, in isolation, part of the conspiracy to oversimplify understanding that I mentioned in the introduction. Once we have looked at problems with structural models, we can go on to consider complementary ways of understanding.

Problems With Structural Models

Though it is not inherent in the concept, structural models are prone to certain problems as a way of giving users effective control of a system.

1. *Learnability:* Since structural models aim at being uniform views of rather complex systems, they themselves tend to be complex. When pushed to cover all behaviors of a system, they can lose their air of coherence with ad hoc elements for dealing with loose ends. For example, though Smalltalk's actor model demands an object destination for each message, there are occasions when, for good reasons, a message must be assumed to be sent to the interpreter, a nonstandard Smalltalk object.

   Because of their compact, tightly interconnected nature, structural models require a good deal of learning before they can be applied to even simple, everyday events. Many of the "little man" model's behaviors are not precisely what one would expect, without a fair amount of coaching, of a little man. For example, a little man (procedure invocation) must "sleep" and call another little man for a subprocedure call. A simpler interpretation which is good enough for most purposes is that the subprocedure "just gets done." Thinking in terms of the model actually complicates understanding early elements of the language. Incremental learnability is sacrificed for the sake of uniformity and completeness.

   These remarks on learnability are more than casual observations. A growing literature on "bugs" makes it clear that acquiring systematic understanding of present day languages is no easy matter. In many ways, it escapes not only children (Kurland & Pea, 1983; Papert, diSessa, Watt, & Weir, 1980), but college students as well (Kahney & Eisenstadt, 1983; Soloway, Ehrlich, Bonar, & Greenspan, 1982). Notions such as recursion fall straightforwardly out of the dynamic structure of many languages (the little man model is sufficient), but they are nonetheless notoriously difficult to learn.

2. *Styles of use:* Structural models, though very good for debugging (that is in cases when one has a given input to the system and needs to afford the time to, step-by-step, trace down an unexpected behavior), are typically rather time consuming to "run." Routine, relatively fluid interaction with the system cannot be expected to occur as a direct result of acquiring a structural model. At the very least, degrees of automation, "compiling" frequently used operations, etc., occur.

   A more fundamental problem is the very kind of knowledge available with a structural model, which makes such models unsuited to certain tasks. Consider: A structural model is almost always unfortunately far from the task of inventing a way to effect some intended result. For example, in planning a program the specification of the goal out of which the plan must arise is typically made only in rough outline, and it is almost always functional, that is, it refers to a conceptual frame outside the privileged structural one. Hence a fundamental gap exists between the functional level of description needed for planning and the structural level. This is quite similar to the gap called "expressive distance" in Chapter 5 by Hutchins, Hollan, and Norman. It is precisely the gap in which the activities of design and engineering usually fall. What follows are some examples of this gap.

Suppose the task is to communicate some information from one procedure to another. One obvious way is to use a variable. It is obvious not because one knows how a variable works, but because one knows a variable can and often is intended to have the effect of "information transmission." In many programming languages, in order to leave some vertical space between two output lines, the programmer has to "print" an empty line (e.g., in Pascal, do "WRITELN"—the write-a-line-of-text command, but without an argument). Not surprisingly, novices must usually be taught this "trick"; it is a potential function not easily seen in the meaning of the command. An example more germane to later discussion is the fact that either a variable or a function might perform precisely the same role in an expression, to provide a needed value. The considerations that dictate whether one chooses to implement that role with one or the other structure may be totally invisible to the functional semantics the programmer attributes to the symbol. A syntax that distinguishes function from variable may therefore be obtrusive and irrelevant to perceived meaning. Designers of a language must face the issue of how much structure they want apparent in the surface form of the language, and whether they can tolerate structural ambiguity from which transparent functional adaptability can more easily be achieved. More pointedly put, the question is how frequently must users think about how it works in order to make it work. I return to this point several times in the section on distributed models.

Learning a command or construct entails learning important side effects of that command that can be exploited to attain particular ends as well as learning a context-free specification of the meaning of the command. It involves learning typical uses of the construct, what one might call teleology ("what it is for"). It involves learning plan fragments, that is, sketches of ways to accomplish things using the construct (such as the counter paradigm for the use of a variable), possibly in paradigmatic combinations with other constructs. All of this functional knowledge is not germane to a structural model, but is clearly part of understanding a system. (See Ehrlich & Soloway, 1983, and Waters, 1984, for work on determining this functional vocabulary, and references to related work.) Though we have concentrated here on planning and related activities, it should be evident that a functional vocabulary is important to other aspects of programming as well, for example, understanding already written programs.

All of this is not to say structural models cannot be of any functional help. If you don't understand much about a system at all, having a structural model is much better than nothing. Indeed, Halasz and Moran (1983) show empirically that having a structural model can be a boon to inventing procedures for accomplishing a new goal with handheld calculators. Tuning (adapting known methods) and even more, radical invention (coming up with significantly nonstandard methods) can benefit greatly from having a structural model since they transcend known functionality. Even memory, for example, reconstructing syntactic or other details of substantially functional ideas ("rational reconstruction") can involve structure. Yet it remains the case that producing even relatively simple programs with little or no functional knowledge would be a tedious and inefficient state of affairs, even if, as is not often the case, the cost of learning the structural model could be discounted.

To summarize, there are two fundamental problems with surrogate models. The first has to do with their tendency toward complexity which makes them slow to "run" and hard to learn. A design heuristic to help deal with this problem is to construct a language deliberately to have a simple structural model by selecting the outline of the model and building structure and syntax around it in such a way as not to lose any "necessary" functionalities. Smalltalk has followed this line, beginning with the root actor and message passing model. But even with simple structure, the second fundamental problem remains. Specifically, the construction of a broad class of functionalities out of a tiny set of structural elements is almost bound to involve great cleverness. While systems designers may be very fond of such cleverness, the novice user is generally less appreciative. Functional understanding is not well supported by a structural model. This side of a system will require specific tutoring, thus the advantage of a small number of universal structures over a larger set more specifically tuned to important functionalities is not clear. A system with a simpler structural model can be at a distinct disadvantage overall.

FUNCTIONAL MODELS

The fact that structural models are removed from application makes them attractive as universal, "mechanistic" ways to understand a system, but such a perspective slights functional understanding. This section explores how to build understanding directly on the basis of functional considerations. As might be expected, the characteristics of functional models turn out to be largely complementary to those of structural models.

To see the advantages and disadvantages of this possibility, consider a case study. Young (1981) details an archetypical functional model. The example is that of an algebraic calculator, one that allows entry in near algebraic format, using parentheses. Young's point was that unlike other forms of calculators (e.g., the stack calculator) it is very difficult to develop a structural model of an algebraic calculator. Instead users must rely on functional models. Consider how one does a simple problem such as "3 + 5 = ?" By typing in 3 + 5, in a way that maps trivially to doing the same thing with paper and pencil, one has set up a context where what should happen is obvious. Pressing = does precisely what one wants: It gives "the answer." "Doing what one wants" is quite a sufficient model of the system for many purposes. The general schematic of functional models is that one has a descriptive frame, in this case doing arithmetic, that includes recognizable objects and actions such as "writing the problem" and "getting the answer." Then the user can understand computational constructs and actions as they function in this frame. "+" is part of writing the problem and "=" means "give me the answer." Functional models might be described simply as rules, e.g., "to get the answer, press '=','" but to think of them in this way ignores the fact that such a rule is memorable precisely because it fits into the previously understood scheme of goals and means, i.e., the one we have for written arithmetic.

Functional models obviously have some defects. Unlike structural models, one cannot expect the general behavior of the system to be evident in a specific context. For example, the state of the system if one were to type 1++ is not constrained by the prototype "writing the problem" model of +. If one expected novices to need to interpret situations equivalent to 1++ (such as understanding some prewritten code) or if the functionality achieved through 1++ were important and not conveniently achieved through other means, one should certainly beware relying on this method of giving users a model of the keystroke +. Functional models provide restricted understanding. The descriptive frame will be weak with respect to structural aspects of the situation that are not usually relevant to the functional frame, e.g., the specific internal state of the calculator after pressing 1+. This knowledge is important for debugging and similar tasks, for example, to know what to do to correct a mistaken + when – was intended. More pointedly, certain combinations of structures—1++ is an example—will be entirely meaningless in the functional frame, even though such a sequence of keystrokes might be not only legal, but useful. With some calculators 1++ defines a constant calculation. This example also points out that if a structure is to serve several functions, a single functional projection often will not be sufficient to allow the user to understand or generate the other functional descriptions.

Functional models provide only a view of part of the system, and that only with respect to a non-universal frame of analysis. Obviously, one needs a repertoire of them; the notion of a single model is tenable structurally but not functionally. With respect to the strengths of a surrogate, this fragmenting of understanding shows weaknesses. On the other hand, from the point of view of incremental learnability, teleology and other important aspects of understanding, functional models can be superior precisely because of their contextual specificity.

DISTRIBUTED MODELS

In this section I describe a kind of learning similar in some ways to that described for a calculator above. But in this case, the model is accumulated through a spectrum of partial understandings—not by virtue of a single functional frame. I refer to models accumulated from multiple, partial explanations as *distributed models*. The notion of a distributed model is derived from ideas I have developed about understanding complex systems in other domains. (See diSessa, 1982, 1983.) These models represent a patchwork collection of pre-existing ideas in the learner, "corrupted" to new ends. They don't appear nearly as "model-like" as structural or functional models. But if "model" is too strong a word to describe them, "understanding" is not. For the kind of learning involved explains some important data on how people come to understand complex systems.

The following example comes from an early stage in learning Logo. It is striking because it shows a quite successful learning sequence that cannot be accounted for in terms of either simple structural or functional models.

Procedure Definition in Logo

Logo beginners almost always start by driving the Logo turtle (a graphics cursor) around with commands like FORWARD 100. It seems certain that elementary school students interpret FORWARD 100 essentially as an abbreviation for an English sentence like "go forward 100 units." The need for input to the command FORWARD is not, therefore, understood structurally but according to the semantic need to complete the sentence, "FORWARD <how far?>." Linguistic and semantic function here precedes and provides a preliminary model of the structure "command plus input."

When students are taught to define their own procedures, the metaphor of teaching the computer how to do a new thing is invoked. One types TO SQUARE :SIDELENGTH followed by the list of commands defining square, as in the following recursive program.

```
TO SQUARE :SIDELENGTH
FORWARD :SIDELENGTH
RIGHT 90
SQUARE :SIDELENGTH
END
```

In structural terms, TO SQUARE :SIDELENGTH is simply part of the syntax for defining a procedure, but the syntax is also intended to support the interpretation of a procedure as a verb. The English infinitive is frequently used definitionally, a function that the Logo procedure definition syntax aims at inheriting. Furthermore, input specification follows the same form as the FORWARD 100 sentence, which is also acceptable English, e.g., "to go far." Abstractly, one sees a problem (teaching a new verb) and a solution (TO SQUARE ...), all of which relies heavily on a knowledge frame, English, that is rather systematically used to help Logo beginners.

> *The Greek meaning of Logo is "word"; this use of natural language in making Logo learnable was quite deliberate on the designers' part. Some indication of the importance of these linguistic considerations comes from teaching Logo to non-English speaking students. In Japanese, verbs usually come at the end of a sentence, and the "FORWARD 100" command form seems harder (personal communication, Leigh Klotz). In Portuguese where it is much less natural to use a word like "square" as a verb, difficulties are encountered (personal communication, Jose Valente). It is also important to Logo that the imperative and infinitive form use the same verb form, "SQUARE" and "TO SQUARE." That this is not the case in many other languages has raised problems in devising effective translations of Logo.*

Not all aspects of the syntax for definition are meaningful within the linguistic perspective or within the functional frame of "teaching a new word." In particular, the use of : deserves attention. In Logo the : (pronounced "dots") denote the value of a variable and are included in the definition syntax to parallel and reinforce the pattern of invocation of procedures with variables as inputs, e.g.:

FORWARD :SIDELENGTH

or more particularly, to parallel recursive call format, e.g.:

SQUARE :SIDELENGTH

in the final line of the above procedure. Another visual metaphor is provided by the pattern of commands making up the definition of SQUARE, which is the same line format one would see on the screen if one just typed them in for direct execution. These consonances are subtle, but contribute to learnability without interfering with other frames for understanding.

The : marker as part of definition syntax has support other than from visually matching the pattern of typical invocation; namely, it has a simple rationalization—to distinguish variable inputs from the procedure name, and to distinguish them in the form one most frequently uses a variable, getting its value. Novices will often respond directly to queries about the definition syntax with such rationalizations: "SIDELENGTH is a variable," or "It's just like when you write SQUARE :SIDELENGTH," (recursive call). Implementations of Logo that changed the syntax to TO SQUARE SIDELENGTH (without the colon) have prompted complaints from novices whose rationalizations were violated.

Being able to depend on distributed models puts us in a markedly different position from what one would expect if a simple structural model accounted for all of learnability, if coherence were measured only in structural terms. This fact is easy to emphasize. Logo is a descendant of Lisp, a language with a uniform, structurally simple control organization and syntax. Essentially all commands are assimilated to the paradigm of function application, where inputs are evaluated to produce a value for the command. In many ways, Logo follows this standard, but procedure definition is a glaring exception.

Assimilation to the function application standard would require SQUARE and SIDELENGTH to be inputs to the function TO. But then one would have to quote SQUARE and SIDELENGTH to denote the fact that they should not be evaluated as part of executing TO SQUARE SIDELENGTH. Thus, one should write something like TO "SQUARE "SIDELENGTH followed by the body of the procedure as a third input. The body itself would have to be represented as a standard compound data object, e.g., a list of lists (the lines of the procedure definition).

All in all, the definition of SQUARE would look like:

TO "SQUARE "SIDELENGTH [[FD :SIDELENGTH] [RT 90] [SQUARE :SIDELENGTH]]

Not only is there loss of template matching to a typical use of the defined object, but there is no problem-specific rationalization for the syntactic markers. The different functions of SQUARE and SIDELENGTH are not marked, and TO is separated by syntactic marks from its close "English" partner, SQUARE. Beginners would need to memorize the syntax with essentially no semantic or experiential support. Of course, for the computer-experienced, the syntax would have a great deal of meaning having to do mainly with the advantages of having very few, context independent structures. Indeed, the argument presented earlier shows that the form of procedure definition above is nearly unique, given strict adherence to function application and the small set of datatypes of Lisp or Logo. Such arguments, probably in fragments, supplement literal remembrance of similar constructs for sophisticated users of languages with strict structural constraints. This would be an example of rational reconstruction, mentioned earlier. But that doesn't help the naive and novice user.

For comparison, we note the functionally opaque, but structurally unexceptional form that is actually provided by many Logos as an advanced version of procedure definition.

```
DEFINE "SQUARE [[SIDELENGTH] [FD :SIDELENGTH] [RT 90]
[SQUARE :SIDELENGTH]]
```

To sum up, the demonstrated learnability of the procedure definition process in Logo is due to its naturalness as a solution to a particular problem when interpreted in a number of frames, each of which partially explains the solution. The list of frames for procedure definition syntax includes:

- A clear functional frame—the problem is making a new procedure; the solution is TO ...
- English
- Visual pattern matching
- Rationalizations

Distributed models such as this one involve learning by prototype in the sense that users learn a construct primarily by example. An instructor often simply shows the student how to do something, much like a parent teaches a word like "dog" to a young child by pointing to one. One does not, at least not initially, expect to provide the user with elaborate, explicit explanations of the details of functional context, why each symbol appears, etc. Instead, rationalizations and other partial understandings provide a backbone of reasonableness that allows the user to remember the form of the construct and understand some of the variation possible.

Special considerations for distributed models. The functional frame has special status for procedure definition because it establishes a meaning for the thing to be modeled, "How do I make a new procedure?" Once the existence and purpose of a computational structure has been established, it is much easier to accumulate rationalizations, and so on.

The visual presentation of procedure definition is also extremely important, but in a different sense. It is the medium supporting many of the distributed understandings. In general, one would expect visual presentation to play that role, replacing abstract reasoning on the basis of structure with visual metaphors, etc.

There are aspects of procedure definition beyond having a prominent problem context that make it particularly apt for using a distributed model. It is learned early on, before structural considerations are likely to interfere. Equally important, the problem solution is frequently enough used that the model will not be dangerously undermined by other experiences and continued learning. After one understands the fundamental structures of Logo, procedure definition is clearly exceptional. Yet because it is used so often, the syntax never comes to feel unnatural. In this respect, one can compare irregular verbs in natural language, which tend to be common words.

In general, one must consider long-term effects of particular functional and distributed models. Some will remain and be integrated as "special case" models. Visual pattern matching is an example. Some will fade away naturally and be replaced where appropriate by structural models. No learner believes Logo is English for very long. But we must be aware that globally destructive misconceptions may be fostered as well as "profitable misconceptions."

Control in Logo

This second example of distributed models seems less rich than procedure definition in terms of modalities; for example, rationalizations play a minimal role. But it is as important to understanding the language. How do children come to understand flow of control in Logo and similar procedural languages?

The text reading model. At the first level of analysis, this is not at all difficult to see. Within natural language, linear structure is used as a default model of the sequence of events being described. (Though it may be difficult to imagine, it could have been otherwise. For example, each statement might be required to include an explicit specification of its location in the sequence.) A sequence of imperatives in the form of a list of instructions is a cliché even closer to a sequential program: "Go to the light. Turn left. Follow Main Street...." It should not be surprising that such a common and powerful model would be spontaneously imported to a somewhat English-like and text-based environment as Logo programming.

The designers of Logo chose to support the *text reading model* without introducing much apparent structure, if any, into the surface form of the language. One can even write several commands on a line without additional syntax.

FORWARD 100 RIGHT 90 CLEARSCREEN

This makes for the ambiguity that computer scientists hate, but it perpetuates the profitable illusion that programs are text telling a story. Compare this to the structurally explicit forms in early Lisps (PROG) and blocks and semicolons of Pascal.

Exactly when and how quickly beginners appropriate the text reading model is an interesting question. That its acquisition is not always automatic is attested to by "recipe programs" that are sometimes produced by young beginners, consisting of a collection of commands with no evident ordering. Even after apparent acquisition, application is sometimes inconsistent. That is characteristic of distributed models, but no more so than of structural ones.

The text reading model typically succeeds some distance into learning about procedure/subprocedure relationship. It is easy enough to understand the use of a word to refer to the list of instructions.

> *This is not necessarily automatic. Children frequently mistakenly connect a procedure with its effect rather than the structural interpretation of a name as a reference to the list of instructions. For example, a child having written the square program shown earlier expects it always to produce a square. But if the turtle is rotated 45° before execution, the program surprisingly produces a diamond, structurally isomorphic, but not implied by the functional model abstracted by the child.*

Problems with the text reading model become apparent with recursion. Reading a text does not imply any new objects, such as a procedure invocation frame equipped with a new set of local variables. The near universal interpretation children adopt is that in a recursive call one starts reading the text again, from the top. Local variables do not have a place in the text reading view of recursion, and a recursive function returning values at each level is nearly impossible to interpret in that frame. Learners may struggle valiantly to find interpretations. But at this stage the text reading model of control has reached its limits of reliable use, and it is time for a structural model.

Control related lexicon. Like many other languages, Logo uses another mode beyond the text reading model to inherit understanding of control from natural language. This mode consists of specific control-related natural language constructions. IF <predicate> THEN <consequent>, and REPEAT are quite successful in their transplantation. IF-THEN's principle failure comes from a slight ambiguity with respect to its following text. In some circumstances, a consequent of the IF may be interpreted as an alternative to the text following the construct, as if it were followed by an ELSE. Then the consequent assumes a "go do this instead" interpretation rather than the nominal and correct "this step is optional, depending on the predicate." Equivalently described, the text following the IF-THEN may be interpreted as an alternative, not as a following command.

Consider the following two programs. In the first case, if the value of X is too large, the intention is to print a warning, and then do the operation ("DO.SOMETHING.WITH X"). In the second case, the intention is to do different operations depending upon the size of X. The first program works as intended. However, the second program has a bug: if X is large, then both operations are done, whereas the student intended only the first to be done.

```
TO DO.SOMETHING.CAREFULLY.WITH :X
IF X > 25 THEN PRINT [WARNING, LARGE X]
DO.SOMETHING.WITH :X
END

TO DO.SOMETHING.CAREFULLY.WITH :X
IF X > 25 THEN DO.SOMETHING.WITH.LARGE :X
DO.SOMETHING.WITH :X
END
```

These branches in the meaning of IF, depending on context, are frames abstracted from common experiences that frequently serve to disambiguate language. Not even a child is likely to ask what to do next at the performance of a play when told, "If a fire breaks out, leave the theater." "Turning on" one meaning or the other helps determine the meaning of succeeding discourse. The common vocabulary of such frames is vital to understand if one wishes to build reliable distributed models. Similar abstractions from experience with the physical world seem to constitute a significant part of physical intuition. And in like manner, it is important to know these in teaching physics (diSessa, 1983).

Solo, a language developed by Mark Eisenstadt (1983) patches this bug at the usual cost of adding explicit structure, loosening the similarity to language. The structure is a new slot to IF that specifies *continue* or *terminate* present context.

WHILE, which does not appear in Logo, is less successful than IF because of its ambiguity about the sequential relationship of predicate to action. See (Soloway, Bonar, & Ehrlich, 1983). Novices will assume the test of predicate occurs whenever necessary to insure its truth, rather than as the first step in the WHILE loop. The following Pascal fragment will add to Y one value of X greater than 5 before stopping, contrary to novice interpretation.

```
WHILE X < 5 DO
     BEGIN
     READ(X);
     Y := Y + X
     END
```

SUMMING UP

Formalists and Hackers: Structure and Function Revisited

I can package much of the preceding discussion into a set of heuristics for designers. We all would like to have simple, complete, uniform mechanistic models available for a computational system. But even if that is achievable, we must avoid the *formalist* bug. We must recognize that function is as important as structure in a language. Inventing methods and understanding the way some prewritten code works are two tasks that require knowing connections between language structure and its use. Function can also offer powerful leverage in coming to understand a language.

On the other side, we must avoid the *hacker bug*, which is in the extreme to allow a structure for each function in the language. Otherwise, we will never be able to trace through the detail needed to debug our programs that are written using guesses as to how something should work. Moreover, we will find aspects of the language hard to remember or to reconstruct without a consistent core.

A course of action that takes advantage of the particular characteristics of these models is to aim for a simple structural model, but to count on functional and distributed models at earlier stages of learning. In particular, the Logo procedure definition example suggests it may be harmless to pick a few important and commonly used constructs to learn primarily through functional and distributed means, even if these break the coherence of the structural model. In order to develop effective distributed models at early stages in learning, we must cultivate making connections to rich, pre-existing areas of human knowledge, such as natural language and spatial and configurational knowledge. Even if early access to the language is gained through functional models, we must expect and accept that certain classes of functionality, particularly those that violate early functional assumptions, become accessible only after full structural understanding is achieved.

Function/structure/function development. I cannot help thinking that the pattern of development I have implied is very general. From function (or through even less systematic means) one gradually develops a sense for the invariances, the structure underlying function, through experiences in a broadening set of contexts. Eventually, a new wave of functionality becomes accessible as this context invariance can play itself out in initially unimaginable contexts and functions.

No doubt this kind of thing happens in the evolution of complex systems like natural language. Categories like *subject* and *predicate* might evolve from useful semantic categories like actor and act. But after the grammar has acquired a life of its own, its structure can be turned around in passive sentences and in other ways to permit refined expression. Humor might have originally served humans only as a sense of violated expectations, but it is not only that now, and it serves a much broader set of functions. I am even more certain that the process occurs in the design of computer systems. Often function motivates a structure that, as it becomes better understood, gives rise to functionalities not even conceived of in its motivation.

Reflecting on the reasons for this pattern brings to light considerations that summarize nicely some of the important general themes of this chapter. The first, functional stage reflects the opportunistic connection to and exploitation of previous knowledge that one must expect if we want learning to progress spontaneously and naturally. The structural stage indicates a system with sufficient internal coherence so as to be understandable in its own terms. Such coherence is also important, in its own way, to learnability and stable remembrance. The final functional stage demonstrates that the power and expressiveness of the system indeed transcends the pre-existing knowledge systems, from which it might inherit a great deal.

Limits of the Analysis

It must be evident that many of the decisions about how to build understandability into a computational system must be made on the basis of the specific model building material, previous knowledge we hope to use, and on what exactly we expect people to do with the computational system. Even the coarse level of analysis developed here has shown great fine-structure. Depending on that fine-structure and other things, great variation in the application of these general considerations will be possible. If, for example, stereotyped use of the language primitives with little programming is assumed, a structural model is nearly superfluous. In some cases, we may be lucky enough to find the right medium so that structure and function can be nearly identified, the ne plus ultra of design. In the language design project for which these modeling considerations were invented (the language is called Boxer: See my Chapter 6) the static structures of the language, data types, etc., are melded to such a high degree with their visual, spatial presentation that structural understanding actually appears easier than functionality. But, more in line with the general move away from structure, we decided to give up even the systematicity of Logo with respect to uniform use of function application. We judged that to be necessary in order to adapt the language transparently to its use, allowing early functional modeling before a structural understanding of mechanism can be developed. (Details may be found in diSessa, 1985.)

Finally, I must disclaim the impression that these kinds of models are natural kinds as far as cognitive structure is concerned. I doubt one could tell one model from the other from their organization in the mind. Distributed models, in particular, obviously encompass multitudes of particular mental processes and configurations. But I hope these models capture typical patterns of the reuse and generation of knowledge. I hope they shake some assumptions about the simplicity of understanding, and the tradeoffs we must make in making a system understandable.

CHAPTER 11

Knowledge-Based Interface Design

WILLIAM MARK

Communication with human beings requires shared understanding of the way the world works—senders hope the receiver's understanding is similar enough to theirs to allow their messages to make sense. Communication with a computer must work the same way. Programmers endow an interactive program with some understanding of some aspect of the world. Users communicate with the program via a set of notations that the program will interpret in the light of its built-in understanding. To the extent the users share an understanding with the program, communication is effective.

User meets programmer at the user interface. With current interface design techniques, this is only a chance meeting. People think of the various aspects of their world in terms of particular sets of actions and objects. Programmers try to design programs that support the actions and objects they believe their user community has in mind (or will learn and find useful). Norman (Chapter 3) calls the programmer's conception the *design model*. The programmer implements the design model by writing functions and data structures that produce the desired behavior for the model's actions and objects at the interface. This interface behavior (along with all supporting documentation) is what Norman calls the *system image*. The user's understanding of the program, the *user's model*, is formed and continually refined through contact with the system image.

A lot can go wrong with the communication process given this setup. The programmer's design model may be severely at odds with the way the user community views that aspect of the world—so that the users have difficulty understanding the program in the context of their *ab initio* beliefs. The design model may be ill-structured or internally inconsistent, making it hard for users to form principles on which to build understanding. The design model may be misrepresented by the system image, so that the users' understanding of the program is askew from the programmer's conception (leading to misunderstandings and inconsistencies with respect to other parts of the program).

From the programmer's point of view, avoiding these problems requires knowing what it takes to produce a well-conceived design model and deliver it in terms of a system image that will lead users to form the appropriate user model. Much of this book is an exploration of these issues. Other chapters discuss two aspects of the programmer's problem:

- Criteria for formulating a design model that makes sense to users and is applicable to their needs (Hutchins, Hollan, & Norman, Chapter 5; Laurel, Chapter 4);

- Ways that the system image affects the formation and evolution of the users' understanding of a program (diSessa, Chapter 10; Lewis, Chapter 8; Lewis & Norman, Chapter 20; Owen, Chapter 9; Riley, Chapter 7).

In this chapter I address the specific issue of *delivering* the design model in the system image: Once a good design model has been formulated, how can the programmer ensure that it is well expressed in the system image?

I first propose a new conceptual model, the *system image model*: a conception of the various actions that could be available at a user interface. It offers a basis from which explicit conceptual models of particular systems can be constructed. I then describe a design methodology that provides an explicit statement of the system image model—one that is independent of any particular program. This independent statement establishes a terminology for dealing with the interface actions and objects that programmers are concerned with. The programmers then use the terms of the system image model to state the parts of their design models that are relevant to user interface behavior. This connects the programmers' conceptual models of what is happening in their programs to a conceptual model of what is happening at the user interface.

In order to deliver this combined system image/design model in the actual system image, the terms of the model must be *linked* to the functions and data structures in the program: To be a delivery mechanism, the model must *state* not what the programmers think should be going on in their programs, but what actually is going on when the program runs.

This business of "stating" and "linking" is made into a viable design methodology by application of some much-studied artificial intelligence technology. The conceptual model is expressed as an explicit knowledge base that enforces rules of soundness governing the use of terms. The knowledge base contains built-in examination and reasoning facilities that programmers can use to explore the consequences of utilizing various terms. The same facilities can be used to ensure that their terms are appropriately linked to the functions and data structures of their programs. The result is an explicit conceptual model of the effect of the design on the system image, grounded in the reality of program execution.

The explicit conceptual model and the knowledge base technology also provide direct benefit to the end user. Terms in the built-in system image model can be prelinked to natural language processing and other reasoning systems for recognizing user descriptions of modeled terms. By making use of these prelinked terms during the modeling process, programmers can automatically deliver powerful understanding and explanation facilities as part of the user interface to their programs.

The design methodology described in this in chapter is based on research for the Consul system (Mark, 1984), a vehicle for experimenting with knowledge-based user interface design for programs in the office automation area. The mechanisms and modeling examples in this chapter are all drawn from Consul research ideas. While I have felt free to "beautify" examples and system capabilities to make presentation easier, the basic principles of what is shown below have actually been implemented.

BUILDING THE EXPLICIT MODEL

The crucial first step in the design methodology is to get programmers to state their conceptual design models explicitly. They will be asked to state their design models in terms of a set of ideas that the system already understands. The idea of expressing something by describing it in known terms is an obvious one—what other way is there? What is less clear is what terms the system should know in order to provide programmers with a useful set for describing their design models.

Programmers' design models are their conceptions of what their programs *do*—that is, what behavior their programs exhibit. Although programmers' conceptions of their programs' behavior may extend to an arbitrary level of detail, the only part that concerns this methodology is the behavior that users can observe. Thus, the useful terms for the system to present to programmers are the ones that describe various kinds of program behavior that users could observe. This is the world of the system image: The system's known terms form a *prebuilt system image model.* For example, since "documents," "electronic mail," "writing," and "forwarding" are objects and actions supported by programs and available to users at the user interface, we would expect to see them in the system image model.

The trick is to build some understanding of these terms into the system, and then to provide a mechanism for allowing programmers to state their design models using these terms. That is, given what the system knows about what programs can do for users, it must be able to solicit from programmers *their* ideas of what *their* programs are doing for users. Programmers must first have a way of determining what the system does and does not know. Then, actually describing a design model requires some sort of dialogue. Ideas must be offered, and the system's understanding of the idea examined and confirmed or rejected. We cannot expect the system to have human-level knowledge, reasoning competence, or dialogue capabilities, but we can define (and, indeed, build) mechanisms for making the model-building task a tractable one (Mark, 1984).

The mechanisms described here are based on research in the knowledge representation area of artificial intelligence (Brachman, Fikes, & Levesque, 1983). Terms are defined by relating them to other known terms in a hierarchical knowledge base. In order to give terms "meaning" (i.e., beyond their relationship with other terms), they are associated with objects and actions that are already understood, or at least whose behavior can be tested and observed. In the methodology described in this chapter, terms are given meaning via their association with the functions and data structures of the programmers' programs: The meaning of each term will be defined by linking it to observable program behavior.

The Prebuilt System Image Model

To make things easier all around, the system should come with as much knowledge as possible—it will be easier to talk to, have a broader notion of how different aspects of the world fit together, and have more built-in answers to the hard problems that every programmer has to solve. Balancing this is the fact that different programmers have problems that require different conceptualizations of the world. The more the system knows, the more they have to disagree with.

As a compromise, the methodology offers a "core factorization of system image knowledge" as a basis. This is meant to be a definition of terms that everyone can agree on *for the sake of discussion*. That is, programmers should be willing to use the system terms for actions like "writing" and "sending" and objects like "documents" and "networks," as long as the system can say what these terms mean, and as long as the programmers can use these terms as building blocks for their own design models.

For example, the Consul system had a system image model relevant to programs in the area of office automation. Part of this core knowledge base—just to give you an idea—is shown in Figures 11.1 and 11.2.[1]

The system is shown knowing things like:

- There are actions and objects in the world (Figure 11.1).
- The "writing" action is a kind of "creating" which results in "information" being embodied in a "document" (Figure 11.2).
- Some objects are "electronic documents," "folders," and "networks." Some of the actions that can be applied to them are "copying," "editing," "deleting," and "browsing." (Figures 11.1 and 11.2).

The core knowledge base also contains inferential knowledge about these terms. For example, Figures 11.1 and 11.2 show the following rules:

- "Electronic documents" can be visualized as "text" if they can be edited by a full screen editor.

[1] The full details of what the system knows about these things cannot be shown here. Also the diagramming formalism (Brachman, 1978) cannot be explained in detail here—but the details are not necessary for an understanding of this chapter. Interested readers may consult the references for a full account of the knowledge representation philosophy and its use in Consul.

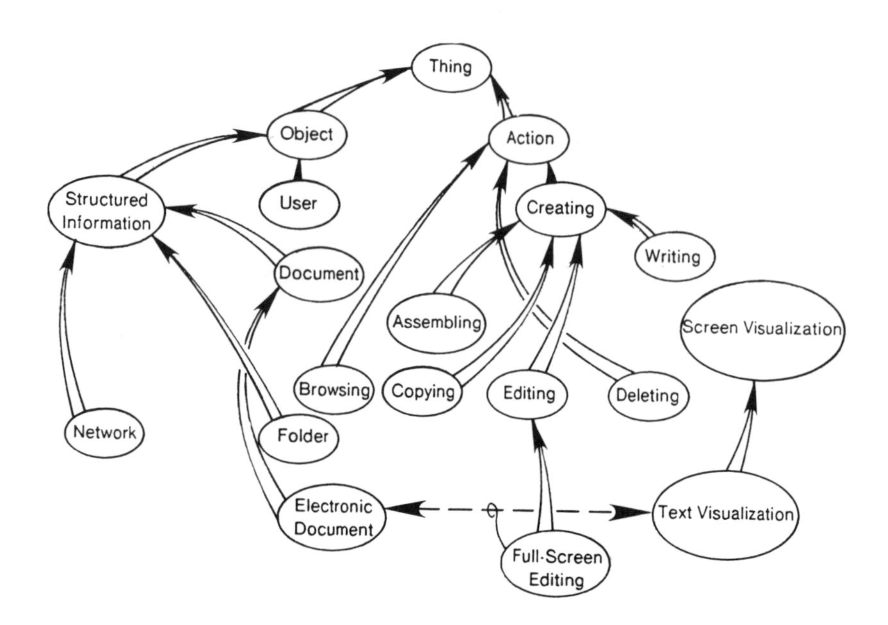

FIGURE 11.1. Part of the core knowledge base. Terms are shown as English words surrounded by ovals. The words are relevant only for presentation; each term is defined in the knowledge base solely by its relationships with other terms. One type of relationship, *kind-of*, signified by a fat arrow, is shown here: e.g., "editing" is a kind of "creating," which is a kind of "action." Terms are used in inference rules, shown as dashed arrows. The inference rule here indicates that something described by the term "electronic document" can also be described by the term "text visualization" under certain circumstances dealing with "full-screen editing." The precise circumstances under which the inference is valid are hard to represent visually, and are omitted in these diagrams.

- "Writing" on a computer can be conceptualized as the combination of two computer actions: one in which an electronic document is made "accessible" to the computer, and another in which a screen visualization of that electronic document is made accessible to the user.

Thus, the system is unwilling to call a computer activity "writing" unless it results in information (presumably in someone's head) becoming accessible to the computer as an electronic document, as well as becoming accessible to the user as a "screen visualization" of that electronic document.

Consul's system image model contained about 400 concepts like the ones shown circled in the diagrams. Different application domains would require a somewhat different set. At the current state of the art

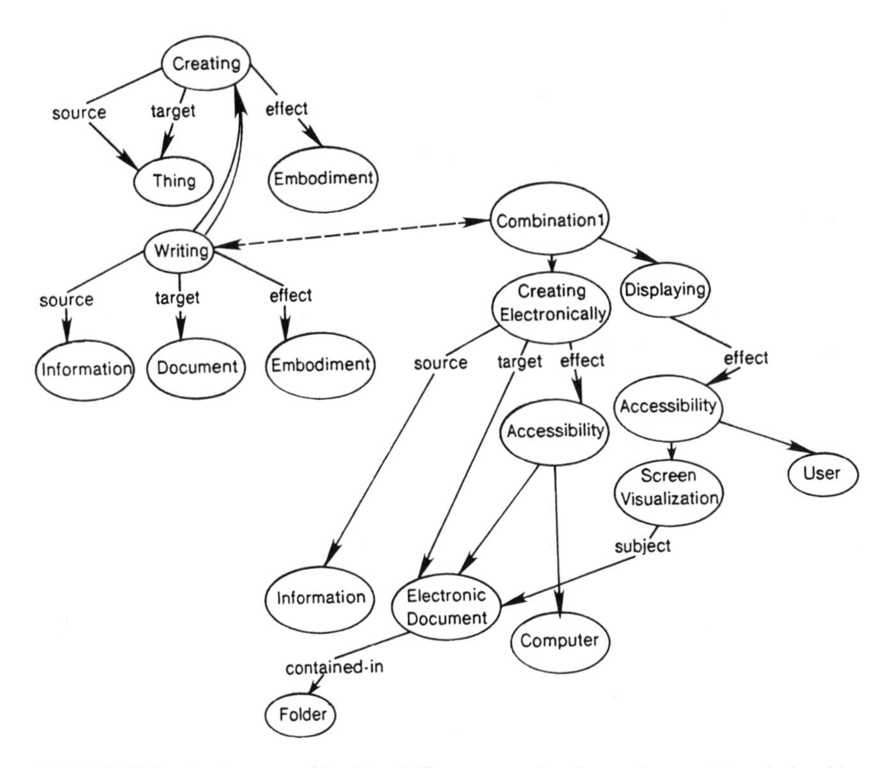

FIGURE 11.2. A close-up of "writing." Terms can also have *characteristic* relationships (shown as solid arrows) with other terms. "Creating" has "source" and "target" characteristics that must be "things," and an "effect" that must be an "embodiment." Characteristic relationships interact with kind-of relationships: The *way* in which a term is a kind of another is defined by its characteristics. "Writing" is a kind of "creating" in so far as it has source, target, and effect characteristics that conform to the requirements of creating. The knowledge base actually represents these relationships in considerably more detail than is shown. For example, the "embodiment" effect of writing and creating really relates the "source" to the "target" (i.e., writing's effect is the embodiment of "information" as a "document"). The inference rule indicates that under some circumstances, something described as "writing" can also be described as a combination of creating and displaying.

in artificial intelligence, deciding what to put in these initial models requires familiarity with the programs in the domain, augmented by some linguistic analysis and introspection on the part of the modelers. Again, I will not discuss in this chapter what should be in such a model. The stress here is on how to use the model to aid in the design methodology once it is built.

Exploring and Explaining

Given that the system has a prebuilt system image in some application domain of interest, the next problem is for programmers to be able to find out what the system already knows about what *they* want to tell it. If the system were another person, the programmers would just start describing and wait for questions (or questioning looks). The system has not reached this level of sophistication; it must put the onus on the programmers to describe things in terms it already understands—but it helps them find and use those terms. The approach is to provide programmers with a combined browser/explainer for exploring the knowledge base in the large, zooming in on details, and asking for English explanations of individual elements.

The system has a highly interconnected knowledge base of hundreds of concepts. Presentation of this information to the user can't rely on simple inspection of this complex data structure; it must be raised above this level in accordance with strategies of how various kinds of concepts should be described to programmers. For example, one strategy for generating English descriptions of a term is by comparison with other terms in the knowledge base. The terms chosen for comparison and the details chosen for actual presentation must depend on whether the term is an action or object, and on other characteristics of the term. Some sample browsing and explanation of the kind offered in the Consul system is shown in Figure 11.3. The programmer has browsed the knowledge base and selected the "electronic document" object and the "deleting" action for explanation (note the difference in the form and content of the two explanations).

Stating a New Design Model

But exploring the knowledge base is not the same as stating a design model. It is unlikely that the prebuilt model will have just the right terms for the programmer's model. The system therefore provides an extension to browsing and explanation that allows programmers to introduce new terms. Programmers use the browser/explainer to choose terms in the knowledge base that are similar to the terms they want to introduce (as a last resort, they can choose "thing," the most general term in the system's knowledge base). The system then provides forms based on the chosen terms for the programmers to modify and augment. The system monitors the form-filling process, confirming or rejecting a programmer's new definitions step-by-step according to what it knows about the chosen term (i.e., its relation to other known terms and the inference rules that apply to it). When the

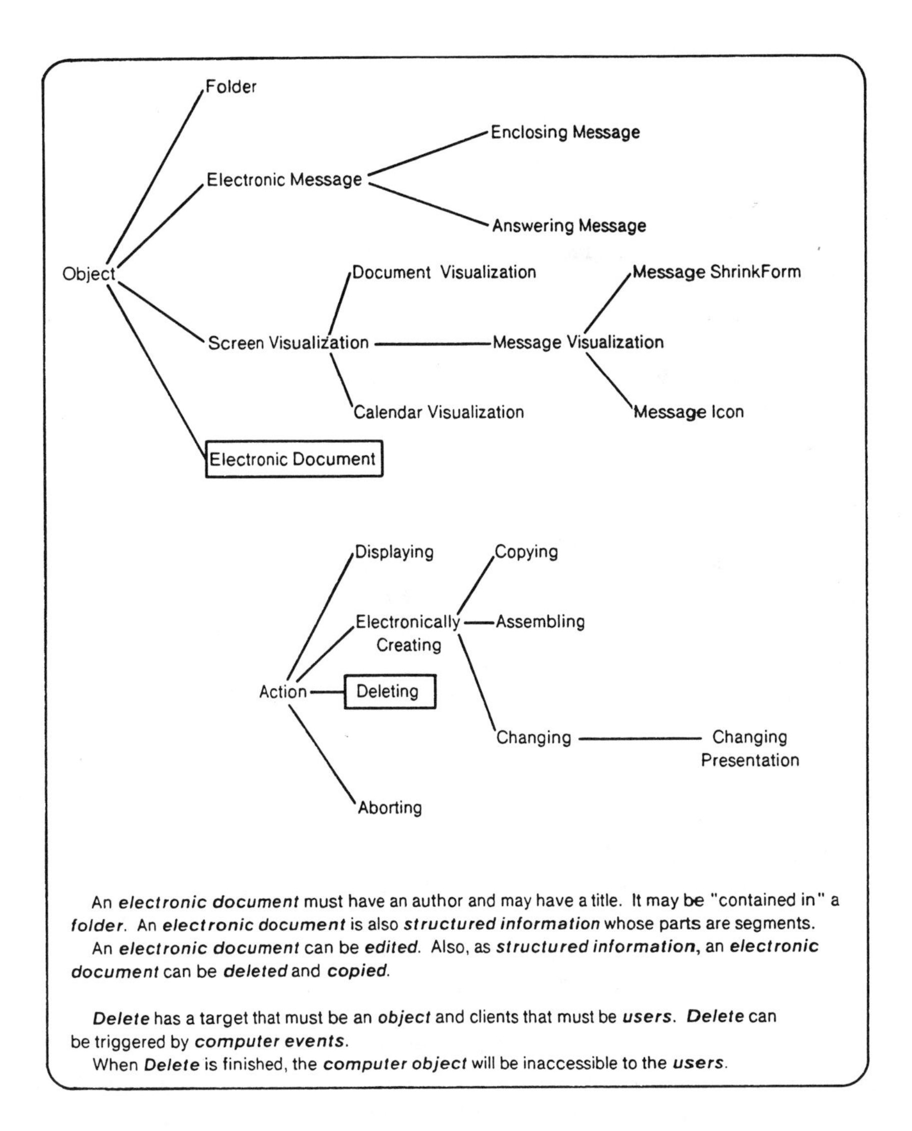

FIGURE 11.3. Examples of browsing and explanation. For browsing, the knowledge base is turned on its side, and kind-of relationships are shown as solid lines (e.g., "Answering Message" is a kind of "Electronic Message"). Any term can be selected (shown by boxing) for explanation. Selecting a term results in a short explanation being displayed. Thus, the figure shows the explanations for the two "boxed" terms: "Electronic Document" and "Deleting." The explanation is generated in the form of a series of English sentences—fairly stilted, but still readable—describing the term in relation to other terms.

programmer is finished, the system translates the information in the form into the representation formalism, and adds the new definition to the knowledge base, automatically relating the new term to known terms according to what was specified in the form. The system can explain to the programmer what these relationships are, and warn of conflicts with existing knowledge.

The system-established relationships for the new term may or may not be what the programmer expected. An advantage of this form of "dialogue" over the human variety is that if programmers disagree with the system's assessment, they can examine the *precise* differences between their new terms and the terms the system knows. But, unlike human dialogue, programmers can't convince the system to "change its mind." For example, a programmer cannot introduce a new kind of electronic document that doesn't have a title, because the system's electronic document term assumes a title. The system can't determine the ramifications of changing some arbitrary aspect of what it knows (i.e., it doesn't know how electronic documents that don't have titles might differ from electronic documents *behaviorally*). The programmer can still introduce a term for "an electronic document, except that it doesn't have a title" by defining the new term as a kind of "structured information," a term more general than "electronic document." In this way, the term will automatically share all characteristics of structured information, but will not conflict with the system's view of electronic documents.

The programmer thus introduces each new idea by extending the prebuilt model, getting more or less direction from the system depending on how close the new idea is to something the system already understands (as one would expect). The system in turn checks each new idea to make sure that it is consistent with what it already knows. This often requires the system to solicit more information from the programmer as it follows necessary connections to other concepts in the knowledge base—a process that continues until all the ramifications of introducing the new idea have been explored. The role of the dialogue is to guide the programmers through these ramifications step-by-step so that they need only make well-understood "local" decisions at each stage. The result of each dialogue is an extension of the prebuilt system image model to include the programmer's design idea.

For example, suppose that a programmer is building a new program to support the writing process. The programmer starts by examining the knowledge base to see what the system already knows about writing, finding a model that requires the creation and display of and electronic document, as shown in Figure 11.2. But our programmer wants "writing" to mean assembling "notecards" into a network based on defined relationships between notecards.[2] The writer is then to arrange and rearrange the network to suit an evolving notion of what the document should say.

The programmer may browse the knowledge base and request explanations in order to understand the system's concepts of writing, electronic creation, display, etc. At some point the programmer requests a form for defining a new writing term. The upper part of Figure 11.4 shows the original form the system provides; the lower part shows the form eventually filled out by the programmer as a result of the dialogue process. The actual dialogue process is variable (the programmer may try various formulations, ask for explanations, etc.), and is at any rate too long to show here. A likely scenario would be that the programmer starts by introducing "assembling notecards" as one of the combination of activities that is to form the design model for "notecard writing." The programmer could either immediately define "assembling notecards" or defer this subdiscussion until later. When the definition is established, the system must be sure that the notion of notecard assembly fits in as part of the combination of creating and displaying that must constitute any model of computerized writing. In knowledge base terms, this means making sure that the "effect" and other characteristics of notecard assembly are compatible with the effect and other characteristics of the built-in "Combination1" concept in Figure 11.4. In this case, the only question is whether a "notecard network" is compatible with the system's concept of an "electronic document."

The system would therefore request a definition of "notecard network," offering a form based on "electronic document" for the programmer to fill out. If the programmer's ideas about notecards are compatible with the system's requirements for electronic documents, the dialogue can simply continue. If the ideas are not compatible, the programmer must either revise the design model of notecards or decide that "notecard-writing" relates to something other than "writing" in the system image model (i.e., perhaps the programmer would be better off to establish notecard-writing as a kind "creating").

All system/programmer dialogues follow this pattern. At each stage, the programmer selects a system concept to build on, and the system provides a form to show the programmer what that entails. Usually the process requires exploring the relationship of the originally selected concept to other concepts in the model. The dialogue therefore appears

---

[2] This is the idea behind the NoteCards system at Xerox PARC (Brown, 1985), on which this example is based (with no obligation to be true to the actual NoteCards system).

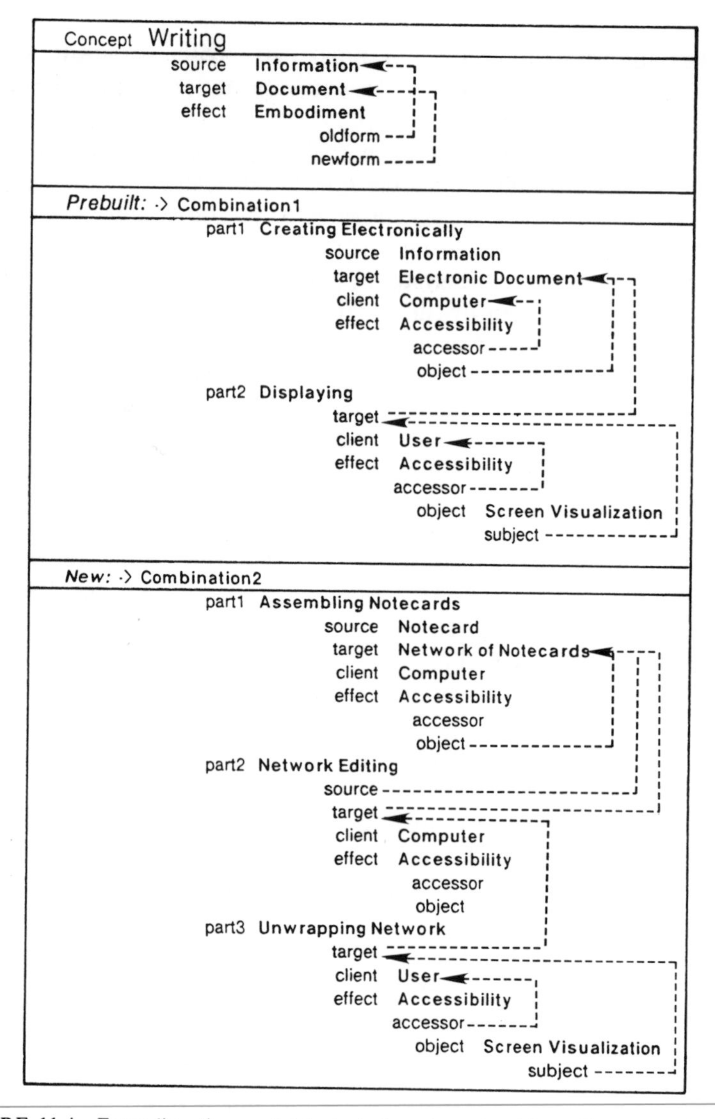

FIGURE 11.4. Extending the model to include notecard-writing. Forms present terms by their characteristics; by modifying these characteristics and (sometimes) adding new ones, programmers can create new terms to suit their purposes. Here, the characteristics of "Writing" and its prebuilt inference rule are shown in the top part of the form. Dashed arrows show required relationships between characteristics (e.g., the target of "Displaying" must be the same as the target of "Creating Electronically"). The programmer has chosen to specify a new inference rule so that something described as "Writing" can also be described as a combination of "Assembling Notecards," "Network Editing," and "Unwrapping Network."

as a continual probing of the ramifications of the programmer's ideas, progressively checking to see if they are compatible with the system's overall knowledge of the subject area. As long as the programmer and the system are on the same wavelength, the dialogue continues and results in the system's model being extended to include the programmer's design (see Figure 11.5). If the programmer cannot agree with the system, it is not a disaster; the programmer simply re-explores the knowledge base to find another (usually more general) concept that *is* compatible with the design. This may in turn require

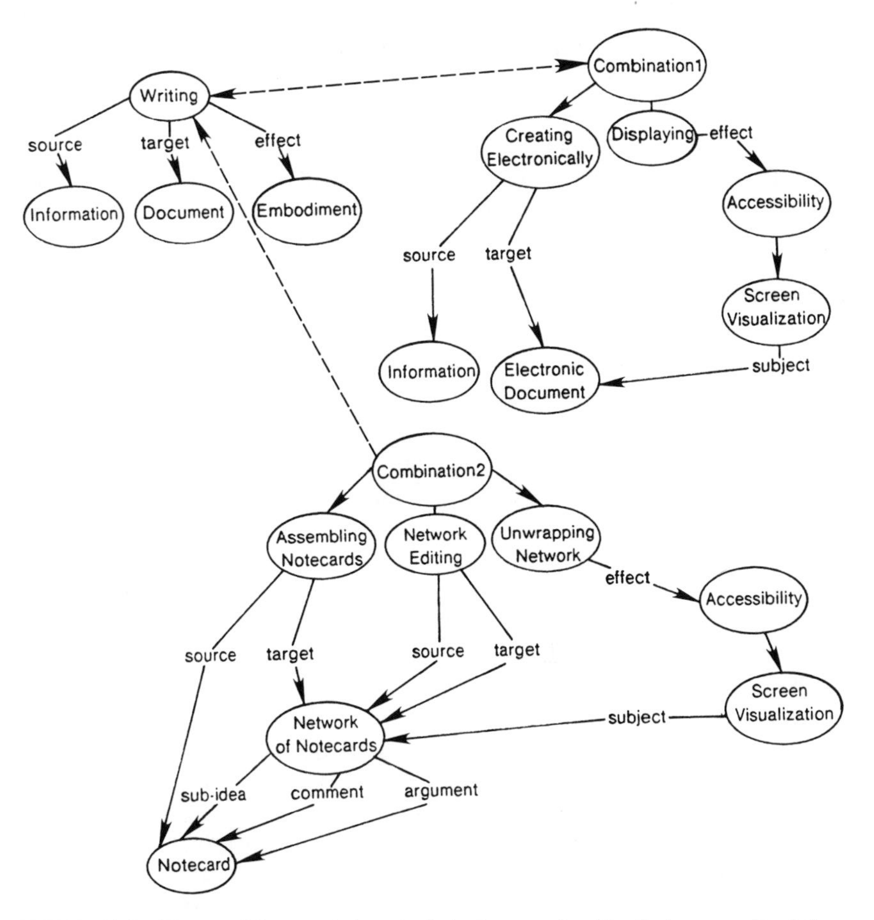

FIGURE 11.5. The model extended to include "notecard-writing." As a result of the programmer's form-filling, the new term "notecard-writing," a.k.a. "Combination2 "in Figure 11.4, has been added to the knowledge base. The inference rule states the conditions under which "Writing" can be viewed as "notecard-writing."

the programmer to change previous selections to more general concepts to achieve overall compatibility, but the process can always have a happy (or at least reasonable) ending: an extension of the system's model to include the new design. The more specific the terms the programmer can find to agree with, the more valuable will be the system's built-in model and reasoning capabilities.

LINKING THE MODEL TO THE PROGRAM

Once a model has been built, it must be made to define the physical interface and the way the rest of the program supports that interface. This design methodology is intended to deliver the conceptual model in the running program. Actual software must be linked to the model such that it implements the programmer's modeling decisions.

The easiest approach would be to prebuild implementations of knowledge base terms. That is, when programmers use an existing term, they buy into an implementation as well. But this is inadequate. First of all, implementation decisions are often based on structuring and efficiency concerns that are hard to generalize (i.e., although the existing knowledge base term may be just right, the included implementation may be all wrong). It will rarely be the case that a general built-in implementation will be adequate; and, unfortunately, the knowledge base technology does not extend to introducing new *implementations* by specialization. Also, different individuals or groups may be working on the implementation; and the model and the program will undoubtedly evolve over time. Prebuilt software is too inflexible to support the needs of individuals or groups over time.

The solution developed in the Consul system (Wilczynski, 1981) was to extend the dialogue process to include incorporation of descriptions of programs and data as well as design model concepts. Just as in building the design model, programmers have a form-filling dialogue with the system until they and the system agree that an implementation is valid. The system's opinion is based on the programmer's previous modeling and implementation decisions, as defined in terms of its prebuilt model and reasoning facilities. As before, the programmers use the browser/explainer to explore the knowledge base, and form-filling to link their implementation constructs to existing terms. The system incorporates these links as inference rules that establish the conditions under which a particular knowledge base term can be viewed as a particular program construct. It will allow or disallow a proposed implementation based on whether it accounts for all of the characteristics of the term and accords with existing inference rules about that term.

A snapshot of the linking process is shown in Figure 11.6. As before, the programmer has used browsing and explanation to focus on the design model term to be implemented, in this case the "notecard" object. The system presents the programmer with a form based on what it knows about notecards. This time, instead of using the form to extend the model, the programmer presents an implementation—the actual datatype that will define this object in the running system.

Programmers can implement a program construct for a term like "notecard" any way they want as long as they account for all the required properties specified in the model. For example, the notecard model requires that notecards have a "file box" relationship with a "folder" object. Any implementation of a notecard *must* include an implementation of this "file box" relationship. What this relationship entails depends on the modeled characteristics of notecards and folders. In this case (as shown in previous model diagrams), since a notecard is

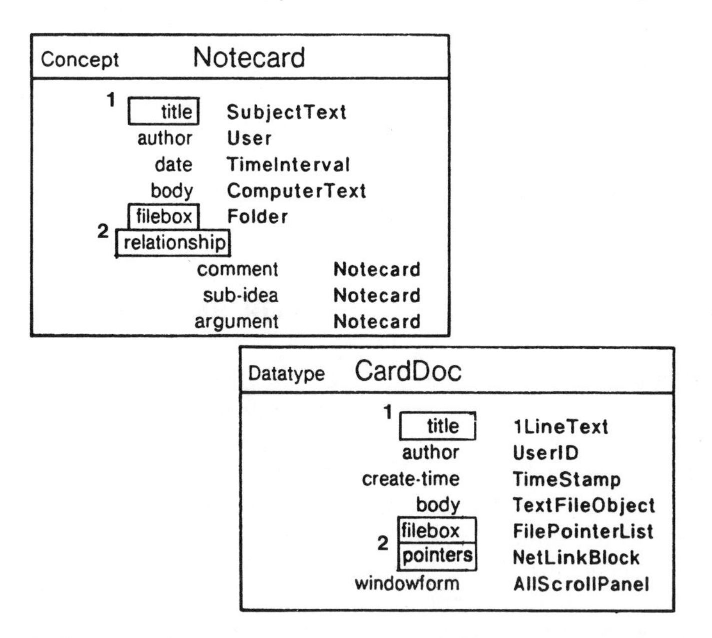

FIGURE 11.6. A term linked to a program construct. The notecard term is linked to an actual datatype by relating elements of the forms describing each. The "title" and "pointers" fields of the datatype have already been associated with the "title" and "relationships" characteristics of the term. The "filebox" field is now being considered, as described in the text. Links are allowed only if the value type for each field can in turn be linked to the term specified in the corresponding characteristic (e.g., "1LineText" must eventually be linked to "SubjectText" if "CardDoc" is to be linked to "Notecard").

a kind of "electronic document," and since electronic documents are necessarily *contained in* folders, the programmer's implementation of notecards must be capable of being "contained in" whatever implementation the programmer chooses for folders.

The system determines whether such a containment relationship exists by using the rest of the programmer's design model and its prebuilt model: in this case, the part of the model that deals with actions that move electronic documents in to and out of folders. The system will probe the programmer until it is sure that the programmer's implementations of these actions take as the appropriate arguments *CardDoc* and the suggested folder implementation, *FilePointerList*. The process continues until each part of the *CardDoc* datatype and all functions that work on it have been checked with respect to the model. The system is simply using its built-in knowledge base, as extended by the programmer, to make the programmer aware of the ramifications of the programmer's own conceptual model.

If the programmer cannot describe some feature of the implementation in a way that the system sees as valid, something is wrong. The design model is somehow inappropriate for the implementation, or vice-versa. The programmer must then modify the design model or the implementation. This is of course the whole intent of the methodology: to influence the software design process by enforcing design model decisions in the implementation.

BENEFITS OF THE APPROACH FOR PROGRAMMERS

Why should programmers go through this term-building and linking process rather than relying on their own internal mechanisms for consistently delivering their designs in terms of software? The answer goes beyond mere "bookkeeping." A crucial feature of any delivery mechanism is that it ensure that similar actions and objects behave similarly, with "similarity" being defined via an appeal to a more abstract conceptualization. This kind of coherency is hard to achieve for any model of significant size. It's even harder for large models built by groups and maintained over long periods of time. It's still harder to achieve across the boundaries of models built by different individuals or groups.

One of the major advantages of having a knowledge base to build models in is that it can provide abstractions to define similarity. It offers a framework in which an individual programmer or groups of programmers can express their designs in terms of a common system image model: a set of abstract concepts of how objects and actions should behave at the user interface. For example, the knowledge base defines "creating" as the action that results in the embodiment of something as a new thing. This is a particular set of assumptions about an aspect of the world: that things come into being as a new form of something else. "Writing" is treated as a kind of creating in which "information" is embodied in document form. Any individual programmer can examine these assumptions since they are expressed as terms in the knowledge base. The system will insist that any model of writing share this abstraction. The abstraction thus defines what the system believes all ideas about writing must have in common: They must be similar in precisely the sense specified by the "writing" concept.

This notion of similarity has consequence through the system's inferential knowledge about the abstract concepts in its knowledge base. One of the things the system knows about writing is that it can be interpreted in the computer world as creating a particular kind of machine-accessible form that can also be made accessible to the user. This further definition of what the system means by writing has direct bearing on the models that can be introduced by individual programmers. Whatever the programmer's notion of writing, it has to share the system's assumption that the activity will result in an electronic medium that both the system and the user will view as an "electronic document"—otherwise it ain't writing.

Stating design models explicitly in terms of prebuilt abstractions therefore allows programmers to sort the features of their designs into *conceptual categories*. This greatly reduces the burden of consistent delivery in the implementation: The programmers can view their problem in terms of implementing groups of similar things, rather than individual objects and actions. The design methodology helps programmers decide which things belong in which groups. Then it helps them ensure that the characteristics of each object and action are embodied in the implementation consistently with respect to the other members of its group.

Besides using the abstractions to define similarity, the system uses them as the anchors of its reasoning knowledge. Any knowledge-based system has to include some fundamental assumptions about how the world works: how change is to be dealt with, how universals like time and location are to be represented, and so on. In addition, the knowledge base must have built-in primitive terms ("information," "accessibility," "containment," "combination," etc.) from which all other concepts are defined. Building in the assumptions and the primitives saves work for individual programmers. In fact, it may well be necessary in order to make modeling—and providing system reasoning facilities—feasible. The crucial saving for individual programmers is that they don't have to start from scratch every time they formulate a design model.

BENEFITS FOR USERS

Dealing with computers means dealing with complexity. Computer users are faced with a tremendously complex set of machine capabilities and limitations. To help them deal with complexity, programmers provide self-contained problem-solving worlds that users can deal with more easily. Designing and implementing these worlds is in itself a complex task, and so far this chapter has been about making the programmers' task more tractable. But does the approach presented here lead to a system that reduces complexity for the *users*, which is after all the real goal?

No matter what anyone does, users are going to form some model of the computer system they are using (cf. Lewis, Chapter 8). Jim Hollan uses the term "shells of competency" to refer to the successive models users build as they become familiar with the system. These shells are artifacts of both user perception and system design. Users react to the complexity of the new environment by mapping some aspects of it into an internal model that they can deal with. System design—especially the characteristics of the user interface—can lead users to form certain models and even change them over time. User interface design can be viewed as a problem in controlling the formation and evolution of these models.

The thesis of this chapter is that the programmers' best approach to this design problem is to create explicit models to help them deliver their designs. But these models also benefit the users directly. These models are in fact what the users will assimilate through the system image. To the extent the models are built in terms of a worked-out framework that provides structure of abstractions the users can share, the users assimilation task is made easier. When programmers build their design models from a foundation of prebuilt abstractions and inferential knowledge, they provide a basic "physics" of the user's world (actions and objects behave according to certain principles). As users' views of a particular program evolve they can be sure that the same physical laws hold.

Grounding the model in the actual software allows users to make dependable assumptions about system behavior; to predict what the system will do even in areas where they have no direct control—to trust the system. There is little hope that a user's model can accurately predict the behavior of something as complex as a large computer program over a wide range of situations unless that model is somehow tied directly to the code. Users rarely have the chance (or desire) to become involved in the program design and implementation process. They must be given concepts that they can relate directly to their experience—the world of the system image—and that are already reliably grounded in the working software.

Finally, users must be able to deal with the program in a language defined by their communication needs and skills. They must know how to describe what they want to do or need to know, whether through physical manipulation of things on the screen or typing in words at the keyboard. The descriptions must make sense to the users and of course to the program as well. The process of making the program understand the users' descriptions can be arbitrarily complex—as long as the users don't have to be aware of the process. They must, however, see the program's understanding process as useful, consistent, and reliable.

The previous section described the benefits to programmers of having abstractions built in to the system's model. These abstractions can be made to directly benefit users as well. An approach explored in the Consul system was to tie input understanding capabilities—for natural language, command, or graphic interaction—directly to the built-in abstract terms of the system image model (Kaczmarek, Mark, & Sondheimer, 1983). For example, a term like "electronic document" was tied to an English understanding system that associated user input phrases like "my memo about overhead rates" or "the message I sent to John yesterday" with the electronic document terms of a programmer's design model. Attached referent-finding reasoners then made the association between these terms and actual data structures in the running program.

When programmers state their design models using system image model terms, they automatically buy into these user input understanding mechanisms as well. The system already knows how to associate English phrases with terms in its prebuilt model. When programmers use prebuilt terms to state their design models, the system can associate English phrases with the programmers' terms as well. When the programmers link their design models to the data structures and functions of their programs, the system can use the links to find the referents for the English phases (e.g., the actual data structure representing "the message I sent to John yesterday").

The methodology can therefore offer sophisticated mechanisms for translating users' input forms into programmers' design model terms, and, through linking, into references to actual functions and data structures.

CONCLUSION

The overall goal of the approach presented in this chapter is to help programmers deliver their designs—not only by reducing the complexity of the delivery process, but also by helping to ensure that the delivered system image provides a good interface for users. The prebuilt system image model provides the programmers with a set of descriptive terms to manipulate—a framework for expressing designs. The methodology helps to make sure that these designs make sense to the programmers by enforcing rules of sound modeling, and by insisting that design concepts be explicitly linked to the actual software that implements them. The methodology also helps to make sure that the designs make sense to the users by insisting that the design concepts participate in a coherent model of the system image, and by linking the design concepts to input forms that are natural to the users. The programs that result implement the programmers' intent: user understandable miniworld within the machine.

SECTION IV

USER ACTIVITIES

The point of the three chapters in this section is that people interleave their activities—and do so in many different ways. Real tasks can take hours, weeks, or even months to perform. As a result, other tasks must of necessity intervene. Some tasks are so important they are allowed to interrupt others. Some tasks require the generation of subtasks, sometimes to satisfy prerequisite conditions, sometimes as corequisites, sometimes simply as relevant, but not necessary components. And then, some tasks are so onerous that people, sometimes deliberately, allow *displacement activities* to take over.

On a broad scale, people and computers often do several things at once: A person can cook dinner and carry on a conversation; a computer can execute multiple processes (time-sharing) while performing input-output operations on its peripherals. However, both computers and humans use an important part of their capacities *linearizing*: carrying on several tasks by switching among them, so that at any instant only one of them is actually being done, even though on a broader scale they are all being done in parallel.

Linearizing carries with it a bookkeeping burden. Any activity carries with it a "context"—all the things that are particular to this instance of the activity. Each time an activity is suspended this context must be stored: Each time it is resumed, the context must be recalled. Thus, linearizing requires support for remembering and recalling. In addition, the fact that an activity is still waiting for completion must itself be remembered and recalled.

To our knowledge, these aspects of real tasks have seldom been discussed, let alone supported (or, in the case of displacement activities, discouraged) by computer systems. The three chapters in this section discuss these aspects of real user activities and suggest several different ways in which computer systems might support the performance of multiple activities. Cypher discusses the situation from the computer's point of view, of how system support could be provided. Miyata and Norman discuss the issues from the psychological point of view. And Reichman shows how the analogy to similar situations in conversation provides useful insight into the problem and shows how analyses of conversational gambits can suggest techniques for allowing people to switch among tasks and, perhaps more important, to resume where things left off with a minumum of disruption.

The three chapters in this section are motivated by the understanding that people usually have many things to do, some of which are related to each other and some which are not. Some tasks take such a long time that other things are also done along the way. Unexpected events occur frequently, interrupting ongoing activities. As a result, normal activities are filled with simultaneous or overlapped tasks, with things deferred until later, or planned tasks awaiting an appropriate time to be done. These factors make behavior complex, even if each activity is relatively simple.

The basic phenomena of interleaved tasks and the basic modes of computer support for people's activity structures are introduced in Chapter 12 by Cypher, *The Structure of Users' Activities*. Cypher approaches the problem from the point of view of the user who is faced with the need to deal with multiple activities that simultaneously compete for attention. He treats the competing activities as interruptions of the ongoing activity and discusses ways to aid the user to transform the simultaneous demands into a linear sequence that can be dealt with one at a time. He discusses the varying ways in which activities may need to share contexts, and establishes in detail the point that people's computer usage is not, in general, in a simple one-to-one relationship with their activities. Good support for activities must therefore allow flexible groupings of processes. He then makes some proposals for activity support in interfaces, particularly in connection with the function of reminding.

The chapter by Miyata and Norman, Chapter 13, supplements the one by Cypher, expanding upon some of the issues and pointing out the relevance to some traditional studies within psychology. Cypher's analysis was based on the activity flow to both the user and the computer. Miyata and Norman take the psychological point of view, centering upon what is going on inside the person. They focus more heavily on humans' abilities to pursue multiple activities and the effects of the relevant cognitive limitations. They thus develop a more detailed analysis of reminding, including the distinction between the signaling component and the descriptive component of a reminder, and hence arrive at proposals for supporting this in interfaces.

Reichman, in Chapter 14, analyzes how users use window systems in an attempt to support their multiple activities. She discusses the developmental trend of computer interfaces leading up to windows that has influenced users to expect more from their systems and to act as if they were having a two-party conversation with the computer. In conversation we switch topics, usually signaling this to the other person by simple clue words such as "well anyway," without recourse to explicit meta-linguistic references to topic. Reichman argues that interfaces need to achieve a parallel smoothness of use and that the lack of such mechanisms in existing systems shows up in various annoying and persistent errors. She then presents a set of contextual, navigational, and visual techniques that, if added to current window managers, would support users' new conceptualizations of the human–machine interface.

Technical Issues

All three chapters depend on the fact that many systems today make a start toward supporting multiple-user activities. The most frequently discussed technique is the use of windows. However windows are not, technically, the heart of the matter.

One important issue is whether one user may have more than one interactive process active at once. Multiple computer jobs are clearly possible, for that is the essence of time-sharing, but with most of today's smaller, personal computers, only one interactive task can be run at a time. Multiple processing is possible on systems such as Berkeley UNIX that permit "suspending," "backgrounding," or "detaching" a process and then "resuming," "foregrounding," or "reattaching" to it later, thus allowing one user at a single terminal to alternate among several interactive processes without terminating and restarting them. The great advantage is that the system context of each process is preserved. Windows add the preservation of the current output state of each process, which can then serve to preserve some of the user's mental context for the associated activity. Windows also serve as a visual reminder of the existence and identity of unfinished activities.

In the world of personal computers, windows face stiff competition from "integrated" systems, systems that achieve by other means the most important aims of a window system—to make it easy, for users whose complex tasks require multiple subactivities, to switch rapidly and easily among multiple parts (programs) of the system, transferring the results of one system into the data structures of another. Whether the system has windows or not, the overall activity will only succeed if the two programs share a common representation with which to pass data back and forth—"integrated" systems stress this interprogram data transfer compatibility. Although windows are a leading technique in supporting multiple user activities, they are neither necessary nor sufficient, and, as discussed in Reichman's chapter, they are not yet supporting users' activities correctly.

From the computer science point of view, the two technical issues a designer must be concerned with in order to support multiple concurrent user activities on one system are, one, allowing the user to move between interactive programs without losing the context built up within each (this is especially important when the two programs are not part of the same user activity), and, two, ensuring that separate programs can be used together to pursue a single activity with an easy transfer of data among them. From the psychological point of view, the technical issues concern support for interruptions, reminders, and resumption of activities, all of which could be viewed as support for the natural and efficient saving and resumption of context. The chapters in this section provide valuable insights into the nature of real activities and into what will be required to support them in a smooth, even if tacit, manner.

CHAPTER 12

The Structure of Users' Activities

ALLEN CYPHER

The exasperated shout, "I can only do one thing at a time!" usually comes from people attempting to do six things at once, at a time when they are only capable of doing five. One way we do five things at once is by interleaving the activities. We dice the chicken while waiting for the water to boil, turn down the heat on the rice while frying the onions, but wait a moment to take out the bread because the white sauce is just starting to thicken.

Whether we are preparing dinner, debugging a program, or getting married, a good part of our mental energies are spent *linearizing*. The many parallel tracks of activities must be organized into a single linear stream of actions to be performed.[1] In a sense we are very skilled at scheduling multiple activities: It is a fundamental process which is a constant part of our mental life. But at the same time, we make a lot of scheduling errors: We let the bread burn while stirring the sauce, or let the sauce get lumpy while removing the bread from the oven.

The field of ergonomics attempts to design objects that take into account the realities of the human body—if a keyboard is too high, the typist's wrists will have to bend and typing will be uncomfortable. In

[1] For a discussion of some of the complexities of multiple activities, see Chapter 13 by Miyata and Norman, especially the section entitled *Multiple Activities: Current and Suspended.*

an analogous manner the field of human–computer interaction attempts to design interfaces that take into account the realities of the human mind. In this chapter, I deal with the reality that people do not simply perform one activity at a time. Program designers put a great deal of effort into allowing users to perform single activities well, but considerably less effort goes into allowing users to arrange those activities. If computer systems are designed so that they actively support and facilitate multiple activities, they will be more comfortable for the user.

CHARTING THE FLOW OF ACTIVITIES

Our computer system provides support for multiple, interleaved activities.[2] In order to get an idea of how frequently users actually do interleave activities, and in order to get a sense of the types of interleavings that occur, I modified our local computer programs so that they would keep a record of each command typed by the user. Whether the commands are to the top-level interpreter or to the text editor or the mail program, they are all collected in the order in which they are typed. Therefore, interleaved activities show up as interleaved sequences of commands. Since it is sometimes difficult to associate commands with the activities which they serve, users may also insert comments explaining their current activities. The Appendix describes the history-collection program in more detail.

Figure 12.1 diagrams the results of one of these historical records. It charts the flow of my activities during one morning of computer use. In the figure, each activity is shown as a box, with interruptions shown as gaps in the shading. Important subactivities are shown as sub-boxes.

It can be seen from the figure that there are numerous cases of interleaved activities. Let's examine some of these cases in detail.

EXAMPLE 1. Three interleaved activities: **read mail ⟶ reposition window ⟶ msg conversation**

By starting at the lower left hand corner of Figure 12.1 and following the arrows, you can see how I began my day with three interleaved activities.

---

[2] The system includes a VAX computer and a network of 15 Sun workstations. All use the Berkeley UNIX operating system and the Sun supports multiple windows.

> I start off the session by **reading my mail** [1]. While I am waiting for that program to start up (it takes about 10 seconds), I type "toolplaces" in a second window — I was annoyed at the default location of the "history" window, and "toolplaces" will help me **reposition the window** [2]. "Toolplaces" takes about 30 seconds to run, so I start up an electronic **msg conversation** [3] by sending off a message to a couple of friends. I then resume the **reposition window** [2] activity: the "toolplaces" program has finished running, and it tells me that the history window is already in its correct location. But that can't be right.... I am confused so I return to **reading my mail** [1].

I temporarily abandoned both the mail program and "toolplaces" because they are slow to start. In the mornings I am impatient with even short delays, but later in the day I still will be impatient when the computer spends several minutes compiling a program. "Waiting for the computer" is a common impetus for interleaving activities.

EXAMPLE 2. External Interruptions: **helping A**

> A short time later, as I am about to start the activity **respond to P's message** [7], I am interrupted by a phone call [8]. User A wants some help with the "fmt" command. I consult my personal database to get the information and then log in to her machine to try it out.
>
> Just as I finish, the message "You have new mail" flashes across my screen. I invoke the mail program and read the message, which continues the **msg conversation** [3] I started earlier. After reading the mail, I have trouble remembering what I intended to do before the interruptions. I eventually recall that it was **respond to P's message** [7].

In this case, two interleavings are prompted by events external to the user. External interruptions can be particularly disruptive since they need not occur at a natural transition point for the user. Computer systems can be helpful in these cases if they are able to provide reorienting information when the user attempts to return to the interrupted activity.

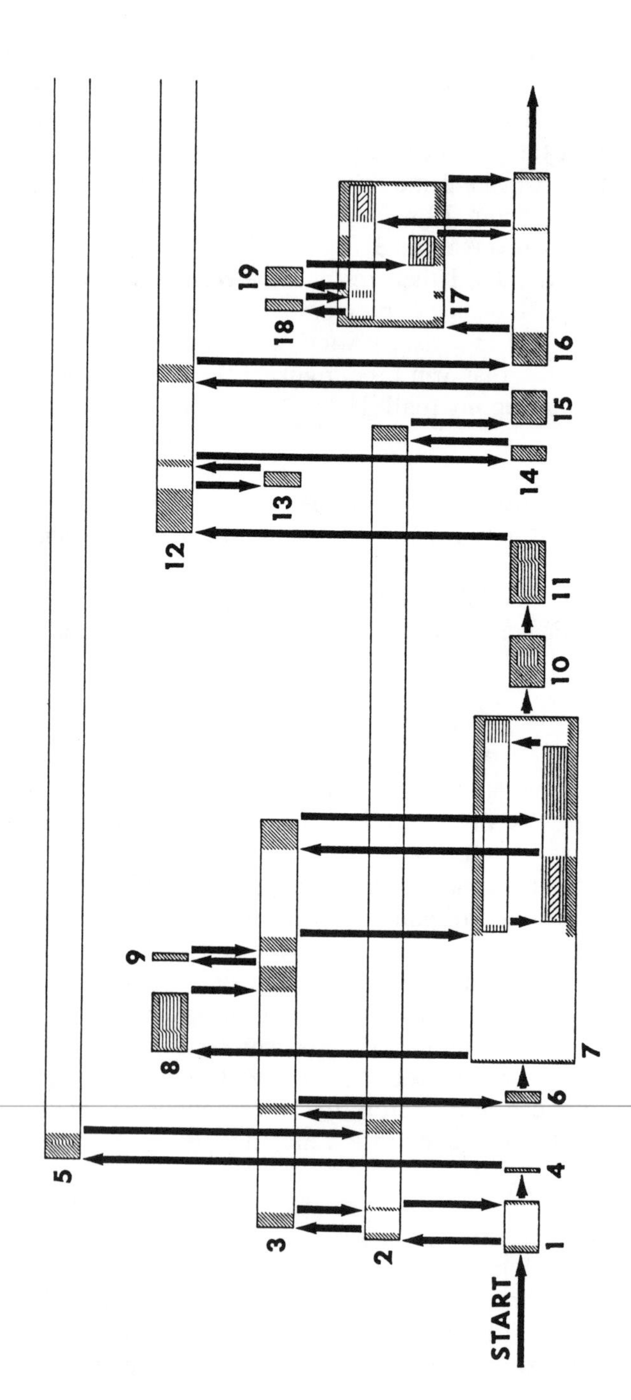
START
1
2
3
4
5
6
7
8
9
10
11
12
13
14
15
16
17
18
19

1. Read Mail
2. Reposition Window
3. Msg Conversation
4. Check Reminders
5. Arrange a Meeting
    - Check Calendar
6. Delete Old Messages
7. Respond to P's Message
    - Send a Reply
    - Set Up a New Account
        - Log In to Remote Computer
8. Help A
    - Find Note About "fmt"
    - Try It Out
9. Delete Outdated Message
10. Mail From Y
    - Find History Programs
11. Fix the Clock
    - Read Documentation
    - Ask for Help
12. Read Over Printouts
13. Look at a Note
14. Play With Windows
15. Read New Mail
16. Make a Note
17. Save It as a Good Example
    - File It
        - Find the Sub-Bin
    - Describe the Example
        - Locate the Text
18. Retitle a Note
19. Make a Main Bin

FIGURE 12.1. This figure charts the flow of activities during one morning of computer use. There are 19 activities, shown as boxes. Shaded areas show when an activity is actually being performed, and arrows show when the user switches between activities. This figure only covers part of one day. Activities {5} and {12} were not completed until later in the day.

The sequence of events during the day can be found by following the arrows. The session starts with **read mail** [1], switches to **reposition window** [2], switches to **msg conversation** [3], returns to **reposition window** [2], and so on.

Subactivities are shown as sub-boxes. The box for **help A** [8] contains two sub-boxes, since this activity involved first "finding a note about the 'fmt' command," and then "trying out the command." Activity [7], **respond to P's message** is more complicated because the subactivities are interleaved. Shortly after the first subactivity of "sending a reply" was started, it was interrupted by the second subactivity of "setting up a new account."

The horizontal axis is scaled by number of commands.

Note how different this case is from the previous one. In example 1, I went out of my way to engage in several activities at once, since I was bored doing one thing at a time. I was willing and able to accept a greater mental load. In the second example, activities are being thrust upon me, obliging me to abandon a simple one-at-a-time sequence. The real-time demands of the tasks at hand impose a mental burden that pushes the limits of my abilities. In this chapter I am not too concerned with users' motivations for interleaving activities; I concentrate instead on ways to support the resulting multiple activities. Chapter 13 by Miyata and Norman focuses on what is going on inside the user's head, and they present a detailed analysis of users' motivations.

EXAMPLE 3. Subactivities: **arranging a meeting**

I want to **arrange a meeting** [5] with R, so I decide to send her a message. I type:

snd ("snd" is our program for sending electronic mail)
To: R
Subject: Meeting
Let's get together to talk about the project.
I'm free at

[remind tomorrow] ("remind" is our calendar program)

1:00 tomorrow.
Is that a good time for you?
-Allen

The "remind" command was typed in a different window. It displayed my schedule for tomorrow so that I could see when I was free.

EXAMPLE 4. Subactivities: **responding to P's message[7]**

I am responding to a message asking me to set up a new account on a remote computer. I start typing a return message saying that I will do it, but instead I just set up the account and report that it is done. This takes longer than I expect because I have trouble linking to the remote computer.

Here the Main Activity "respond to the message" contains two Subactivities: the Subactivity "send a reply," and the Subactivity "set up the account," which contains a Sub-subactivity "link to the remote computer." You can see in the flowchart that "send a reply" is temporarily interrupted by "set up the account." Incidentally, this entire activity is interrupted by new mail which continues the **msg conversation** [3].

User Activities

When we design programs, we think about the sorts of things that users will want to do and then we build tools for carrying out those tasks. Implicit in this approach is the belief that there will be a fairly good match between *computer programs* and *user activities.* But how well does this assumption hold up in practice? Among the activities discussed so far are "reading the morning mail," "arranging a meeting," "carrying on an electronic conversation," and "helping another user with a command." The programs used by these activities include the "mail program," the "calendar program," and the "personal database" program. So in this list, the only case where an activity and a program are matched is in reading mail.

Single Activities and Multiple Programs

One way for a mismatch to occur between activities and programs is for a single activity to call upon more than one program. The activity of **arranging a meeting**, for example, involved two programs: mail and calendar. The ramifications of this simple fact are far-reaching and have a dramatic effect on how a user interacts with a computer system.

A system which is oblivious to the use of several programs for a single activity places all of the burden of program management on the user: in the case of **arranging a meeting**, I would have had to abort my message when I realized that I did not know when I was free tomorrow. Then, after consulting the calendar program, I would have then had to call the mail program anew and retype the message.

Saving context. At issue here is the fact that when one engages in an activity, one builds up a *context.* My context in "arranging a meeting" included a partially typed message. When you are engrossed in any challenging activity, you build up a rich mental environment that is populated with current problems, attendant confusions, and potential solutions. Even a momentary interruption will cause this elaborate *mental context* to collapse. Complex activities on a computer can lead to equally elaborate *computer contexts*. A half hour with an interactive debugger, painstakingly stepping through deeply nested code, can produce a context which could be reconstructed only by repeating the half-hour procedure. With proper design, these elaborate computer contexts can be interrupted without collapse.

Freezing programs. Computer systems can provide support for interruptions by saving the complete state of a program. The Berkeley UNIX "Stopped Jobs" facility is an example of this approach. It allows any process to be frozen at any time, thereby freeing the computer for some other process. When a process is frozen, its entire state, or context, is remembered so that it can be resumed precisely where it was left off. Most window systems (but not, for instance, the Macintosh) allow for multiple interactive processes for a single user. It is often said that windows support multiple activities, but in fact it is the underlying "multiple user processes" that enable multiple activities. Windows are a particularly good representation for the user.

In the "three interleaved activities" example, I interrupted the activity of reading my mail. Part of the state or context of the mail program is a variable which specifies the "current message," the message that I am currently reading. This context was saved when I interrupted this activity. When I resumed reading my mail, I typed the "next message" command, and the mail program knew where to continue.

To be precise, a "stopped jobs" facility supports multiple *programs*, not multiple *activities*. This distinction is significant in those cases where activities do not correspond perfectly with programs. Suppose in the **arrange a meeting** example that I suddenly want to interrupt this entire activity while I am consulting the calendar. I could freeze the calendar program, but there is no sense in which the system is aware that my mail program is also associated with this activity. What I want to do is to freeze my *activity*, not just one of the programs associated with that activity. But an "activity" is not a part of the computer system's vocabulary—it only knows about programs. When faced with the fact that the system does not think in our terms, we just do our best to align our activities with the units that are available. So we tend to identify an activity with the program that it is currently using. As a consequence, if I digress from the **arrange a meeting** activity for a long time, I may return to the calendar program and forget that it is associated with the previous mail program. And if I attempt to hide this activity by closing the calendar's window, the window for the mail program will remain open. In the context of activity management, then, the idea behind "user centered system design" is to allow users to treat their activities as manipulable units.

Windows. Windows divide one screen into multiple virtual screens, each behaving like a complete screen. Windows are a further step in saving a suspended context. Not only are all internal variables saved, but the screen image that is presented to the user is saved as well. One of the considerable advantages of windows over standard terminals is that this saved image can be present on the screen even while the user is engaged in an interrupting activity. It is hard to imagine a more effective reminder of the interrupted task than its image on the screen.

Windows improve the user's vocabulary by adding a concept that is one level above that of programs. By keeping all of the programs for a single activity inside a single window, the user can indeed treat those programs as a single unit. In practice, though, windows are often used to make the multiple programs of a single activity available simultaneously. This is common, because when two programs are being used within the same activity, it is likely that the user will want to pass data between them, see what they are both displaying, and interleave commands to them. In fact, in the **arrange a meeting** example, I used a second window to run the calendar program. This example suggests that window systems can better support activity management if they allow the user to link related windows and treat them as a manipulable unit. So, for instance, the "close" and "open" functions would close and open all of the windows together. This is discussed further in Chapter 14 by Reichman.

We can go one step beyond the use of multiple windows in support of an activity and consider using multiple screens. By a "screen" I mean a collection of windows that fill the computer's screen. The Symbolics Lisp Machine demonstrates the potential of this approach. This machine is organized around a few general-purpose tools which are integrated so that they can all be used in concert to perform an activity (typically the development of a program). In the process of debugging, the user may have one screen which contains a window for source code, a window for giving commands (the "lisp listener" and "debugger"), and a window for the running program's interface (the "application"). A second screen may contain the file system editor, a third screen the "flavor examiner," and a fourth screen the data-structure "inspector." When one screen is active, there is no visible trace of the windows associated with the other screens. In normal use, then, the user will hop amongst screens in the course of performing a single activity.

Multiple Activities and Single Programs

The other way for a mismatch to occur between activities and programs is for more than one activity to call upon a single program. This lack of correspondence between activities and programs also leads to activity-management problems. The potential difficulty with sharing a program is that the two activities will each want to establish their own contexts, but since the program only has one set of context variables, the contexts will clash. The one activity's setting of a context variable will get clobbered each time the user switches to the other activity.

Context Clashes

In the "three interleaved activities" example, the first activity (**read mail** [1]) and the third activity (**msg conversation** [3]) used the same program (the "mail" program), so there was the potential for a clash. The main context variable in the mail program is the "current message" variable. Commands to "display message X" or "display next message" make use of this variable. In the "read mail" activity, I read messages 1 and 2 by using the "display message #1" and "display next message" commands. This set the "current message" variable to 1 and then to 2. Next, I **repositioned the history window** [2] and then went to the **msg conversation** activity. Suppose that in that activity I had chosen to "read the message from J," which happens to be message number 8. This would have set the "current message" variable to 8. Had I done this, when I returned to the original **read mail** activity and asked to "display next message," I would have been shown message number 9 instead of the desired message number 3. I was lucky in the real example because I chose to "send a message to J" rather than "read the message from J," and sending does not affect the "current message" variable.

As another example of context clashes, consider multiple-buffer editors such as Emacs. A "multiple-buffer" editor allows the user to edit several different files from within a single instantiation of the editor program. The editor handles this by maintaining a full complement of context variables for each of the files in question. This feature can be used effectively for single activities involving multiple files. As such, it is a good example of a design which anticipates the needs of the user. If you decide to change the word "brother" to the word "sibling" throughout a set of six files, you can do this in a single operation. However, this feature can also be used to perform multiple activities within a single editor. This opens up the possibility of the word "sibling" cropping up unexpectedly in files belonging to an unrelated activity that just happened to be in the way.

While-I'm-At-It Activities

A third example of context clashes occurs with "While-I'm-At-It" activities. A common example of "While-I'm-At-It" activities occurs when you are looking at a listing of files and happen to notice an outdated one—it is convenient to just delete the file right away and then continue on with whatever you were doing. Technically, you have interleaved an activity ("delete file" in this example) that is completely unrelated to your primary activity except for the fact that it occurs within the same context or environment as the primary activity. That is, the two activities share exactly the same context but their goals have nothing in common: I refer to such activities as "While-I'm-At-It" activities. In Figure 12.1 there is an example of a "While-I'm-At-It" activity (**delete outdated message** [9]) which occurs as a quick digression from **msg conversation** [3].

Handling context clashes. It is valuable for program designers to try to anticipate the typical "While-I'm-At-It" activities for their application. For these activities, it is possible to resolve the potential context clash by ensuring that the "While-I'm-At-It" activity does not change the setting of any context variables. For instance, consider the "current message" variable in the mail program. In order to read my first three messages, I would give the commands "display message #1," "display next message," and "display next message." These commands set the "current message" variable to 1, 2, and 3, respectively. But suppose I interrupt my reading after the second message, and delete message number 8. Now what will happen when I give the "display next message" command—will I see the desired message number 3, or will I instead see message number 9? It would have been natural and consistent to make the "delete message" command change the value of the "current message" variable to the message after the deleted one. Fortunately, the designer was more clever: The "delete message" command leaves the "current message" variable unchanged, so in fact I will see the desired message number 3.

The success of this design approach—of allowing for simple operations which will not alter the user's overall context—is dependent upon the fact that "While-I'm-At-It" activities tend to be short and self-contained. When unanticipated difficulties arise and the digression becomes more protracted, the user may regret not having performed it as a separate activity.

But what about the more complicated types of context clashes? How can we handle the case of **msg conversation** clashing with **read mail**? A partial solution is to design modularity into the program. The program is segmented into several subparts, each with its own context variables. The idea is to group together commands whose use typically coincides with a common user subactivity. Then if multiple activities call upon different subparts of the program, they will be able to use the same program without clashing. For instance, the mail program has a subpart for sending mail which is distinct from the subpart for reading mail. This accounts for the fact that **msg conversation** and **read mail** shared the same program without clashing. However, it is inevitable that two activities will eventually want to access the same subpart of a program, so segmenting cannot be considered a complete solution to the problem. A more involved **msg conversation** would eventually have caused the value of the "current message" variable to be changed.

The most direct solution to the problem of clashing contexts is to call up separate instantiations of the program for the different activities. This way, the user has a complete, independent set of context variables for each activity. This solution is commonly used with text editors. Which file is being edited is such an important context variable that a user who wants to edit two unrelated files will generally choose to invoke two separate instances of the editor.

One drawback to separate instantiations is that there is a time penalty for the user in starting up a new instance of the program. A more interesting drawback is that the two instances will have *no* context in common. This hardly seems a problem, since I have been assiduously attempting to avoid shared context. But in the next section on "Related Activities" I discuss situations where some sharing is important.

A fourth solution is to design activity management explicitly into the program. In this approach, the program eliminates clashes by keeping separate copies of its context variables for each different activity. When some shared context is desired, some of the variables can be shared, and separate copies can be made for the others.

I wrote a program called Notepad which adopts this approach. It has special "Interrupt" and "Resume" commands for switching amongst activities. Because it is so easy for users to engage in multiple activities within Notepad, the history flowcharts show considerably more interleaving when it is used. An example can be seen in the latter part of Figure 12.1 (activities [16]–[19]). Notepad falls short of the goal of true activity management because it only remembers a single note for each frozen activity. This is quite analogous to the example where UNIX would freeze the calendar program but be unaware that the mail program was part of the same activity. An improved Notepad would allow groups of notes to be manipulated as a single unit. Another way of saying this is that there needs to be a context variable called "current activity."

When activity management is explicitly designed into a program, it yields the added benefit that its features are available for managing the subactivities of any activity which uses that program. Consider a complaint that I have heard about our mail system: In the middle of composing a message, the user realizes that a copy of the message should be sent to a third party. The mail program has a "compose" command (it is called "input" mode) for composing messages, and a "cc:" command for sending copies. But the "cc:" command cannot be used while in "input" mode—that is, the "composition" subactivity cannot be interrupted by the "cc:" subactivity. So the user waits until the text is complete, intending then to add the cc line. But the user forgets and sends the message, forgetting to use the "cc:" command. A similar frustration occurs when a user is composing an answer to a message and finds it impossible to view the message that is being answered.

Explicit activity management within the program could remedy these situations by allowing the subprograms to be individually interruptible. The program would manage its *subprograms* in the same way an operating system manages its *programs*.

Related Activities

My focus so far has been on interleaving activities that are not related to each other. For instance, the activity of **arrange a meeting** [3] had nothing to do with the activity of **read mail** [1]. In situations of this sort it is perfectly reasonable to assume that the user is unconcerned with the one activity while busy with the other. If the previous activity remains on the screen, it is only as a reminder of its existence; it has no bearing on the execution of the latter activity. In contrast, there are numerous important situations where the interleaved activities are related to each other. The most common source of related activities is *subactivities.* Consider the situation where you are answering a message and you need to take a look at the message that is being answered. The subactivity of "looking at the message" is related to the "answering the message" activity: You are looking at the message in order to help you carry out your answering activity.

Simultaneous Interaction

In cases of *related* activities, the user (a) wants both of the activities to be visible simultaneously; (b) wants to be able to interleave commands to the activities; and (c) wants to be able to pass data back and forth between the activities. We can summarize this by saying that users desire "simultaneous interaction" with related activities.

How can systems accomplish this simultaneous interaction? Making both activities visible and receptive to commands can be achieved by using windows, and a limited degree of data-passing can be achieved through "cut and paste" facilities. But full data-passing, which honors the contrasting data types of different programs, is a complex problem requiring a sophisticated programming effort (See Chapter 11 by Mark). The **arrange a meeting** [5] activity illustrates the need for data-passing. In that example, I consulted a calendar as a subactivity and used a second window so that the calendar would be visible while I continued composing my message in the mail program. If the information displayed by the calendar program had been complex, I might have wanted to copy its display into my message.

Shared Context

In the **arrange a meeting** example, it was necessary for the system to provide special support in order to facilitate "simultaneous interaction." The reason for this is that the subactivity used a different program from the main activity. In cases where subactivities use the same program, there is no difficulty in viewing both activities, giving them commands, and passing data between them. In fact, since there are no such problems, it is unlikely that the user will even be aware that multiple activities are present. However, related activities which share the same program have their own special problem: The issue of "context clashes" which I discussed earlier is now complicated by the need for a certain amount of "shared context." This problem is best explained by an example. Suppose I am busily fixing a bug in a program, using the text editor to modify the code. I run into a problem so I send a message to a colleague for help. While waiting for the reply, I decide to work on another bug in the same program. This is a situation where two subactivities are interleaved. The special twist presented by situations of this sort is that the subactivities have some context which is shared and some which is unshared. The shared context is that they are both modifying the same file. The unshared part is that each subactivity will want to establish its own markers into the text, its own set of strings that are saved in buffers, and its own independent command history. If I just continue to use the same editor, I will mix the markers, buffers, and history of the former subactivity with those of the current one. On the other hand, if I start up a new instantiation of the editor, the two editors will develop clashing versions of the shared file. What I really want to do is to start up a new editor which inherits the shared context of the two subactivities and which creates its own context for the unshared part.

The Emacs editor has a facility which approximates this: it is called a "recursive editing level," and it has the effect of invoking a new instance of the editor inside of the original instance. The new instance shares some context, such as the file and the location pointers, but it maintains its own command history. Deciding which parts of the context to share and which to separate is a complicated question which must be tailored to the expected needs of the application. The recursive editing level in Emacs is mainly used within long, complicated commands, so that the user can make a few changes without having to abort the complicated command.

The principle of recursive instantiations forms the basis for another highly successful program, the "break" package, which is found in many LISP systems. The "break" package allows a user to nest subactivities (and sub-subactivities) within other activities in an attempt to resolve subproblems that crop up while working on other, larger, problems. Each subactivity inherits the complete context of its parent activity.

REMINDING

Users engaged in many different activities can appreciate some assistance in keeping track of those activities. If some of the cognitive burden of keeping track of activities is removed, users can concentrate more on actually performing the tasks, instead of spending their energies on remembering and scheduling those tasks. Chapter 13 by Miyata and Norman discusses ways for assisting users in remembering their activities.

Windows Are Not the Whole Solution

Windows and icons provide one way to remind users of interrupted activities. They can be quite effective—up to the point where the activities become so numerous that they clutter the screen. At this point, it may prove more effective to replace the jumble of overlapping windows with a more concise list of titles or descriptors of the activities.

Another way (besides clutter) that visual reminders lose their effectiveness is when they are not easily associated with their activities. In the same way that activities need not correspond one-to-one with programs, it is also true that activities need not correspond perfectly with windows. When there is no visible unit on the screen that is uniquely associated with an activity, there will be no direct way to be reminded of that activity. For instance, the display of a calendar program may not serve as a useful reminder for an activity that the user conceives of primarily as sending a message. Likewise, if a window is being used by three different activities at once, it cannot effectively remind the user of all of those activities.

Given the shortcomings of windows as reminders, it is interesting to note how far the Symbolics Lisp Machine has moved away from the simple use of windows as reminders. One reason for this is that it makes such extensive use of windows within a single activity that most of its window operations (moving, reshaping, etc.) are commonly enlisted to serve the current activity. In order to compensate for this, there are special predefined keys for retrieving particular windows (e.g., the "Lisp Listener" window or the "Flavor Examiner" window) or the "previous" window. This obviates the need for a window to be visible in order to select it.

Reorienting

Once you have successfully located the activity you want to resume, you may still be in need of some assistance from the computer: If you are returning to an activity after a lengthy digression, you may have difficulty remembering where you left off. It is generally the case that you need more information to resume an activity than you need when you are engrossed in performing that activity. For this reason, it is valuable if systems can provide additional visual and contextual cues to reorient the user who is trying to resume an activity. Text editors, for instance, maintain a considerable amount of hidden information: There are buffers containing text to be copied or moved, and pointers to important locations in the file. On returning to an editing session, then, it may be very helpful if the editor lets you display the contents of these buffers and pointers. Also, a "movie" of earlier screen images and a listing of the commands given prior to the digression may help you to remember where you left off.

User-Centered Activity Management

I have emphasized the distinction between a user's *activities* and a computer's *programs*. The interfaces of traditional operating systems take the computer's programs to be central and leave it to the users to manage their activities. What if the interface took the users' activities to be central? This means that the top-level units that mediate the users' interaction with the computer would be activities rather than programs. Of course, the users would still have to deal with computer programs, but the programs would now occupy the second tier of conceptual units. The interaction would be "user-centered."

Let me describe a session on a hypothetical system with this sort of "activity manager." The interface will maintain an "activity list" of all of the activities which I have started but not completed. In addition, it will have available a list of recent activities which have been completed. The reason for including a list of completed activities is that the history lists which I have examined contain a surprising number of cases where an activity that was presumed to be complete is "resurrected" some time later. A glance at Figure 12.1 will convince you that **reposition window[2]** is an example of a resurrected activity, and that **arrange a meeting**[5] was resurrected later that day. This list of completed activities will serve a function similar to an "undelete" command.

An activity manager that keeps a list of all interrupted activities opens up the possibility of maintaining activities across sessions with the computer (Bannon, Cypher, Greenspan, & Monty, 1983). The activity of writing a program, or a paper, may span several weeks, and it would be convenient for the system to keep track of your progress on such an activity and the associated computer context. There are also activities which one wants to perform once a day, or once a week. An activity manager could insert reminders for these tasks into the activity list. Furthermore, such an activity list could be conveniently used as a "todo" list—a place to jot reminders about tasks that need to be done in the near future. The "displacement activities", which are discussed in Chapter 17 by Owen, also have a place in the activity list. These activities are special, in that the user ordinarily wants them hidden from sight. But at those times when the user has nothing special to do—or actively wants to avoid some particular chore—this list of diversions could be consulted.

When I log on to my computer at the start of the day, it displays the screen just as I left it when I logged off. The activity list reminds me of things I have to do, and the windows for my current activity are already open. I choose to read my new mail first, so I temporarily hide away the current activity. Since all of its windows are linked, they all disappear at once, leaving me with an uncluttered screen. I read my mail and move two of the messages to my activity list. They are automatically assigned their own instantiations of the mail program which will handle any future correspondence on these topics. Finished with my mail, I push the "Resume" key. I need not recall what the original activity of the day was, as the activity manager has remembered this for me.

A few minutes later, the "debug program W" item in my activity list begins to flash. I select this activity and see that my colleague has answered my query. She has figured out how to fix this troublesome bug, so I make the change and recompile the program. Without waiting for it to finish, I return to my original activity and complete it shortly thereafter. I then decide to make a modification to a paper I am writing, but suddenly the phone rings, so I jot "clarify X" on my activity list and grab the phone. The caller informs me of a terrible bug in program Y, so I turn to my table of "standard layouts" and select a "debugger."

Because certain common activities (like debugging) always require a particular set of programs, and because users become accustomed to certain ways of laying these programs out across various windows and screens, the activity manager maintains a table of a user's "standard layouts." So when I select the "debugger" layout, a collection of useful tools arrange themselves on my screen.

I soon tire of the debugging chore and I look for a change of pace. I see several choices in my activity list, but I am unable to recall what they refer to. So I ask for more information, and I am treated to miniature displays of the various activities. For some activities I am shown a miniature screen, for others, a particularly representative window. For yet another activity, a short descriptive paragraph that I wrote months ago appears. Chapter 16 by Draper suggests various ways for allowing the user to choose the amount and type of information that is displayed for each activity.

The Notepad Program

The Notepad program implements some of these ideas about activity management. Like the ThinkTank program for personal computers, it is designed as a tool for "thought-dumping"—the process of quickly jotting down a flood of fleeting ideas. Thought-dumping places a premium on the ability to record an idea with the minimum of interference. In Notepad, notes are organized by assigning titles to them and filing them in bins. These operations can be postponed arbitrarily so that they will not interfere with the process of jotting down ideas. In addition, the process of writing a note can be interrupted at any time by the Title and File commands.

Notepad has explicit activity management commands. If the current idea is suddenly interrupted by a new idea, the user gives the "Interrupt" command. This puts the current idea on an "Interrupted Notes" stack and supplies a blank page for the new note. The stack is constantly displayed to the user and serves as a reminder of interrupted ideas. When the new note is complete, the user gives the "Resume" command to resume the interrupted idea.

The use of a stack means that Notepad embeds activities; resuming activities in an arbitrary order is not supported as well. Notepad supports postponing with a special "Jot" command, which allows the user to jot down a short reminder about a new idea without having to leave the current idea. This gets the idea out on the table and allows Notepad to remind the user to pursue it.

Problems With Activity Management

An activity manager can take over some of the user's burden of managing activities, but it is restricted by what it knows about the activities. Many user activities are performed only partly on the computer. Various subactivities and meta-activities may be performed by talking to a colleague or by simply remembering them. This can cause significant problems for an activity manager because there are arbitrary holes in its picture of the activity. If the user consults an on-line manual, the computer can remember this activity when it is interrupted. But if the user consults a printed manual, or a local expert, the manager will be unaware of the activity. Similarly, a user who interrupts the activity of debugging a program may be unable to remember where to pick up the task once it is resumed, because the interrupted activity was some complicated reasoning process that was totally within the user's head.

ACKNOWLEDGMENTS

I would like to thank Autumn Chapman for designing the original activity flowcharts, and Steve Draper and Dave Owen for providing many helpful comments. I would also like to thank Don Betts for designing and drawing Figure 12.1. The impetus for the study of activity structures came from the Activity Structures working group, consisting of Lissa Monty, Liam Bannon, Steve Greenspan, and myself. This early work was published as "Evaluation and Analysis of Users' Activity Organization" (Bannon, Cypher, Greenspan, & Monty, 1983). This group developed many new ideas about creating a complete working environment for each user activity. I am particularly indebted to Lissa Monty for her ideas about gathering history list data, and to Liam Bannon and Steve Greenspan for their ideas about workspaces.

APPENDIX: COLLECTING HISTORY DATA

Activity flowcharts are prepared from an "annotated history list"—a record of all of the commands that the user types during the computer session. The history list records the linear stream of commands that the user performs to carry out his or her activities. If activities are performed one at a time, the commands will cluster into neat, separate packets, whereas an interleaved activity will have its commands sprinkled throughout the historical record. I refer to these lists as "annotated" because users can add explanations of what they are doing, as they are doing it.

The history list used in this chapter was recorded from a Sun Workstation, that has a bit-mapped display and a window system. It uses Berkeley UNIX. Because this system uses both windows and the Stopped Jobs facility, it provides considerable support for multiple user activities. Had these data been collected from a personal computer, I would expect the results to be quite different. The software currently available on personal computers is generally quite unsupportive of multiple activities.

To record these data, the major interactive programs on the system—the top-level (shell) interface, the text editor, the mail program, and the note program—were modified to send a parsed copy of every command to a central collection program. The collector receives commands from all of the windows in whatever order they are performed, so that embedded activities show up as embedded commands. A segment of a history list is shown in Figure 12.2.

This collection technique has an advantage over the standard technique of recording keystrokes in that it lets each program parse the input string into a meaningful command. However, it is important to note the limitations of this history-list data. Only the user's commands are recorded; there is no record of the computer's responses or of computer initiatives (such as "you have new mail"). This unfortunately reinforces the stereotypical view that "the user commands, the computer performs." It makes it difficult to appreciate situations where the user and the computer are working together, like co-routines, and situations where the computer is initiating activities.

Another problem is that the history list includes only those parts of the activity which actually take place on the computer. For instance, consulting another person is an unrecorded activity, so social interactions of the sort discussed in Bannon's Chapter 21 are missing from the record. The user can mitigate this problem somewhat by making annotations about relevant unrecorded activities.

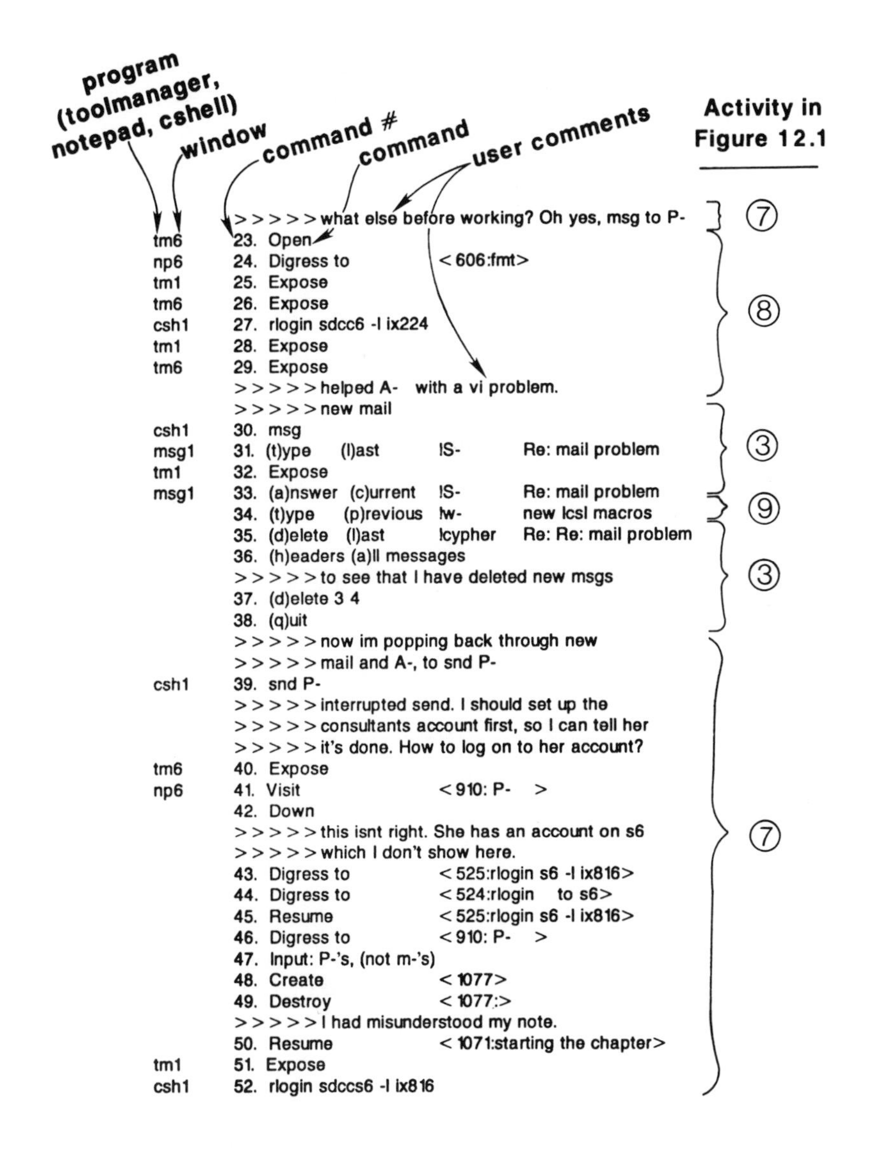

FIGURE 12.2. A segment from the annotated history list for the session shown in Figure 12.1.

CHAPTER **13**

Psychological Issues in Support of Multiple Activities

YOSHIRO MIYATA and DONALD A. NORMAN

That people engage in multiple activities at the same time should not be surprising. Nonetheless, little is known in contemporary psychology about either the phenomena that result or of the underlying mental structures. In this chapter we present a quick tour of the psychological theory relevant to the understanding of people's actions and of multiple activities. Consider this an "approximate theory" that is sufficiently accurate and relevant for the analysis of the basic phenomena. We conclude with a discussion of how the theoretical ideas can be applied for system support of multiple activities.

THE PSYCHOLOGY OF MULTIPLE ACTIVITIES

There are three major areas of psychological studies relevant to the study of multiple activities:

- Studies of memory, especially studies of short-term and working memory and of the organization of knowledge in long-term memory.

- Studies of attention, including studies of "controlled" (conscious) and "automatic" (subconscious) behavior, studies of "simultaneous attention" (how much can be done at the same time), and studies of "selective attention" (what activities interfere with one another).

- Studies of action, how people actually accomplish tasks.

Although these topics constitute three distinct areas of study within psychology, we believe the three to be intimately related. In particular, the theoretical tools of memory schemas and activation values underlie all three. These topics are all under active study and there is no general resolution of the appropriate theoretical mechanisms. But for the analyses that follow, we need only an approximate model, and for this purpose, much is known.

Memory

A reasonable approximate model of the Human Information Processing system divides processing structure into conscious and subconscious operations and memory into two classes of structure, short- and long-term memory (STM and LTM). In this approximate model, we treat working memory and STM as the same structures, and in the chapter we primarily refer to working memory. For current purposes, all that matters about STM is to recognize its limited capacity—for example, as described by the "5-slot model of memory" in Norman's Chapter 3, working memory can be thought of as having a capacity for only 5 items at any one time. Long-term memory (LTM) consists of organized knowledge units, called schemas, that structure knowledge and also contain the procedural information necessary to control actions. Several aspects of LTM are important, including the difficulty of acquiring new information and the problems and issues in retrieval of information, once acquired. An important aspect of memory is "reminding," the manner by which one event may cause retrieval of the memory for another. (See Card, Moran, & Newell, 1983, for further discussion of approximate models of memory; see Norman & Bobrow, 1979 and Schank, 1982, for a discussion of reminding and memory retrieval.)

Two Control Systems: Conscious and Subconscious

The large body of psychological research on attention allows us to develop a two system approximation to the control of behavior. One system, conscious control, has limited resources, especially that of STM or working memory. In general, the resources required to do any particular task subtract from those required for the simultaneous conduct of other tasks. In practice, these resource limits are severe enough that we can assume that only a single task can be under conscious control at any one time. The other system, subconscious control, seems to develop specialized procedures for tasks that are relatively independent of one another. As a result, we can treat subconsciously performed tasks as resource unlimited, so that several can be done simultaneously (as long as they do not require joint use of the limbs or sensory organs). Only well-learned, routine tasks can be done subconsciously. Note that the subconscious control system does not appear to use STM.

> *The conscious limitations do not preclude a person from simultaneous conduct of subconscious or automatized activities. Indeed, skilled people make use of automated actions to allow themselves to do several things at the same time. A good typist thinks of other things while typing: A skilled computer user plans ahead for several activities while doing the first. Again, although the physical limits of our limbs and sense organs are real, skilled practitioners (such as circus performers) can do things we never would have thought possible. As Buxton shows in Chapter 15, proper development of input devices allows us to extend the number of operations we can perform simultaneously.*
>
> *Note too that the definition of "task" changes as skill develops. Thus, a beginning piano player treats control of the two hands as separate tasks, but the accomplished pianist does not. We return to all these issues in the discussion of "backgrounded" activity.*

Conscious control is used primarily in four situations:

- When the task to be performed is novel or ill-learned.
- When the task is perceived to be especially critical, difficult, or dangerous.
- When there is a need to override the automatic control, either to cause actions that would not otherwise take place or to inhibit those which would take place (but are not desired).
- When there is a need to resolve conflict among schemas (or activities), especially when an ongoing activity is interrupted by demands from a different activity.

These properties of conscious control are discussed in more detail by Norman and Shallice (1986). Norman and Shallice assume that conscious control occurs by increasing or inhibiting the current activation values of schemas.

Planning

A primary need for planning comes about because of what Cypher (Chapter 12) calls "linearization": the act of taking the several activities that might be competing for attention and arranging to do them in sequence. That is, if we can do only one task at a time, then when many need to be done, they must be arranged in a linear sequence. Planning needs to occur for other reasons, however. Some tasks must wait until prerequisite conditions are satisfied. Planning helps discover these situations and arrange the order of task execution so that prerequisites are satisfied. Some tasks may have critical timing requirements: Again, the planning stage helps arrange matters appropriately. Finally, some tasks should be postponed (even if one wishes to do them) when more important work should be done instead, and some unpleasant tasks should be done, even if one would rather not.

One of the major difficulties in planning comes about because of the limitations of human processing resources. The limits on conscious resources and working memory capacity mean that in-depth planning is often not possible without external aid. Moreover, at the start of an activity, many different ideas are often jumbled together. The problem is that concentration upon one tends to cause the others to be forgotten. The primary planning aids are those that support memory. In addition, various retrieval aids are required to allow further work on the ideas and to act as reminders, to make sure that once generated, plans are not forgotten. We treat these aspects under the discussion of reminders.

Interruptions

Interruptions can be both external and internal. External interruptions result from events in the environment. Internal interruptions come from our own thought processes—new ideas that draw attention from the current activity. Interruptions introduce new tasks on top of the ongoing activity, often unexpectedly. As a result, conflicts arise. Because of a person's limited processing and memory capacity, one suspends work on current activity at the risk of losing track of the current activity by failing to resume the work where it was interrupted.

But taking the time to make the current activity recoverable runs the risk of losing the new idea.

There is a conflict in processing between the need to concentrate upon one thing in order to give it full processing capability and the need to be alert for unexpected, but relevant and important, thoughts and external events. As a result, the human information processing system seems to exhibit conflicting properties: continual concentration and continual distraction. The first property moves the system toward operation as a single purpose, dedicated processor, primarily attending to one task and ignoring other events: We call this *task-driven processing*. The other property moves the system toward operation as a responsive processor, continually changing its activities to reflect new thoughts and ideas or to respond to events in the environment: We call this *interrupt-driven processing*. When people are deeply engrossed in a book or movie, they are task-driven. When people are in a job that requires constant interaction with others, whether by telephone or in person, they are apt to be interrupt-driven.

Task-driven processing. In a task-driven state, people are so occupied by the processing of the ongoing task that there is apt to be an effective decrease in sensitivity to events external to the activity. This lack of sensitivity to external events has been widely studied by psychologists interested in "attention." A typical experimental paradigm investigates a person's ability to process information in a secondary task while concentrating upon a primary task. These studies show that there is a severe limitation in the degree to which information is processed from the secondary activity. Although people remain aware of the existence of activity around them while concentrating upon a task (while under task-driven processing), they can do only minimal processing of this external activity, not enough to draw meanings and implications. Thus, while deeply engrossed in a book, people can be quite unaware of questions addressed to them. People do note the presence of activity, but not the content: Signals intrude, but meanings are only minimally extracted, if at all. (A reasonable review of these phenomena, although a little old now, is Norman, 1976. See also Shiffrin, 1986).

The implications of these findings for our analyses are that if people become too engrossed in the task to which they are paying conscious attention, they will not process other events that occur on the computer screen. Note that they will detect gross signals, but not their meaning. Thus, even when people are heavily engrossed in task-driven situations, they will notice abrupt sensory signals such as flashes of light or auditory tones. Unfortunately, the type of signals needed to attract the attention of people heavily engaged in task-driven activities are also the kinds of signals that are most intrusive and annoying to those who are not in this state, for they draw away processing resources from the main task.

Interrupt-driven processing. The phrase "interrupt-driven" processing refers to the situation in which people are especially sensitive to extraneous events, easily distracted by extraneous thoughts and external signals. The normal working environment is full of potential interruptions: phone calls, talking, requests for help from colleagues, and so on—unpredictable and often irresistible interruptions. Internal interruptions occur as a natural result of thought, as new ideas and new topics get suggested by the processing for the current topic. The result can be captured by some other thought, perhaps unrelated to the task that is supposedly being performed. It seems more difficult to maintain attention on a topic that is purely internal than on one which has external support.

Whether a person is in a state of task- or interrupt-driven processing is a function of both the person and the activity. Differences among individuals play a role: Some people are more distractable than others, some are more easily controlled by task-driven structures, others are more distractable by extraneous events or thoughts. The amount of external activity clearly is relevant: It is obviously difficult to maintain a task-driven state in the presence of external events irrelevant to the main task. Task-driven processing continues when then processing is dominated by the schemas relevant to the activity. The result is that there are few resources available for other activities. Interrupt-driven processing occurs when the activity does not have much structure, or if the external support for the activity is such that schemas are not always kept activated. The result is that there will be idle resources that will tend to get used for extraneous activities.

Multiple Activities: Current and Suspended

We now turn to an analysis of the different types of multiple activities. First, activities are either currently controlling actions or are suspended: Call the first case "current activities," the second "suspended activities." Second, there are two forms of current activities, those that are in the foreground of conscious attention ("foregrounded activities") and those that are not ("backgrounded activities"). And finally, there are two forms of backgrounded activities: external and internal. The result is three major classes of activities:

I. Foregrounded Activities:
A current activity under conscious control.

II. Backgrounded Activities:
There are two classes of backgrounded activities: external and internal. Externally backgrounded activities are those done by some other agency. Internally backgrounded activities represent ongoing activities under "automatic" or subconscious control.

III. Suspended Activities.

Current activities. We distinguish among several different types of ongoing, current activities, divided into two major categories. One category is reserved for the primary activity, the activity that is the focus of conscious attention: We call this the "foregrounded" activity. The other category is used for ongoing, active tasks that receive little or no conscious attention: We call this the "backgrounded" activity.

> *Thus, in writing a paper (or a program), there are at least two tasks going on: One is the development of the ideas, the other the act of typing them onto the computer keyboard. The development of the ideas should be the foregrounded activity and the act of typing the backgrounded activity. This holds only for skilled typists: The act of typing is backgrounded; typing takes place simultaneously with the development of the ideas and without interfering with them. For nonskilled typists, this is not possible. They must focus so much attention upon the typing that it becomes the foregrounded activity, interrupting and suspending the development of the ideas. This leads to severe disruption of the task of idea development, and, as a result, many nonskilled typists cannot compose at the keyboard, but prefer other means of composition, one where the translation of thoughts to symbols is more automatic (for example, dictation or handwriting).*

Backgrounded activities. Backgrounded activities result whenever a task is performed "automatically," without conscious supervision, thus allowing other activities to be done at the same time. There are two classes of backgrounded activities: *external* and *internal.*

Externally backgrounded activities.

The phrase "externally backgrounded" comes from the field of computer science and the development of computer systems that allow a program to be run in "background": performed by the computer as an independent job, without interaction or supervision by the user (and often at a lower priority than "foregrounded" jobs). We generalize this concept to refer to any task that is being performed by an external system, such as another person or a computer, without requiring supervision. The important point is that when a task is backgrounded, it no longer requires conscious attention and other activities can be started. Examples in the domain of computers are frequent. Thus, in writing a paper, the getting the "hardcopy" printout of the final product is often "externally backgrounded." In similar way, compiling and loading a long program is "externally backgrounded." This allows the user to start another job before these backgrounded tasks are completed. Often it is desired to resume the backgrounded task as a foregrounded activity when the external system has completed the tasks. But if no appropriate signal or reminder is presented to the user signaling the completion, there may be problems in recalling that a task was backgrounded and, hence, a failure to resume.

Internally backgrounded activities.

We generalize the notion of "backgrounding" still further for the category of "internally backgrounded" tasks. Here, we refer to the situation where the person continues the task, but in situations where it is so well practiced and learned that performance can take place with minimum conscious control, thus allowing other things to take place at the same time. Thus, skilled practitioners can do one task (e.g., talk or type) while their main attention is devoted to something else. Although this kind of "automatic" or "backgrounded" performance only takes place with highly skilled, well-practiced behavior, it is still a common, frequent occurrence.

Errors, especially "slips," are likely to occur when attention is not focused upon the current task. Thus, "capture errors" are most likely to occur in "internally backgrounded" activities. An expert typist intending to copy a limited part of a written page is apt to end up having copied the whole page: The more distractions or other activities present during the action, the more likely the task will be internally backgrounded, and the more likely the error.

Suspended activities. Activities that are not current are called "suspended." The schemas relevant to the activity remain activated, but do not control performance. Activities can be suspended for a variety of reasons, including both external and internal interruptions, boredom or fatigue with the task, the need to delay until some prerequisite condition has been satisfied, or even from a judgment that things are not going well and that delay would help. Some of the issues confronting the suspension or postponement of activities are discussed by Cypher in his chapter. Whatever the reason for the suspension, the result is a task awaiting the appropriate time to resume execution. The critical issue with suspended activities, of course, is how, when, and whether they will be resumed.

Reminding

Reminding is required if suspended activities are to be resumed at the appropriate time or place. The reminder, therefore, is a signal that indicates that a suspended or backgrounded task still exists or that it is ready for further processing. From the practical point of view, reminders help the person remember what is to be done. From the theoretical point of view, the problem is to reactivate relevant schemas and to re-establish whatever information is needed in STM or the environment so that the appropriate trigger conditions will be satisfied. It is good to keep in mind that although reminders are valuable in re-establishing a planned or suspended activity, they often interrupt ongoing activity.

Reminders as signals and as descriptions. There are two aspects to a reminder:

1. *Reminder as signal*: to indicate that something is to be remembered;

2. *Reminder as description*: to aid in retrieving what was to be remembered.

These two aspects of reminders can be quite independent: A reminder can succeed at one aspect while failing at the other. For example, a cooking timer is a common aid in preparation of a meal. But the ringing of the timer acts only as a signal that some externally backgrounded activity has been completed. The timer is nondescriptive, and it is up to the cook to retrieve what activity is being referred to. A list of activities to be performed during the day provides a good description of each individual activity, but if the list is hidden from view, it acts as a poor signal: The list may not be examined because it was never noticed.

For a reminder to be effective in specifying the activity, it must act as a memory retrieval cue. The reminder can be thought of as a partial *description* of the to-be-remembered materials (Norman & Bobrow, 1979). The relevant factor is specificity of the cue. Sometimes the cue itself specifies the task. Dirty dishes in the kitchen sink serve as excellent descriptions of the task to be done—wash the dishes—but they are apt to have poor signaling qualities if they are hidden from casual view. A boiling tea-kettle is intermediate: The sound of a tea-kettle boiling is an effective signal, but it may not always be a good description of the activity. A telephone call from a friend reminding of a dinner appointment serves both as a good signal and a good description. Messages on display screens can be effective reminder cues. A light, a nondescriptive message or icon, or the common mnemonic of tying a string around the finger all provide only very partial descriptions of the items to be remembered, oftentimes requiring deliberate and difficult retrieval processes.

> *An example of a reminder that acts as a good signal but a poor description comes from the job control features of the Berkeley distribution of UNIX. On our computer system, attempts to execute the "logout" command are sometimes rewarded with the message:* "There are stopped jobs." *This message is a reminder that some tasks have been suspended but it gives no indication of what those tasks are. (It also gives no indication of the possible courses of action, and beginning users have reported feeling frustrated by the presence of a reminder and the failure of the "logout" command to work, combined with a lack of information about the cause of the problem and no hint as to what they are supposed to do about it.)*

For a reminder to be a good signal, it must be conspicuous. This can be accomplished in several ways. The literature on selective attention indicates that simple sensory signals can be especially effective, being noted even when a person is deeply engaged in activity (flashes of light for a person engaged in auditory task: a tone or bell for a person engaged in a nonauditory activity). Discrepancy is especially relevant: An event that is unusual or not expected acts as a good signal. A frequent or common event does not. An alarm is effective only in an environment in which the sound of the alarm is infrequent. People seem especially sensitive to change and to violation of expectancy. When a discrepancy is noted, the resulting processing can interrupt ongoing activity, drawing attention and processing resources to the discrepant part of the environment. All this suggests that a cue is most effective when it is discrepant from one's ongoing expectation. Tying a string on a finger is an effective cue only if there is not usually a string around the finger. Display of a message or signal light is effective only if there is not usually a message or light in that area.

SYSTEM SUPPORT FOR MULTIPLE ACTIVITIES

Several different aspects of multiple activities require support. Cypher has discussed a number of the issues in his chapter: Here we expand upon the notions, with special emphasis on support for reminding (resumption of tasks following interruptions and suspensions). First, we discuss support for transitions between one activity to another focusing on two aspects: suspension of activity and reminding of activity. In addition, we discuss some aspects of support during execution of an activity, especially with regards to the execution of simultaneous activities.

Support for Suspensions of Activities

What happens when the decision is made to suspend current tasks? This usually takes place when some internal or external event occurs that makes it desirable to change the task currently foregrounded: See the description of these interrupting events in Cypher's chapter. This means that a new task is foregrounded and the previously current task becomes backgrounded or suspended. If the change occurs at the conclusion of the current task or at a natural breaking point, then there is probably no difficulty.

Unpredictable interruptions are likely to occur at potentially disruptive times. As a result, support for suspensions of activities has three aspects. First, the system should be designed so that it is easy to suspend an activity when this is desired, without interfering either with memory for the current task or with the thoughts relevant to the interrupting task. The suspension should not require much activity, or else thoughts in working memory will tend to be lost, thus interfering with the activities to be performed on the interrupting activity. Second, sufficient information should be saved with the suspended task so that when the activity is resumed, it can be continued where it left off (recovering the active thoughts is the hard part). Third, a reminding structure should be established so that the user does not forget that the task is still unfinished. Making the suspension easy to accomplish (the first point) while simultaneously saving enough context and unfinished ideas to allow smooth resumption (the second point) are somewhat opposing requirements: yet another tradeoff to worry about.

> *Cypher's* Notepad *program (Chapter 12) addresses all three aspects of the suspension of activities. It is easy to interrupt a task—a single keystroke will do it.* Notepad *saves all context and continually displays (and thereby reminds one of) the titles of the tasks that have been interrupted. The user can review the list of unfinished and interrupted tasks, and the program provides a simple means to step back through them. Completed tasks do not appear on the list. This procedure also provides the user with a way to deal quickly with potentially disrupting events. The user can decide to interrupt the main task only briefly enough to "jot down" a simple reminder of the interrupting event, then to resume the main task. This avoids major disruption by the new event while still noting its existence.*

Support for Reminding

When an activity is suspended, it needs to be resumed at a later moment when time is appropriate, when some prerequisite conditions are met, or when the user is free. The task-driven aspect of processing suggests the need for reminders to overcome the limits on memory and processing capability. The problem is that focusing upon one activity makes a person unresponsive to other events, including reminders of the activities that have been suspended or backgrounded. As a result, once an activity has been started, it is possible (likely) to forget about other things that should be done. If an action is to be started at some specified time or when a specified condition arises, the intended activity is likely not to be done if the person is engaged in some other activity at the time.

What would an ideal reminder look like? It should:

1. Inform the user when conditions are ready for resumption of a suspended or backgrounded activity.

2. Remind the user when something has to be done immediately.

3. Not distract from the current activity.

4. Continuous or periodically list activities that have been suspended or backgrounded.

5. Help resumption of an activity by retrieving the exact previous state of the activity and making it available to the user.

When should reminders occur? The problem with reminders is that they are also interrupts: One task's reminder is another task's interruption. Suppose that there is some control over when a reminder is presented to the user. When would be the ideal time to do so? What are the natural breaking points in activities? Deciding when to remind is a very sensitive problem. The importance and relevance of the conditions are important factors, but how can these be determined by the system? If the user is producing new ideas constantly, or if the user is working on a very complicated process like programming and has to keep a lot of information in working memory, it is not a good idea to interrupt the train of thought. An experienced assistant knows when to interrupt the boss or not. How can this expertise be captured?

In general, the major factor that determines whether a reminder should interrupt the current activity is the relative importance of the two activities. But even if the current activity is less important than the interrupting one, it is still necessary to examine the state of the current activity before interrupting: There are some states at which interruption would be very disruptive, and some where it would not disrupt as much. (We discuss this factor in the next section—the stage analysis of user activities.)

Sometimes, it is possible to have the user specify which activities can (or should) interrupt the current activity. In extreme cases, people lock their doors, turn off their telephones, and put up "do not disturb" signs. Some systems allow the user to control whether or not system messages or announcements of arriving mail will be permitted to intrude. It is often desirable to take into account the importance of the interrupting activities. Thus, although irrelevant messages might not be wanted, if a task has been externally backgrounded, the message that reports its completion is a desired interruption. Similarly, if in the writing of a paper, information is sought from some other person, then that person should be allowed to interrupt, but only for the relevant topic.

Reminders and interrupts provide a large set of tradeoffs among options. One problem is to determine how priorities can be established without excess effort by the user. Some means have to be devised for handling interrupts and reminders that do occur, but that are withheld from the user: Interruptions that are not important should be put off until the user finishes the current activity, but they should never be discarded. The two separate aspects of reminders can be handled separately: reminders as signals and reminders as description (the message part of the reminder). Different kinds of reminders might be allowed to use different levels of signals, from covering the whole screen, overwriting whatever is there (the most obtrusive, attention-getting form) to a subtle indicator in one corner of the screen indicating that reminders are awaiting user-initiated action.

Interruptions and the stage analysis of user activities. The analysis of seven stages of user activity (in Norman's Chapter 3) suggests that an interruption would be least disruptive if it occurred between the completion of the last stage—evaluation—and the formation of a new goal or intention. However, it is also clear that some points of interruption between stages should be less disruptive than others. In particular, interruptions where memory load is high should be disruptive, whereas interruptions where load is low (perhaps because of the reliance on external cues) should be less disruptive. This suggests that interruptions should be most disruptive while in the planning or evaluation stages: i.e., during the formation of the intention and the development of the action, and during the interpretation and evaluation of the outcome. Interruptions should be least disruptive at the juncture between execution and evaluation, where there is maximum use of external information. It is clear that an interrupting message should not be presented in the stages of execution or perception, stages where the user is directly interacting with the system.

> *Of course, these observations are not easy to follow because the stages where disruption is apt to be most serious consist of mental activities (planning, interpreting, evaluation), and these activities are usually not visible to the system. However, a good procedure might be to present relatively unobtrusive reminders to the user just following the completion of an action. If timed right, the reminder will not disrupt the*

ongoing task, but will be present and visible when the user completes the cycle and is starting to think of the next sequence of activities. The visibility of the reminder has to be carefully selected so that it does not disrupt if it comes on during evaluation, but yet is noticeable when the user completes the tasks. Think of the user buried in the task, unwilling to be interrupted, but every so often finishing a cycle and "coming up for air," quickly breaking from the task and taking a quick look around. The reminder should only be noticeable during that "breathing spell."

Memory Aids. The old folk saying, *Out of sight, out of mind* makes a good slogan for designers. The slogan speaks directly to the data-driven aspect of physical reminders: Something that is physically present keeps its memory schema in an activated state through data-driven activation. This means that one way of reminding is to keep visible the activity that is to be remembered. In computer systems, this could be implemented in a variety of ways, but the most common today is through the use of windows, lists, menus, and icons. (Windows are important enough that we give them their own special section.) Keeping things to be remembered constantly present has several advantages and disadvantages.

The advantage of continual presence is continual reminding. The deficits are distraction and clutter, as well as potential loss of working space. Keeping piles of work to be accomplished on a desk diminishes the usefulness of the desktop as a working space: The same is true of the computer screen. The problems of distraction are real. Being reminded of the important tasks that remain undone can have serious implications: It can demoralize, it can cause rapid switching among ideas. (This leads to the human state analogous to the computer system state of "thrashing": So many resources are used in contemplating all the tasks yet to be done, that there are no resources left to do the task.)

Making reminders nonvisible avoids all the deficits of visibility, but also avoids the virtue as well: Out of sight, out of mind. A compromise solution is possible. Reminders can be simplified, text can be abbreviated, windows "closed" or "shrunken." A group of descriptive reminders could be replaced with a simple flag or note that informs (reminds) the user that more specific reminders are awaiting attention. Then, when the user is ready to deal with the interrupts, the flag or note could be expanded into its set of more complete, more detailed reminders.

There is an important set of issues here that should be explored: The interaction of stress and performance. When under high stress, human performance deteriorates. The conscious resources available decrease. There tends to be a focusing upon one task to the exclusion of others, even if the task is not directly relevant to the problem (and sometimes even when execution of the task makes the difficulties more severe). This is a special problem with the ability to deal with interruptions. The ability to perform backgrounded jobs and to plan decreases. Working memory capacity seems to decrease. The more stressful the situation, the more disruptive interruptions become, which increases the stress.

Support for Concurrent Activities

One point that should be emphasized is that many complex tasks require simultaneous, concurrent activities. Thus, as we write this paper, it is often useful to have an outline of the entire paper as we write any particular section—an outline that changes as the writing continues reflecting the ongoing writing activity. Moreover, because this paper is so closely related to Cypher's chapter, we need a copy of that chapter in front of us as we write. Similarly, we need to refer to other sources of information, such as reference lists, or working notes. Finally, because this is a joint activity, at times we need to interrupt the writing activities in order to see if the other person is logged on to the computer network or if mail relevant to the writing of the paper has been received (while somehow avoiding the distraction of the other messages that would be discovered, irrelevant to the writing activity, but probably a more attractive pursuit of time).

So far, the best way we know to provide this kind of support is to use multiple displays. This can be done either through multiple terminals or by using windows on a terminal that has a sufficiently large screen to allow simultaneous display of the output of several ongoing programs. The problem with this solution is to avoid the surplus of riches: Some of the displays may be useful in the conduct of the activity, but they also act as dangerous lures, leading one to stray from the unpleasant duties of the current task.

The problem of distraction by other, more attractive messages or thoughts that are not relevant to the current activity is a pervasive difficulty. It leads to the types of interruptions Cypher called "While-I'm-At-It." The problem is exacerbated by the manner in which information is usually displayed: No attempt is made to suppress irrelevancies. This characteristic is common to many information systems: It is especially problematic in the use of dictionaries, atlases, and encyclopedias. It happens mentally as well: The attempt to think of one topic often leads to thoughts of others, at first related to the item of interest, but possibly leading to thoughts far astray. The extraneous retrievals can be both beneficial and disruptive. But the important point is that it would be useful to have a better control over the tradeoff: To be able to choose to be diverted or not, depending on the context.

This paper was written with two different kinds of terminals, each of which supported a different style of support for concurrent activities. One system, the *SUN Workstation*, provides for a large, bit-mapped display with multiple windows, allowing for simultaneous display of a number of components (see Figure 13.1). Thus, a typical session on the *SUN* has one window devoted to the text, one to an outline and another to an outline of Cypher's paper. In addition, other windows are used for other, suspended or backgrounded activities.

Most of the work, however, was done on 24-line by 80-character terminals. Here the support for concurrent activities came primarily through the "jobs" facility of Berkeley UNIX, so that although only the foregrounded activity was visible on the screen, it could be suspended immediately by the typing of a single character (control-Z) and a previously suspended program made visible by the typing of a relatively short command (*fg %N*, where *N* is its "job control" number). Still, this is disruptive and it prevents simultaneous viewing of different components of the task. As a result, work on these normal terminals had to rely more on external support, such as printed copies of the papers and outlines. In many ways this was more convenient, for external reminders were easier to read and scan than even the multiple windows on the SUN, but in other ways it was less convenient, for the paper was static and did not reflect changes that took place over the course of a work session.

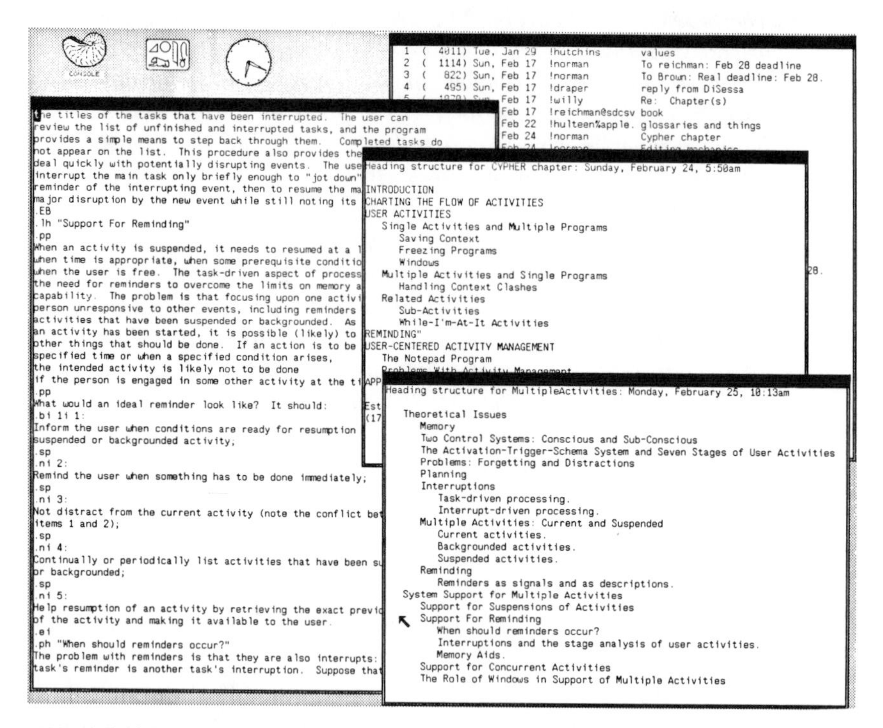

FIGURE 13.1. A typical screen on the *SUN Workstation* being used in the writing of this paper. The large window on the left supports the foregrounded activity: the writing of the text. Other windows provide other information related to main task, such as the outline of this and Cypher's chapter (the two medium sized, overlapping windows on the right). Other windows provide support for and reminders of suspended and backgrounded tasks. Thus, one window shows the time of day (as a clockface), another shows mail received by the message system. Several icons represent closed windows, acting as "signals" of other activities that have been suspended. Finally, and not visible, an alarm clock program runs in background that will signal at the requested time, and (one hopes) cause the user to stop work and get on to some other activity.

How well a system supports multiple activities can depend on the details of the hardware as much as the software support. Thus, we found dramatically different support was provided by the very same computer system, terminal, and software as communication rate (Baud rate) was changed. Thus, the job-control facility of Berkeley UNIX works reasonably well to support multiple activities when the terminal is in the office, operating at 9600 Baud—it takes about 2 seconds to display a full screen. However, when working at home, at 1200 Baud, it takes as much as 16 seconds to display the screen. As a result, the job control facility was not used much to bounce back and forth between jobs: To leave the paper in order to check one item on an outline (saved in a suspended job) and then return could take as long as 32 seconds—a long time to spend simply waiting. When one of us switched to a 2400 Baud modem at home, the difference was dramatic: It cut display times in half, and the result was a substantial increase in the amount of switching back and forth among activities.

The Role of Windows in Support of Multiple Activities

Windows were originally designed as explicit supports for the conduct of multiple activities. Some of these aspects have been forgotten in the current craze for window systems. Moreover, there has not developed any systematic body of knowledge about the properties of windows, either as reminders or just as useful interface structures.

Window systems make possible the display of considerable information for each of the multiple activities that are currently active, subject to limitations on the size of the screen and the memory space allotted to handing the screen map. All sorts of reminders can be presented on the screens because a major portion of them are continually visible. Windows themselves can serve as reminders of the existence of the activities contained within them. Life, however, is a series of tradeoffs, and window systems are no exception.

Windows distract. Too much extraneous information on the screen distracts from the main task. It distracts both by drawing attention to subsidiary, irrelevant activities and also by cluttering the visual appearance, thus increasing the difficulty of finding the point of interest. People are easily distracted, especially as work on a long, tedious (but necessary) task drags on and attention momentarily wanders from the work. At that point, it is *not* desirable to have a subsidiary window that contains an ongoing game display, or a list of new messages received by computer mail, or partially completed other tasks. In this case the window system, is much *too* effective as a reminder. A slogan summarizes the tradeoffs inherent in all reminding schemes (not just those involving windows): *One task's reminder is another task's distractor.*

Organization of windows. Major discussions in the window-design community take place on the desirability of the various ways of organizing windows: Should they be neatly placed on the screen, or should they be allowed to be put wherever they are handy? Should the user have control, or should the system provide neat partitioning of the screen space in some predetermined manner? Should cleanup be done automatically, or only when requested in some predetermined manner?

Much of this controversy is outside the scope of this chapter. Some of it is unsolvable, both because different tradeoffs occur in different circumstances and also because different people have different preferences and work habits. However, we do wish to remind of one important principle: *exploit spatial location.* People can use space as effective memory aids. This allows reminding to take place simply by the use of markers in space, with minimum or litle need for verbal labels or special icons. Even where icons or full text is available, spatial location acts as powerful cues to the contents without the need for the disruptive processing required to interpret the icons or text. But space can serve as an effective reminding tool if it is consistent. This means that windows, icons, or other reminders should have fixed positions, and that each time the computer system is used, the same positions are always used for the same information. Automatic systems that restructure the screen each time the system is called can be as disruptive as the well-meaning cleaning person who "tidies up" a person's private possessions, desktop, or bureau drawer.

ACKNOWLEDGMENTS

We thank Allen Cypher and Steve Draper for their comments on the manuscript.

CHAPTER 14

Communication Paradigms for a Window System

RACHEL REICHMAN (ADAR)

A BRIEF DIGRESSION

To understand the normative procedures that users might be bringing with them to the human-computer interaction, consider the idea that the person and computer are having a two-way conversation. Let's start by looking at some of the structures and procedures used in everyday discourse to see if there are analogs for them in the human-machine interaction forum.

As an example, consider the following piece of taped dialogue that I presented to two groups of subjects. The dialogue is composed of two stories. The two groups heard identical stories except that for one group the last utterance of the dialogue was preceded by the phrase "well anyway." The groups' interpretations were diametrically opposed. The group with the "well anyway" said the last utterance referred to the first story; the group without the "well anyway" said it referred to the second story.

K: Uh, I got a terrible story for you. Poor Sue.
R: Yeah
K: You know she's been having trouble with her car. So she brought it into the garage,
R: Yeah
K: And she thought, you know, it would be $200, not that expensive,
R: Uh huh
K: First they told her that it was just this little plug in back of the engine which was just a minor part but was gonna cost a lot anyway because it was hard to get at.
R: Uh huh
K: And then she went in today, and, oh boy, they gave her this list.
R: You're kidding. Do you want to know something? I mean, I hate to interrupt you but my girlfriend had the exact same story happen.
K: Oh God, you're kidding.
R: No, it's like really strange. My friend, uh Jane, brought her tape recorder into Fedco,
K: Uh huh
R: Right? And, you know, she only brought it in to get cleaned
K: Uh huh
R: And then she was going in to pick it up and she calls them and says, "Hey look, is my tape recorder all ready?"
K: Yeah
R: And they go, "Well actually, you know, your reverse isn't working very well and your recording isn't working real well.
K: Oh, wow
R: It's just like outrageous.
K: That's terrible.
R: I know.

Now, read the next sentence:

K: Well anyway, she brings it in for one thing and they tell her ten others.

To whom does "she" refer? In my experiment, some people heard the sentence above, others got the same utterance by the same speaker without the "Well anyway" preface. 84% of the people who got the "Well anyway" said "she" referred to Sue, the girl in the first story. Only 24% of those who didn't get the "Well anyway" said "she" referred to Sue. The results are statistically significant.

"Well anyway" worked as a marker to indicate the return to the preceding unfinished story about Sue and her car. "Well anyway" is a conventionalized marker for the expected return to a topic interrupted by something closely related to it (e.g., one whose generalization is the same as the interrupting topic). In discourse, whether or not a potential ambiguity exists, we use markers of this form. These markers help coparticipants structure an ongoing exchange in the same way.

A feature enabling coherent discourse is a set of conversational rules that conversants in a given language community share. These rules provide speakers with mechanisms to indicate to coparticipants context shifts. They also dictate what reference forms are unambiguous.

In the preceding excerpt, R has performed the conventionalized move of an interruption. The effect of an interruption move is to put the interrupted context in an open status assignment. An open context does not govern development of the digression context but it does create certain constraints on further development: One expects a return to the interrupted context on completion of the digression.

Now, among the shared rules of discourse that conversants have is the rule that pronouns like "he" and "she" and deictics like "this" and "these" can only be used to refer to entities in a context which is (a) active, and (b) has not previously been closed. During R's discussion in the digression context, K's initial topic was in a nonclosed open activity context. On return to it, one can use the pronoun "she" to refer to Sue, a high-focused element of this topic. On the other hand, given no return, the topic remains in the digression context and "she" is taken to refer to Jane, the high-focused element of topic 2.

But, how is it that 84% of the people knew that a return had taken place and that topic 1 was not earlier closed by topic 2?

This is where the conventionalization of discourse activities plays an important role. It is a conventional move in discourse to perform interruptions. It is also a conventionalized discourse expectation that on completion of the interruption the speaker will return to the initiating subject of the interruption. "Well anyway" is a marker of such resumptions. One of our internal reflections of this expectation is not to give the interrupted context a closed status assignment.

Now if we hypothesize that a person and a computer are having a two-party conversation, let's see what, if any, analogous conventions have support in that domain.

COMMUNICATION WITH COMPUTERS

Early communication procedures between people and computers were quite limited, linear, and constrained. With the advent of full screen displays and multiple window facilities, however, the interaction begins to approximate our everyday type of interactions among people. These interactions rely on a set of conventions that participants of the community share. They enable the participants to structure their reality and interactions similarly.

A basic feature of everyday interaction conventions are methods to partition an ongoing exchange into delimited contexts and to track, as an interaction proceeds, the currently relevant context. To participate in an interaction, individuals must be able to build compatible context models. In everyday interaction, individuals are able to switch adeptly and easily between contexts. The amazing thing about person-person interaction is that participants usually seem to be able to follow the different twists and turns that coparticipants engage in. They are able to do so without having to be told exactly what the coparticipant is about to do. This ability stems from the shared conventions that the individuals are using.

These conventions specify a standard set of relations that contexts can have with one another. Associated with each relation are also rules on how to establish it in a given environment and the effect of its establishment on preceding contexts.

Window systems. What do they afford us? What is the relation between context modeling and interacting with a window system? Why do I claim that with the advent of window systems our interactions with the computer begins to approximate our everyday interactions between people?

First and foremost, windows provide us with a visual display of contextualization. Let's begin with an example of an individual's use of a particular computer and window system, a debugging session on the Xerox 1108 computer (the Dandelion) as described in Pavel, Card, and Farrell (1983).

The user's initial display consisted of three windows (see windows 1–3 in Figure 14.1): a window in the upper right hand corner displaying a clock; a window in the lower right hand corner displaying a graphic design; and a teletype-style window in the upper left hand corner in which the user could type commands to the computer and on which the computer would type messages back.

In the example described, the user began by typing in the teletype window the name of a Lisp function which was to be executed. The system responded with an output window and waited for the user to

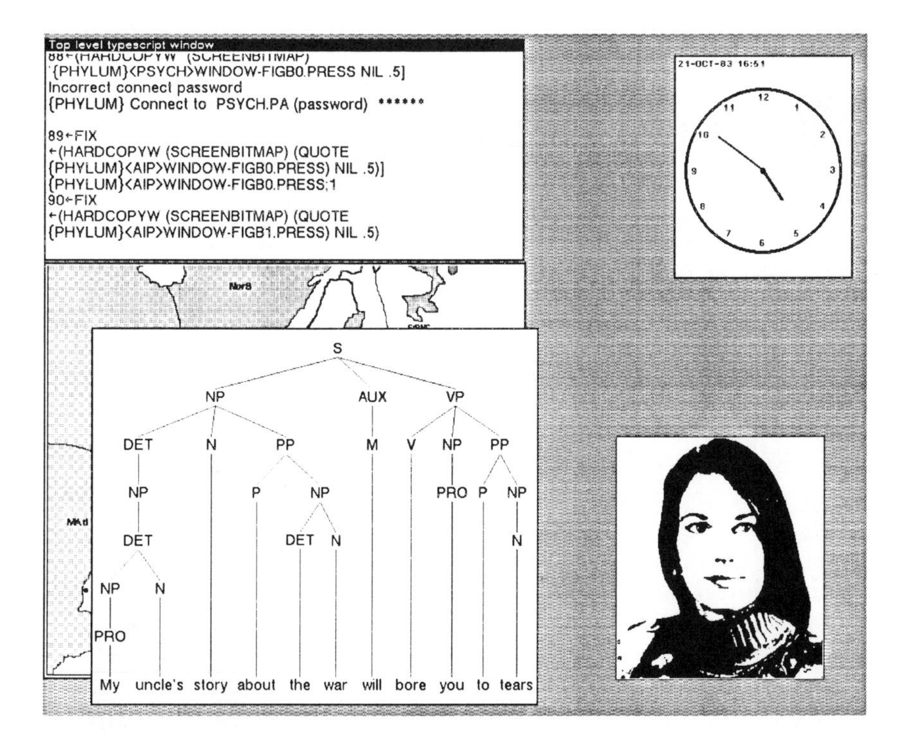

FIGURE 14.1. A multiple window configuration.

indicate where this window should be placed on the screen and how large it should be. After receiving the user's response by mouse input, the system placed the output window in the desired destination and placed the output of the function, a parse tree, into this window (see window 4 in Figure 14.1).

The user then pressed an interrupt key while manipulating arcs of the parse tree. This caused the system to create and place a new window display on the screen, window 5 in Figure 14.2. Windows 6–8 then appeared sequentially afterwards as the user asked for further details about the routines involved in creating and manipulating the initial parse tree in window 4. In window 6, a sequential list of all routines executed thus far was given. After the user clicked on one of these routines, Cursorposition, the system responded with window 7 which contains a list of the variables and their bindings for tracking the cursor. Next, the user clicked on the variable 'ND' in window 7 and the system responded with window 8 where it listed the component

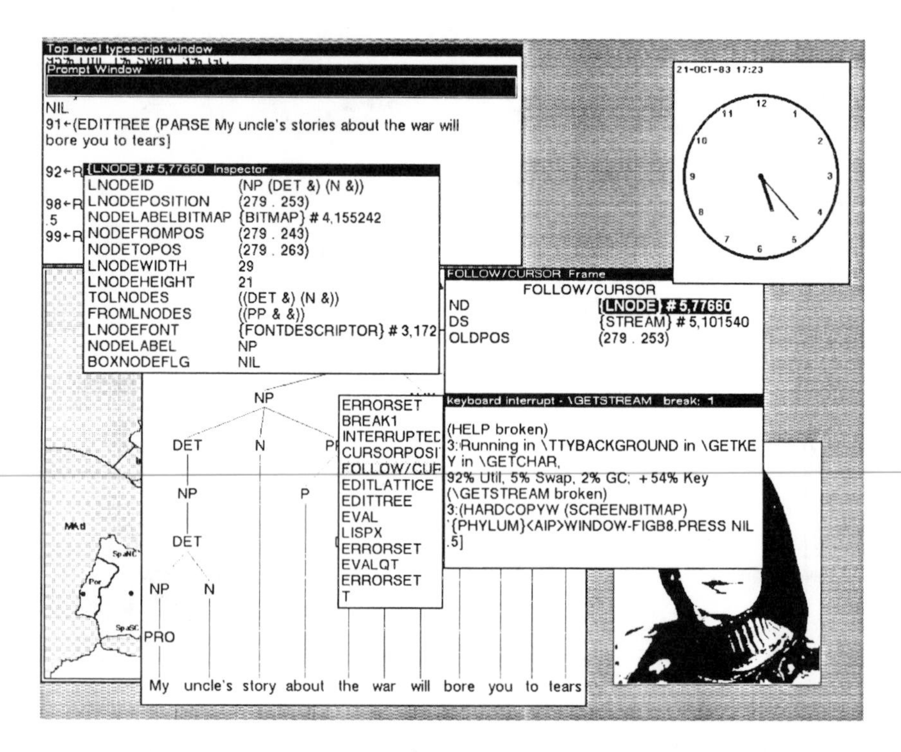

FIGURE 14.2. A multiple window configuration with overlays.

parts of the 'ND' data structure.

As discussed in the Cypher and Miyata and Norman chapters in this section, we have here an example of a user engaging in multiple activities within a single session. In this example, each activity has been put into a separate window and as such each window constitutes a separate context. The contexts, however, are highly interrelated. For example, the contents of window 8 are a "further elaboration" of an item in window 7; the same relation holds between the contexts in windows 6 and 7. While all these windows are simultaneously visible to the user, a user basically interacts with one window at a time. This window is called the *active* window in most window-management systems. Any input typed by the user is received by this active window. All other windows are considered nonactive.

In everyday interaction, specifically in language, people are always dealing with multiple contexts in different activity statuses. In language there is only one active context and depending on the relation of this context to preceding contexts, these preceding contexts are assigned different status assignments. Depending on the particular activity status that a preceding context is in, participants have different means of returning to it. I will illustrate that similar navigation schemes can be used to good effect by window management systems.

While window-system managers on different systems vary greatly in their degree of functionality, most systems do not support the type of implicit context navigation and tracking used by people in their daily interactions. In particular, there is no differentiation in activity status between the different nonactive windows in the environment. As a result, users are not provided with differing access schemas to these different windows. However, because the window systems give visual evidence of contextualization, users often assume that the conventions of contextualization used in everyday interaction are also supported by these systems. While such support would incur an overhead cost it seems the right direction to go. In this chapter, I'll look more closely at (a) the notion of context in computer usage; (b) the correspondences between context in computer usage and in language; (c) how context tracking is supported in language; and (d) some possible schemas that could be used to have context tracking and identification supported in our window system interfaces.

Let's begin by briefly tracing through human–machine interfaces over the years. The focus of this trace is to highlight those aspects of change that have led users to expect more from systems, and in particular to expect some type of analogous support for multiple contexts as found in their everyday systems.

THREE FORMS OF INTERACTION

Indirect Interaction

Using a computer, whether via a window system or a simple screen, is a communication process. The screen is the visual interface to the computer. It echos back to us the commands being issued to the computer and it reflects the computer's responses to these commands.

In the history of computing, the shift from batch processing to interactive computing brought along with it major software changes. A most important change was a new conceptual image about the nature of the human-machine interaction. One might even claim that before the transition there was not a human-machine interaction to speak of at all. The input to the computer was both physically and conceptually distinct from any of the workings of the computer. Entering the data or program into the computer was like passing a note to a friend who would not answer until some time later. In that medium, there was not a direct, real time, interaction between the user and computer. User, interface, and computer, were distinct.

But when we shifted from batch-processing to interactive programming (on-line computer interaction), no longer did we have to wait minutes (or even days) to get some response; we could type on the computer terminal and get an immediate response. At this stage communication was beginning. But there was some time delay between input and output. The teletype was a paper medium and software written for this interface was governed by that fact. For example, the basic commands of editors written for teletype interaction manipulated lines and the words and characters of these lines. As on paper, one proceeded line-by-line through the text file. Separate instructions were needed to edit the file and to see the effects of that edit on the file.

The teletype software fit its medium. But, it left its mark. It was clear to all users that a barrier, an interface, existed between them and the computer. Part of this intermediary feeling between user and computer stemmed from the nonsharing of information between the user, the terminal, and the computer. The user could not see what the computer was doing. The interface and computer were separate.

Limited Direct Interaction

In the next stage, a constellation of hardware changes made it possible to support a full screen display so that text editors could display an entire screenful at a time. It now also was possible to show the results of any operation immediately and automatically.

It took awhile for the new software, taking advantage of the new medium, to be written. It took even longer for users to change their working habits to suit the new medium. For years, users familiar with line editors continued to use them on the new displays. For years, most people, even those using the screen editors, first wrote their files longhand and edited them on hard copy. Some still do. It often takes quite awhile for the power of new technologies to be utilized. We will see that the same phenomenon is happening with our window systems of today.

Although an improvement over the earlier systems, the full screen terminal and accompanying software changes still had made the barrier between user and computer self-evident. While communication between user and computer was now more direct as a result of their sharing a fuller and richer environment, the communication was still limited and stilted. A predominant feature of the stiltedness, again, was driven by the medium of interaction: At any point, the user could only be actively engaged in a single process, the one visually represented to the user by the single-screen medium. But peoples' tasks on the computer are rarely limited to a single process or to a single file at a time. People switch between activities midstream and are used to doing so from their everyday interactions.

Thus, though the interaction between user and computer within a single activity was more direct and natural than in previous mediums, the overall environment was not. Users were still limited to a single activity at a time. The unnaturalness of the overall interaction is somewhat as if you had a friend name George to whom you could talk about chemistry in a chemistry lab and about food in a restaurant and for whom the twain never meet. In fact, you would need a third language to get George to go from the chemistry lab to the restaurant. Now, because George was never seen simultaneously in both environments, and in any single environment George showed no evidence of ever having been, or ever being capable of being in the other environment, you got the feeling that at any time you were only communicating with a part of George. You never had the whole George. You were not even sure there was a whole George.

Some single-stream systems do provide some form of midstream switching. However, they do so in a very limited and constrained fashion. They also incur overhead costs. In the Berkeley UNIX operating system, for example, one can suspend a process thereby relinquishing the computer and screen to another process. The system automatically assigns the suspended process a job number. To return to the last suspended job one types a single command; to return to other suspended processes, the user first gets a list of job numbers and then specifies the desired job number command. Again, the interaction is unusual. It is like always having to leave a current discourse topic explicitly in conversation, metadescribe the different topics that were talked about, and then use one of these metadescriptions to return to an old topic. Metadirectives break the natural flow of thought. In discourse, simple linguistic *clue words* like "But" and "Well, anyway" are usually used instead (Reichman, 1978). These clue words do not shift the level of discourse and thus only minimally interrupt flow content.

Direct Interaction

The most recent technological change provides support for windows—dividing one screen into multiple virtual screens, each behaving like a complete screen. Within each window one can invoke a different process. One can also have nonprocess windows; for example, one can write a program that will create a window and which will send graphic and/or text output to it.

Window systems make it easy to move among different processes. The user simply uses the mouse to move the cursor from the window that contains the current process to the window that contains the new process (some systems require clicking as well). During the stay in the second (or third, or fourth, or fifth, etc.) process, the user's state of activity in earlier processes is remembered and often visible. In addition, in some systems, a limited amount of intercommunication between processes is allowed (for example, on the Symbolics 3600 some shared knowledge space exists between a Lisp listener and an editor window).

Multiwindow display screen systems, then, give a major change. No longer is the computer-interaction world composed of one interactive process at a time. Now it is composed of a whole set of interactive processes, many of which are simultaneously visible to the user.

This greatly enriches a user's environment. No longer do users have to interrupt their flow of thought in order to shift to a related activity or to return to a previous one. Related contexts are often simultaneously visible and can be accessed and referenced in subsequent engagement. Leaving or returning to activities does not require explicit naming and/or freezing. Mouse shifting and clicking language used within activities is now used to affect the transitions as well. This gets rid of our strange friend George who needed a separate control language to shift between activities and, who had an entirely distinct communication procedure for whatever part of the universe he was in currently. The interaction becomes more natural and supportive of the way we naturally work.

THE ALIENATED WINDOW

Something is still lacking. As in earlier systems, our roots and metaphors for the new interface is driving our software development and use of the new technology. An important metaphor that underlies many window systems is the "desk metaphor" (Malone, 1983). Window displays can be thought of as computerized desks with each window being like a separate piece of paper on one's desk. Some systems require "neat desks" while others allow overlapping windows and "messy desks." In either case, like papers on a desk, the windows are pretty lonely. What organization or interrelationship exists between the contents of these computerized papers, are, in the main, in the mind of their authors. In the desk medium, writers may impose some kind of organization for their papers; for example, by constructing one pile of papers "to be taken care of immediately" and another pile "to be taken care of in the next few weeks." The organizations, however, are in the mind of the writers who must somehow remember them. There can be other kind of relationships: consider a bookkeeping ledger. On page 1 there might be a list of current sales, and on page 5 a current inventory. The person in charge of this ledger understands that an entry on page 1 necessitates an update on page 5. But the pages are isolated: They do not know of any relationships.

Now why has this metaphor come into prominence? Once again, we must look at the roots of the technology. The parent of the multiwindow system was the single screen display. In the same way that the limits of the teletype interface strongly influenced and directed the users' and developers' initial use of the full screen displays, so too the single screen display seems to be limiting our uses of the window technology. A lot of the software encourages users to view their interaction with these systems as comparable to being logged into a number of separate single screen computer terminals at the same time. But surely this is not a new and ultimate use of the technology. It does not capture or effectively use the possible contextualization features of windows at all. Rather, again, users are working in impoverished and isolated environments.

Current window systems mainly assume that each window (process) is independent of the other. This radically undermines the power and utility of such systems. It is in direct conflict with how users view their interaction. Communication postulates are violated and errors result. In language, for example, we do not randomly start up different topics of conversation. There is a coherence to our communication. Similarly in the window domain, we are not randomly creating twenty unrelated windows and engaging simultaneously in twenty unrelated tasks. Often in fact, new tasks are started because they are required by existing tasks. As illustrated in the example cited in the introduction, in the window-system environment, there are often interconnections among the different windows in our environment, though these interconnections are not currently visibly depicted or supported by most window managers. The interconnections, however, direct a person's interaction with this environment. As in the natural language forum, these relations set up their own expectations about proper communication procedures.

In the language forum, different discourse expectations result from the way a conversation progresses. It is crucial that each participant be able to follow the twists and turns of the discourse and for each to understand how a current piece of discourse is related to the prior discourse. This enables them to build compatible models of the interaction. Conversants use subtle signaling devices to achieve this. Depending on the particular type of twist and turn, different devices are used and expected. Conversational gambits stand in strong contrast to window display systems. In window systems, transitions are treated equally and identical signaling procedures are used to make the transitions. For the users, however, transitions are not functionally identical and as in any ordinary communication system, they will not use identical devices to make the different transitions.

Tracking the current discourse topic is one of the major tasks that a person performs as a conversation progresses. "Oh, we've left that topic and are onto a new one. Okay." Similarly, in a multiwindow display system, a user has to know which window is currently active. Many users on differing systems are getting into trouble here. For example, one of the main complaints users have of Texas Instruments' menu-based Naturalink natural language system is "Which menu in which window am I suppose to be making a choice from now?" On Lisp machines users have the same problems. They have two different editor windows on the display. They think they're typing into one file in window B, and meanwhile their typing goes into a different file in window A. "What?" "Oh darn" are the exclamations.

Why are users having so much trouble? One reason is simply mechanical. Most of these systems mark the active window by a little blinking pointer or a highlighted title bar. These visual indicators are too limited. The real reason, however, is conceptual. Users' use windows as a visual reflection of the set of overlapping and interleaved activities that they are engaged in. And they expect the computer to be doing the same. "Given what I've done so far and what I'm typing in now, it only makes sense for this input to be going to window B and not to window A. How come my computer isn't following?" Or, "I'm staring at window B, obviously that's where the input should go. Why isn't the computer following?"

The problem is that while the user sees and knows the interrelations the computer does not. In tracing through earlier interfaces, I claimed that the lack of shared contextual knowledge between user and computer made the interface obvious and unnatural. A similar lack still persists. There are no markers or underlying representations in the computer for window/activity interrelation. Currently, users do not have the means to indicate to the computer that they will, for example, be leaving for the moment activity 1 to pursue activity 2, and that they'll be going back to activity 1 shortly. Nor can they return to preceding contexts by means of specifying a functional and semantic relation. But it is in these terms that they view the flow of their interaction. How can users get the computer to have the same expectations that their discourse partners have? In current systems they can't. There aren't any standard mechanisms for users to indicate to the computer why they're leaving a current window and for how long they will be gone. This lack causes problems.

For example, to change between windows on a Symbolics 3600, one must move the mouse and click. Two actions. These two actions are required no matter what the cause of a shift may be. Now, in actual practice, people seem to do okay when there is no relation between the windows they're shifting between. Where, however, an obvious relation is involved (i.e., obvious to the user), more often than not, users would move the mouse but forget to click. Why? Well, they were using a clue-word type of, not a "Getting back to X" type of metadescriptive return. In language, some circumstances require metadescriptions, others do not. On current systems, we have no such differentiation. There is no relation between a user's visual display, the functional interrelations between the windows of this display, and the mouse and menu commands to shift between the different windows-processes on the display.

CONTEXT, COMPUTER APPLICATIONS, AND WINDOWS

Windows are used to contextualize activities. Let me give a few examples of how context is or is not currently being supported by systems.

Programming Examples

A main feature of context support is supporting the relations between objects within a single or multiple context(s). Spreadsheets such as

Visicalc, Lotus, and Multiplan, provide such support. In a spreadsheet, one's domain is decomposed into a number of cells. Cells may be linked to one another, and an update in one cell can ripple through to all other cells to which it is connected. Multiplan, for example, allows one to name both individual cells and groups of cells. In Multiplan, one can specify that the value of some cell is an algebraic function of some of these other named cells. On updating a cell, Multiplan then automatically updates the value of all cells whose value function references the cell being updated.

This type of object dependence is also found in some general purpose systems. For example, the Loops system from Xerox supports object dependence by allowing users to define some objects as *active values* (Bobrow & Stefik, 1982). Users may then specify special functions that will get executed whenever the object's value is either referenced or changed. These functions can be quite powerful and can represent internal constraints between the different attributes of a single object (e.g., change the type of plane to Boeing 777, then change its passenger capacity to 200). The functions, however, can also reference attributes of other objects and in this way we can model dependencies between different objects as well.

Supporting context entails two things: knowing when things should be interpreted together and knowing when they should be interpreted separately. Take a relational data base system, for example. A user has a set of files, each containing different types of information. One file may contain the names and addresses of manufacturers; another file may contain, for each part used in a particular industry, its name, size, color, and manufacturer. The two files form individual contexts, though interrelationships do exist between them. Sometimes users will access one context at a time, at other times they'll access them jointly. For example, asking for the address of a manufacturer who supplies part X requires accessing both files simultaneously.

Now, in designing our new computer interfaces, using the virtues of the multiwindowed displays, we'd like our underlying programming and display software to simultaneously capture how these files are separate yet linked. For example, asking to display these two files might cause two separate windows to be displayed with arcs linking the overlapping fields. Currently, no system that I know of provides this type of visual reflection of interconnection.

But what about underlying support? Here, too, we have a lack. For example, consider dBASE III, a popular commercial relational data base system. While it allows users to work with multiple files, it does not support the different ways that users may or may not be interconnecting their accesses to these files. Just as in the current window systems,

"A switch is a switch is a switch." So, for example, suppose that a user is using some screen editing format file, F, for a file R, and the user then switches to work on a file S. On return to the R file, the user has to respecify that format file F should be used for R. A typical response is: "What does my intervening work on S have to do with the format I'm using for R?" Nothing. But the system still has to be told.

There is an even worse situation in dBASE III. dBASE III allows users to explicitly associate different work areas with different files. This strongly suggests context sensitivity to the user. (See Chapter 9 by Owen and Chapter 7 by Riley on user models.) It's like giving users separate windows for their different files. Users may think, "Okay, now using the separate work areas I won't have the above problem." Wrong. The same problem happens. For example, just as users can set up formats for files, they can also set up indices for them. Once again, the minute a user switches to another file, even if this file is in a different work area, on return to the first file, users have to respecify their index for it. Why is this happening? Simple. There is no principle or understanding of context in these systems.

Similar situations occur even in the most sophisticated computer environments. This actually should not surprise us since most commercial software is based on laboratory work. Both system designers and researchers are governed by the old roots. For example, consider a favored "serious programmer's" on-line editor. Emacs supplies users with an 'UNDO' command which will undo the effects of a user's last command to the system. However, as in our dBASE III examples, the Emacs command is not context sensitive. Thus, if a user first performs actions A–C in Buffer 1, then switches to Buffer 2 and performs actions D–F, and then returns to Buffer 1 and does an UNDO, F will be undone, not C. But, in Buffer 1, action F is meaningless; it should not even exist. Again, no notion of real context.

In contrast, let's go back to the Loops system. Here, we do find some basic principles of structure and context underlying the design. A basic primitive in Loops is the *knowledge base*. A knowledge base is built up as a sequence of *layers*, where each layer specifies changes made to the preceding layer. Within a single knowledge base, there exists, then, memory of the progressive changes made to one's file. And at any time one may retrieve any layer that one wants. Knowledge bases themselves are partitioned into *environments*. An environment provides a lookup table that is used to get the value for some referenced object. Depending on the currently defined dominances, a reference to an object will return its value from one of the different knowledge bases. This way again we get the model of an object's referent being context dependent.

Context switching and interleaving are basic features of human interaction, whether in ordinary everyday conversation, using computer databases, main frame editors, or personal Lisp machine facilities. Few systems, however, provide adequate underlying programming support for the type of multiple activity interleaving that we need.

Visual Display Examples

Having shown that many different computing environments need underlying programming support for context identification, I'll now talk about some of the visual needs as well. I'll give two examples (see Figure 14.3.) The first I take from a fairly sophisticated and well-developed system on the Symbolics 3600 (Gross & Payne, 1984). All items on the right hand side of Figure 14.3 are mouse sensitive and clicking on anyone of them results in either a function execution or a pop-up menu of additional choices. In Figure 14.3 we see an overlay window in which one such function was run.

The Symbolics 3600 treats all windows as independent from one another and one is arbitrarily allowed to select and de-select windows. We earlier showed that this was a positive feature of the system. It modeled some aspects of users' multiple activity structures. It gave them an easy mechanism to do what they naturally did in everyday life,

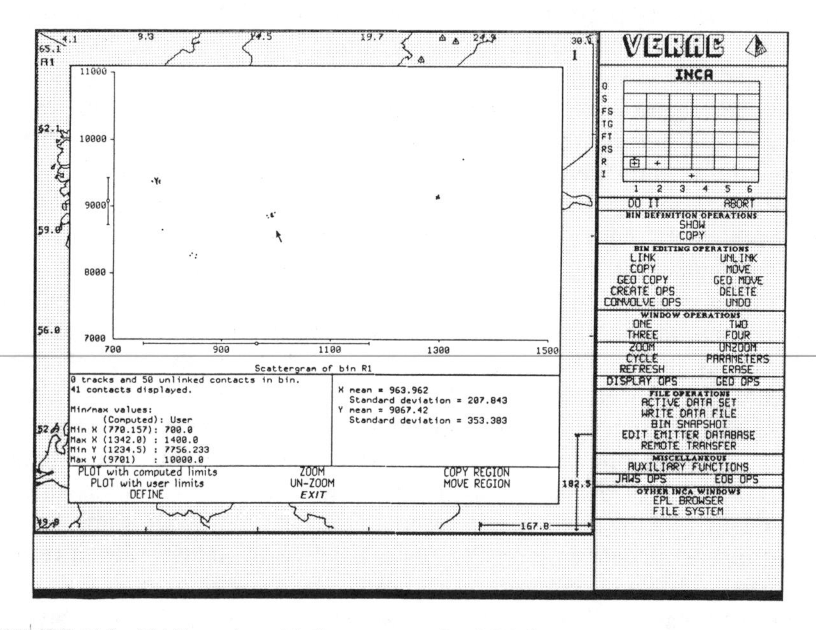

FIGURE 14.3. MAP overlay with Scattergram of cell R1 (background shading added).

in both discourse and other cognitive transactions. Here's the pitfall: *Not all windows are functionally independent.*

The function in Figure 14.3 was put in its own overlay window to get maximal use of the screen display. Because the Symbolics window manager does not support a window being separate and yet dynamic and dependent, the dependency between the two windows in Figure 14.3 is not visually apparent. Using the single shift mechanism, then, users would simply leave the overlay window to reactivate the main process. There was trouble: There was an interference between the different functions associated with these functionally interrelated subprocesses.

Current systems, in general, provide little or no support for object and context dependence. Our language of communication in multi-window systems should include primitives for specifying the interrelation between the different contexts that we're setting up in our different windows. A major aspect of the semantics of a window is that it delineates a context that has possible dependencies and connections to other contexts. But, our current systems mainly see windows from a syntactic viewpoint. Their nonsupport and nonreflection of dependency is reminiscent of early AI planning systems that decomposed a problem into a set of *noninteracting* subgoals (e.g., STRIPS, Fikes & Nilsson, 1971; ABSTRIPS, Sacerdoti, 1974). It was only in later planning systems that the presumption of noninteraction was discarded (e.g., NOAH, Sacerdoti, 1975). It is time to do the same for windows.

The problem of context visualization and support extends beyond just our windows. The problem is systemic to all graphical objects on our display. Basically, we need a visual constraint language for display objects, whether they are entire window contexts or particular entities within these contexts.

For example, on the top rightmost portion of the screen in Figure 14.3 you see a matrix of small cells. These cells can either be empty or have some data in them. If a user clicks on one of these cells a pop-up menu of possible operations will appear. It turns out however that some operations are only legal if at least one cell in a preceding row has some data in it. This constraint, however, is not easily represented graphically and it is not supported by the mouse, i.e., all cells are mousable even if in the current environment there are no legal operations for it. We need a language in which to display these types of constraints to a user. Just looking at the matrix, no notion of interrelation is evident. I believe the same language that can be used to reflect window dependencies can and should be used to reflect the dependencies between individual objects on our screens.

Most users who use a window system figure out for themselves ways to get their displays to support the underlying structure and organization of their work. They often begin a session by setting up their display to support the situations they usually get into. In the terminology of Pavel, Card, and Farrell (1983), users preset their window *working set*. I'll give a personal example from the Symbolics 3600.

I usually begin a session by setting up the particular window structure on my screen shown here:

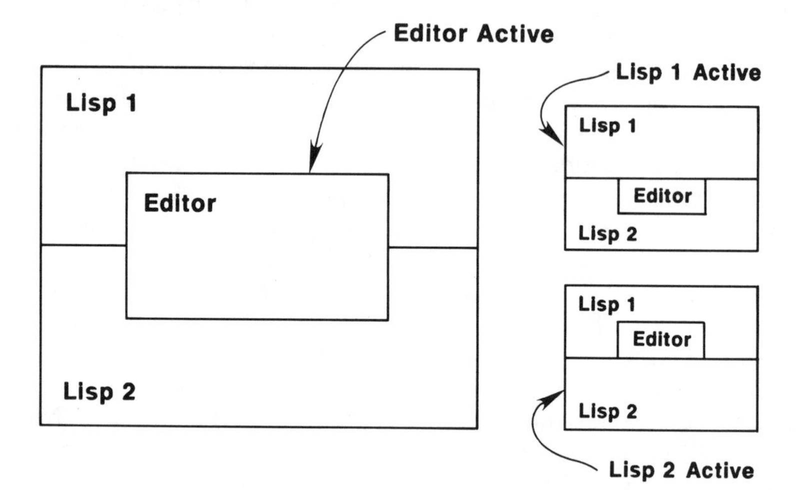

Since I only use one Lisp window at a time this must look pretty weird. It is, but it was a quick way for me to get the window manager to support what I needed. I used this setup for my Execute-Fix cycle (i.e., execute a function, get a bug, go back and fix the function). Why? Well, when I first began to work on the Symbolics I used to use its Split screen command for this cycle getting the display shown below:

Editor

Lisp

I had two problems with this. First, it didn't give me enough room for my Editor window. Second, I kept forgetting to click in my transitions between the Lisp and Editor windows. In the first configuration, I got more room for my Editor window and I also couldn't get errors on my transitions. I *had* to click since without the click neither Lisp window was completely *visibly* available. It wasn't a matter of some tiny blinker blinking or not. It was self-evident, the window was being obscured. Similarly, in going from one of the Lisp windows to the Editor window, I was *forced* to click because without the click only half of my Editor window was visible. Again, the display strongly reflected my constraints. It forestalled errors and kept things at the nondistractive automatic way of doing things.

Now, you still say, "But why did she need two underlay windows?" Simple. If my underlay were a single Lisp window, clicking on it would cause the Editor window to disappear. Then I had two problems: First, I either had to (a) search through a menu to get the Editor window back after the Lisp portion of my Execute-Fix cycle, or (b) do a lot of burying and exposing of windows, or (c) use control character commands to get the Editor window back. All three took me to the metadirective level. Second, I couldn't see any part of my code as Lisp was executing. This, of course, was a needless impoverishment of my environment. So I had to use a second underlay window on the display even though I never actually activated it.

While different systems have designed their screen managers differently, with some providing more or less support for contextualization, none so far provides users with a set of conventions for differentiating between context switches. No set of standard relations have been defined for the human–computer interaction and, as such, these interactions do not support contextualized navigation through distinct and related activities. Let's turn now to the language forum where such support exists.

THE CONVERSATION METAPHOR

Remember the opening dialog of this chapter. We could conceptually analyze the window-system errors that we saw of a person's not clicking on a return to an Editor window after getting a bug in a Lisp window as the equivalent of using a pronoun or deictic reference. The earlier activity of writing is not closed by the latter activity of testing and executing the file. Since we can expect resumption of the editing activity (either further defining of the function, fixing it, or closing the editor process itself) it seems natural to expect minimal signaling. Isn't mouse movement enough?

In discourse, depending on the type of activity switch, we get different suspension statuses and corresponding different means of returning to the different activities. For example, where in fact one activity closes another, on return to the closed activity, when we first re-enter this context, we cannot (do not) use pronouns to refer to things in the previously closed context. There, a click would not be so readily forgotten. It is mimetic to what we do in everyday interaction. As an example, consider the following taped dialogue between a therapist A and a patient B.

> B: We could briefly discuss something, my mother—you see, I don't really want to because I don't really want to sit and talk about her here, you know, in a way I'm talking poorly of her, I guess.
>
> (APPROXIMATELY THIRTY MINUTES OF TALK)
>
> B: I think in a way that's what she does to me, and I don't like it. So I try not to do it to her.
> [SHE refers to B's mother]
>
> A: But you said you have some feelings about bringing up this whole topic of what goes on between you and your mother. You said because it was negative?

In A's last utterance, B's mother is referred to by the full descriptor "your mother" rather than by a pronoun. B had just referred to her mother with a pronoun in the preceding utterance. Why the difference?

The answer is *context dependence*. A's utterance refers to B's earlier topic of discourse, B's not feeling comfortable about discussing her mother with A. This topic was discussed briefly and closed earlier on in the discourse. Therefore, in referring to an object of this context, full descriptors are called for, even though the same entity was just referred to with a pronoun in the preceding, now closed, context.

In our window-system environment, a closed context corresponds to a process within a window which is completed and for which we have no particular expectations or requirements of a return. For example, let's say you had the file system window open on your screen. You get the name of a file you want to look at. You then create a configuration of a Lisp window and an Editor window which you overlay over the file system window. Once you get the filename you are looking for, the activity which prompted opening the file window is completed. There's no particular reason to think it will be returned to. Whether or not a user physically closes the window or not, functionally, its status is closed.

I believe that windows with a closed status should not be on a current display. Having them there is just distracting and wasteful of the limited screen real estate. Reactivating them should require a nonpronoun reference. In fact, just as in discourse, here some type of full descriptor and metalevel description could be required. For example, for closed windows we might require users to access a pop-up menu which cited all previously created windows. Users would then have to search this list and click on the appropriate name to get the closed window back in their environment. Alternatively, we could require users to bury and expose all intervening windows. Both procedures are metalevel activities and are analogous to what is required for these contexts in everyday discourse.

A differentiation between the status of different contexts/activities is important in our window system. For example, where one returns to an Editor buffer after getting a Lisp error, the Lisp environment is in a *controlling* status. Why? It is the reason for our being in the Editor window: It controls and governs what we'll be doing in the Editor buffer: We could bet 100 to 1 that after completion of the edit we're going to be back in the Lisp window rerunning the function. Thus, for both controlling and open contexts (like the first story in the introductory dialogue) we have an expectation of return. Open contexts however do not govern development of a current context: They don't have to remain on the display while the interrupting activity is executed; controlling contexts in contrast do and thus should be left on the display automatically by the system.

On the other hand, let's say that a user returns from a Lisp window to an Editor window in order to fix a Lisp function and then decides to look at an earlier version of this code where in fact this particular function worked all right. Another Editor window is called overlaying part of the screen. The user looks at the second file and accordingly fixes up the function in the first Editor window. Now, the user returns to the Lisp window. While the user is back in the Lisp window, the first Editor window is in a controlling status. It has a reciprocal relation to the Lisp window that the Lisp window has to it. The user is in the Lisp window to test a part of the Editor window. But, now, the second Editor window has a *generating* status and it should be left on the screen. If there's still a problem with the Lisp function the user will go back to the first Editor window which in turn *might, but very well may not,* need the second Editor window again.

In general, window managers, embedded in their single display roots and desk metaphor, do not support these types of status differentiations. A context is either active or nonactive and differential access to them, in general, is not supported via functional or structural markers.

RE-EXAMINING THE CONVERSATION METAPHOR

I hypothesized that the ease and naturalness of multiple window systems to support multiple activities caused users to transfer the expectations and procedures they used in daily interactions to their interaction with the computer. But interacting with the window display system differs from ordinary conversation in a number of crucial ways. These features are (a) eye contact, (b) time, and (c) control. Ordinary conversations move quickly and who gets control of the floor is a competitive and negotiated process (Polyani, 1978; Sacks, Schegloff, & Jefferson, 1974). In contrast, in the window system world, it is the user who basically always has the floor and is in control. Users can interrupt a computer session to go out for lunch, then come back and continue from where they left off. The computer won't notice the gap. In an ordinary conversation, of course, by the time you returned from lunch, your co-conversant would be long gone. At this stage, the interaction between computer and user totally changes. In fact, perhaps the metaphor of a conversational partner should be replaced with the metaphor of a patient assistant.

The distinction in metaphor is crucial in deciding what type of system support to supply. In particular, the reason ordinary conversations work so well is that conversants signal to each other their context shifts and the types of shifts they're taking. A sensitivity to the models of others is an intrinsic feature of communication. Albeit this sensitivity is mainly subconscious, automatic, and built into our conventionalized rules of how we talk, it is nevertheless there. Speakers do not stop and say to themselves, "Why am I bothering to signal to the other person what I'm doing." With the computer, however, it is a different story. Currently, the multiwindow computer allows a user to shift tasks with no overhead costs. Treating the computer as a conversational partner, incurring extra overhead costs, may be contrary to a user's attitude towards the computer. What was subconscious and easy in one system becomes conscious, taxing, and a burden in another. This is especially true if the user cannot see the benefit in the added communication and takes the attitude of "Why am I telling this tool what I'm doing?"

> *An interesting dialectical feature to this attitude is when the computer does not do what the user wants and the user cannot understand what went wrong. To a novice or intermediate user, at these times, it is the computer who seems to be in total control and not vice-versa.*

Usually, there is no apparent payoff. We cannot effectively converse with another person if we aren't both tracking the discourse and getting the same underlying partitioning of topics and context. But most often we can interact quite effectively with the computer even if it doesn't know how we got to where we are and why we are doing what we're doing. The payoff is basically for trouble spots and extras that we'd like the computer to support. So why bother if the nature of the interaction doesn't require it?

The Assistant Metaphor

Interaction with an assistant (or set of assistants) is a dynamic and multifaceted process. For example, in telling an assistant to perform a task, often it is necessary to explain how to do related subtasks. Moreover, you might explain several unrelated tasks at the same time. We expect an assistant to understand the varied interrelations among the different tasks and to perform accordingly.

Alternatively, a set of assistants may form task hierarchies. For example, one may be an assistant of another or two assistants may be responsible for different phases of a single task. The varied possibilities affect the expected outcomes and nature of interaction. For example, if in your presence I tell your assistant something, I will expect you to be aware of the contents of that communication.

Thus, the assistant metaphor shares with the conversational metaphor the basic shifting and sharing of information requirements of new window systems. It is a better metaphor, however, because of the control aspect of the user's interaction with the computer and window system. In studying the use of window systems I have been struck by the number of mistakes users make after a pause of interaction. For example, if a user spends some time looking something up in a manual or talking to someone else, more often than not, upon returning to the display screen, the user will have a new set of activities in mind. At this point, the user begins by looking at the new place on the screen, ignoring what was previously done. The user seems to assume that the window system has kept up.

The conversational signals and procedures that we use with assistants is much more relaxed than with a second conversational party.

We expect assistants to be at our service. We expect them to know our needs and to have an updated mental model for each person they assist. Thus, even if your secretary works for ten different people, if on Monday you dictate a letter, on Wednesday you can ask if "the letter" is finished. You need not disambiguate the reference to "the letter" though the secretary may have since typed four different letters for four different people.

SUPPORTING MULTIWINDOW INTERACTION AND COMMUNICATION

Who's Going to Tell Who What?

We are currently at a crossroads, faced with the following dilemma. If users see their interaction with the computer as one of boss and assistant, then, they will be reluctant to take the time and effort to mark and/or explain to the computer their activity structures. They will continue to forget about, and be insensitive to, the computer's model of what's going on. They will use eye movement and expect the computer to keep up. Certainly, they will not take too much time to ensure that it has done so.

What can we do? Well, perhaps we want a very intelligent computer assistant. This assistant, like our assistants in everyday life, will anticipate our actions and will try to complete its task without having to ask us for more information. If necessary, it will make inferences and guesses on its own. Why? To uphold the prevailing slogan of such interactions, "Bother the boss as little as possible."

> *Not bothering the boss, however, will of course entail hiding and keeping secret how we figure out what it is we think the boss is doing. While this goes against the grain of much of what is being advocated in this book, one must admit that in real life this seems to work out pretty well.*

But now we have a problem. Most users of computers know that the computer is treating their actions as independent. As Cypher points out, common user interruptions are "while-I'm-at-it actions." They have experienced arbitrarily shifting between different tasks and suffered no negative consequences. For example, I recently got a second drive for my Macintosh. I was debugging a Basic program, which first created a file, F, of names and addresses. The program allowed the user to update the records of F via the screen display. In debugging the program, I wanted to look at the contents of F before it got updated. However, since the Basic program created F as a text output file, I could only look at it using the Macintosh editor program, MacWrite. My MacWrite was on another disk, so I inserted this other disk into my second drive. But now, all of a sudden and quite unexpectedly, in the middle of this debugging exercise, I realized I could test the effect of the two drives on the Macintosh window system. Most of the software that I had used in the Macintosh did not support multiple windows. "But what," I thought, "if the two files to be displayed come from different drives?" I therefore momentarily stopped debugging and issued another open file command; this time for a file on the MacWrite drive. Alas, the window for F closed and I had to reopen it as my next move. How disappointing. But, my immediate response really was, "It's a good thing the computer can't follow my actions or else it would think I was totally incoherent."

> *Assuming coherence, the system would have decided initially that I didn't really want F. But then I asked immediately for F again. Weird. Or assuming nonrelatedness, the computer wouldn't understand why I first ask to open something, then close it by asking to open something else, and then close that by asking to open the first thing again.*

Sensors that track eye movements would allow construction of a smart assistant. We could probably create a pretty complete grammar for most user action sequences. Associating a standardized set of effects with these actions would enable us to track a user's shifting context structures. Contexts could automatically be closed where appropriate. To re-enter them, perhaps, we'd require a user to go through a window menu of closed windows where reference by name would be necessary to reactivate the closed context (as in discourse). In contrast, where moves like "Edit for bug in Lisp," were detected, the system would put the Lisp window in a controlling status and a mere shift of the mouse to that window would be sufficient to reselect it. (See Reichman, 1985, for an example of such a grammar for computerized conversational move tracking.)

However, changes based mainly on eye movements and mouse shifts without clicks would chance a lot of "unintentional" effects. I am therefore not too wild about such a mechanism. People do use automatic signaling devices in ordinary discourse. They do do some signaling to assistants. I want the freedom to go off on random digressions. We don't want to risk the computer making all these inferences about our actions, right or wrong, and keep them hidden from us.

> *To inform us of the inferences and tracking structure would defeat the whole point of the boss–assistant metaphor. It would also keep taking us to the metalevel of interaction. This would be a direct violation of having this process be mimetic of everyday interaction and having it be easy and nondistractive.*

A set of user commands is warranted. Assuming such a set exists, what will the effects of these commands be on our displays? How will the display reflect the dependencies and underlying organization of the interaction?

What Will the Display Look Like?

To support our needs:

1. There should be underlying support to define individual contexts and the relations between these contexts and the objects they contain.

2. These dependencies and interrelationships should be made explicit on the visual display.

3. There should be a separate language that allows users to note the types of activity shifts being made.

4. There should be a correspondence between contexts and windows. The semantic import of a window should be that it constitutes a context. Linking contexts then becomes the equivalent of linking the visual reflection of these contexts, that is, windows.

5. The constraints derived from a context's status and interrelationship with other windows should be self-evident to the user.

6. Navigational schemas should be provided to users so that they can navigate back to old contexts via reference to a functional relation that this preceding context has with a current one.

Our displays should reflect status assignments. I have illustrated that a minimum of four categories are required:

1. Active
2. Controlling
3. Generating
4. Closed

At first I thought of using color to reflect status; for example, yellow for active and blue for controlling. I dismissed this idea however: The associations are arbitrary and nonmimetic to what they're suppose to represent. On the other hand, we are accustomed to red for **stop** and green for **go**. Perhaps, then, red could be used for closed windows and green for an active one. A better use of color, however, seems to be for window grouping. What if we used, for example, different shades of a single color for all windows of a particular working relationship, and different shades of another color for windows in a different working set. Then, in each working set, the active window would be given the palest shade of that color; the controlling window the next palest color; all generating windows a deeper shade of that color; and closed windows of that set, the deepest shade of all. That the windows were related would be evident from their all sharing a same color. That they had different status assignments would also be visually reflected: The darker the shade, the further in the background.

For a similar effect, we could use different black and white pattern backgrounds. We have, in fact, done that manually in this book for a clear presentation of Figure 14.3. An active window would be clear white; the controlling window would have interspersed dots so that it was still quite readable; the generating window would have an even denser overlay of dots making its contents more difficult to read; and the closed window would either be removed from the screen or given a really dense overlay. Here, the status is pretty clear, "Gosh, I can hardly read the contents of this window. Of course it's in the background."

Bit patterns, however, don't reflect the semantic groupings of windows as well as color. To see where semantic grouping would be useful, consider the overlay windows labeled 5, 6, 7, and 8 in Figure 14.2 presented earlier. These four overlay windows were created for purposes of investigating a same function. They therefore form a single working set. When the investigation activity is completed, a window manager should remove and close all four windows together. While they are on the screen, the fact that they are so interconnected should be reflected to the user. A same color for all four would be good here. Perhaps we need both color and bit patterns simultaneously.

In addition to visually representing to the user that a relation exists, we want the particular relation to be self-evident as well. For this I suggest a taxonomy of context relations. I would have named arcs linking the windows in an environment and allow users to mouse these names to return to a linked window of that relation.

> *Alternatively, one could even have a voice-recognition box that would recognize the relations involved and the accompanying clue words of these relations. As users shift between activities they are in all likelihood saying these words to themselves in any case. Voicing them would therefore be totally nondistractive and automatic as it is in ordinary discourse.*

Only windows related to the current development of the active window would be visible at any given time. However, the named arcs to the nonvisible windows would be visible from the active window and the user could then return to these windows as well. To return to windows not related to a current one, a user would have to go through the fuller metadescriptive process of listing all closed windows through a system menu. In addition to effecting a return, such transitions would also effect automatic status reassignments and color reshadings to the other windows in the environment. This would be comparable to our ability in language to refer to a coconversant's preceding utterances by referring to its functional relation in the discourse, e.g., "Your challenge of my remarks was absurd," and, it would similarly capture the structural effects of such a reference on one's model of the discourse context.

Some systems have bits and pieces of some of these supports. For example, in the KEE system environment old messages in the system's output window are overlayed with dots to reflect their noncurrency (KEE, 1985). This same system, however, does not support returning to previous windows based on a semantic or functional relation to a current one. In diSessa's Boxer system (Chapter 10), when users perform a subactivity, Boxer opens up an embedded window for this activity. This visual embedding reflects the underlying subordinate relation between the two activities.

In addition to representing window linkages, it is equally important to be able to represent that objects within these windows may be interconnected and dependent on each other. For example, in Figure 14.3, which illustrates the Symbolics system, there are small dotted triangles in the underlay window (see, for example, triangles between 19.7 and 24.9 in the upper portion of the underlay). The underlay display is a map, and the triangles represent the data contained in cell R1 about these locations. This correlation between the circles and cell R1 is visually represented on the display by "R1" appearing on the topmost left hand corner of the map display and a "•" appearing in cell R1. While here some visual indication of the connection is represented, a user must read and scan these separate parts of the screen to hook things up. What if the triangles and associated R1 cell were both colored the same color in this context? Here, the association would be self-evident.

For this same figure, we earlier discussed the existing dependencies between the rows of the matrix. Perhaps we should have a row grayed out if no operations are presently available for it? This would be like the Macintosh's interface which grays out those operations that are inapplicable in a given environment. Perhaps we should have the different rows organized in some color scheme to reflect dependence? Perhaps, explicit links between different rows and different cells would be warranted?

These are just some of the things that I think a visual display should be able to represent. They are all features of context dependence and interrelation. The obvious analog to a window is a context. The obvious analog to switching among windows is switching between different contexts. These analogs need support.

In conclusion, to begin to use the potential of the window technology, its single screen and desk origins should be left behind. Instead, we should look at person-person communication and activity forums in natural settings to see how multiple activities are usually supported, and to understand the everyday conventions that people bring to bear in their use of the new computer systems. We can already conclude some things: A set of activity interrelationships should be defined. A set of object interrelationships should be defined. We should have a dynamic, easy, and nondistractive means for users to communicate these types of interrelationships to the computer. The computer should then visually reflect these relations back to users and provide them with a set of varying navigation mechanisms which are derived from, and are based on, the different types of relations involved.

SECTION V

TOWARD A PRAGMATICS OF HUMAN-MACHINE COMMUNICATION

In linguistics it is traditional to separate the analysis of language into semantics and syntax. Those who study language, however, have long noted interesting phenomena that do not fall neatly under these divisions (and which are less amenable to formal analyses). These additional subject areas are known by various names, but the term "pragmatics" seems most common. Pragmatics addresses issues such as why the response "Yes" is not appropriate as an answer to the question "Can you pass the salt?" and why the response "I'm not wearing a watch" can be appropriate as an answer to the question "What is the time?" Analogous issues arise in human–computer communication, and so we suggest labeling them with the same term: *pragmatics*.[1]

The two chapters in this section, in effect, deal with pragmatics,

[1] The term "pragmatics" has been proposed before in this connection (see Buxton, 1983) but the idea seems not to have yet gained a foothold.

although neither mentions the term. Other chapters too are connected, most notably the previous chapter, Reichman's "Communication Paradigms for Window Systems," with its concern for how a person might signal shifts between activities smoothly and naturally. A characteristic of pragmatic phenomena is that, although they often are concerned with "large-scale" structures such as user activities or conversation, they are typically expressed in "low-level" things such as mouse clicks, or words such as "but" and "hello." Don't be fooled: "But" and "hello" may seem nonentities syntactically and have insignificant meaning, but their role in structuring a conversation is crucial.

In Chapter 15, Buxton demonstrates that critical attention to the details of the input mechanisms are essential if we are to optimize interface design. He points out that current systems are designed with a paucity of input devices—often only a keyboard—thereby severely restricting their effectiveness. He goes on to discuss how even apparently small variations within a class of input device changes their effectiveness for different tasks considerably. The chapter might at first appear to be about the mechanics of input devices, but it deals with more than that. The chapter is really about the pragmatics, about subtle features that can seriously affect the nature of the interaction.

Buxton demonstrates quite convincingly that too little attention has been placed on the varieties of input devices (existing and potential) and on the critical details of their operation. He draws upon his considerable expertise as a professional musician (especially electronic music) to argue that we should realize that people have five major output mechanisms which could therefore use five input devices: two hands, two feet, and voice. There is no reason why we could not use all five simultaneously, and he suggests that at least we should develop systems that encourage use of the two hands simultaneously, each on a different form of device. (We could perhaps add a sixth—eye movements—but effective eye movement sensors are not yet practical.) Buxton restricts himself to simple kinds of inputs. He does not discuss the important variety of output devices that are possible, and the paucity of experimentation with sound as output (not speech—sound) is much to be regretted (see, for example, Bly, 1982). Buxton is fully aware of the power of audio signals as data, he deliberately restricted the topics in this chapter to make his points: (a) We have seriously neglected the details and pragmatics of input devices; (b) details do matter; and (c) we can do better, much better.

In Chapter 16, Draper begins by introducing the term "inter-referential I/O" as a generalization of the ways in which the inputs and outputs to a system can refer to each other. This allows for a common language and history between a user and a system, allowing both to refer to the same concepts readily and easily. These are critical aspects of the "first-personness" concept of Laurel (Chapter 4) and the Direct Engagement notion of Hutchins, Hollan, and Norman (Chapter 5). Draper suggests that systematic support for inter-referential I/O could have important benefits. First, it could provide better tools for support of user-machine dialogue. Second, it should allow for error correction and recovery through cooperative interaction, by allowing the user to elicit clarifications and explanations. These are points taken up by other chapters: The ability to interact gracefully with the system to take care of error is an important component of the discussion by Lewis and Norman in Chapter 20 and in the discussion of "repair" by Brown in Chapter 22.

Draper's chapter, like that by Buxton, again points out the interaction between apparently low-level details and high-level concepts. On one hand, the concept of inter-referential I/O is a low-level suggestion (keep the records needed to interpret pointing to objects on the screen and other references to previous I/O) but, on the other hand, it adds up to a transformation of the whole quality of the interaction and supports repair to the dialogue and smooth transitions between "levels."

These chapters can be seen as steps towards a pragmatics of human machine interaction. The notion however cries out for further exploration: There may be a whole perspective (not just hints) waiting to emerge on the "pragmatics" or "ethnomethodology" of human-computer communication that would explain what are now isolated, curious observations.

CHAPTER 15

There's More to Interaction Than Meets the Eye: Some Issues in Manual Input

WILLIAM BUXTON

Imagine a time far into the future, when all knowledge about our civilization has been lost. Imagine further, that in the course of planting a garden, a fully stocked computer store from the 1980s was unearthed, and that all of the equipment and software was in working order. Now, based on this find, consider what a physical anthropologist might conclude about the physiology of the humans of our era? My best guess is that we would be pictured as having a well-developed eye, a long right arm, a small left arm, uniform-length fingers and a "low-fi" ear. But the dominating characteristics would be the prevalence of our visual system over our poorly developed manual dexterity.

Obviously, such conclusions do not accurately describe humans of the twentieth century. But they would be perfectly warranted based on the available information. Today's systems have severe shortcomings when it comes to matching the physical characteristics of their operators. Admittedly, in recent years there has been a great improvement in matching computer output to the human visual system. We see this in the improved use of visual communication through typography, color, animation, and iconic interfacing. Hence, our speculative future anthropologist would be correct in assuming that we had fairly well-developed (albeit monocular) vision.

In our example, it is with the human's effectors (arms, legs, hands, etc.) that the greatest distortion occurs. Quite simply, when compared to other human-operated machinery (such as the automobile), todays computer systems make extremely poor use of the potential of the human's sensory and motor systems. The controls on the average user's shower are probably better human-engineered than those of the computer on which far more time is spent. There are a number of reasons for this situation. Most of them are understandable, but none of them should be acceptable.

My thesis is that we can achieve user interfaces that are more natural, easier to learn, easier to use, and less prone to error if we pay more attention to the "body language" of human-computer dialogues. I believe that the quality of human input can be greatly improved through the use of *appropriate gestures.* In order to achieve such benefits, however, we must learn to match human physiology, skills, and expectations with our systems' physical ergonomics, control structures, and functional organization.

In this chapter I look at manual input with the hope of developing a better understanding of how we can better tailor input structures to fit the human operator.

A FEW WORDS ON APPROACH

Due to constraints on space, I restrict myself to the discussion of manual input. I do so fully realizing that most of what I say can be applied to other parts of the body, and I hope that the discussion will encourage the reader to explore other types of transducers.

> *Just consider the use of the feet in sewing, driving an automobile, or in playing the pipe organ. Now compare this to your average computer system. The feet are totally ignored despite the fact that most users have them, and furthermore, have well-developed motor skills in their use.*

I resist the temptation to discuss new and exotic technologies. I want to stick with devices that are real and available, since we haven't come close to using the full potential of those that we already have.

Finally, my approach is somewhat cavalier. I will leap from example to example, and just touch on a few of the relevant points. In the process, it is almost certain that readers will be able to come up with examples counter to my own, and situations where what I say does not apply. *But these contradictions strengthen my argument!* Input is complex, and deserves great attention to detail: more than it generally gets.

That the grain of my analysis is still not fine enough just emphasizes how much more we need to understand.

> *Managing input is so complex that it is unlikely that we will ever totally understand it. No matter how good our theories are, we will probably always have to test designs through actual implementations and prototyping. The consequence of this for the designer is that prototyping tools (software and hardware) must be developed and considered as part of the basic environment.*

THE IMPORTANCE OF THE TRANSDUCER

When we discuss user interfaces, consideration of the physical transducers too often comes last, or near last. And yet, the physical properties of the system are those with which the user has the first and most direct contact. This is not just an issue of comfort. Different devices have different properties, and lend themselves to different things. And if gestures are as important as I believe, then we must pay careful attention to the transducers to which we assign them.

An important concept in modern interactive systems is the notion of *device independence.* The idea is that input devices fall into generic classes of what are known as *virtual devices,* such as "locators" and "valuators." Dialogues are described in terms of these virtual devices. The objective is to permit the easy substitution of one physical device for another of the same class. One benefit in this is that it facilitates experimentation (with the hopeful consequence of finding the best among the alternatives). The danger, however, is that one can be easily lulled into believing that the technical interchangeability of these devices extends to usability. Wrong! It is always important to keep in mind that even devices within a class have various idiosyncrasies. It is often these very idiosyncratic differences that determine the appropriateness of a device for a given context. So, device independence is a useful concept, but only when additional considerations are made when making choices.

Example 1: The Isometric Joystick

An "isometric joystick" is a joystick whose handle does not move when it is pushed. Rather, its shaft senses how hard you are pushing it, and in what direction. It is, therefore, a pressure-sensitive device. Two isometric joysticks are shown in Figure 15.1. They are both made by

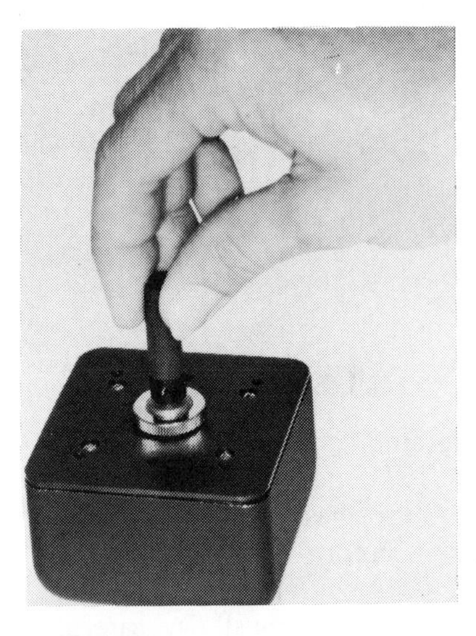

FIGURE 15.1. Two isometric joysticks. (Measurement Systems, Inc.)

the same manufacturer. They cost about the same, and are electronically identical. In fact, they are plug compatible. How they differ is in their size, the muscle groups that they consequently employ, and the amount of force required to get a given output.

> *Remember, people generally discuss joysticks vs. mice or trackballs. Here we are not only comparing joysticks against joysticks, we are comparing one isometric joystick to another.*

When should one be used rather than the other? The answer obviously depends on the context. What can be said is that their differences may often be more significant than their similarities. In the absence of one of the pair, it may be better to utilize a completely different type of transducer (such as a mouse) than to use the other isometric joystick.

Example 2: Joystick vs. Trackball

Let's take an example in which subtle idiosyncratic differences have a strong effect on the appropriateness of the device for a particular transaction. In this example we look at two different devices. One is the springloaded joystick shown in Figure 15.2A. In many ways, it is very similar to the isometric joysticks seen in the previous example. It is made by the same manufacturer, and it is plug-compatible with respect to the X/Y values that it transmits. However, this new joystick moves when it is pushed, and (as a result of spring action) returns to the center position when released. In addition, it has a third dimension of control accessible by manipulating the self-returning, spring-loaded rotary pot mounted on the top of the shaft.

Rather than contrasting this to the joysticks of the previous example (which would, in fact, be a useful exercise), let us compare it to the 3-D trackball shown in Figure 15.2B. (A 3-D trackball is a trackball constructed so as to enable us to sense clockwise and counter-clockwise "twisting" of the ball as well as the amount that it has been "rolled" in the horizontal and vertical directions.)

This trackball is plug-compatible with the 3-D joystick, costs about the same, has the same "footprint" (consumes the same amount of desk space), and utilizes the same major muscle groups. It has a great deal in common with the 3-D joystick of Figure 15.2A. In many ways the the joystick in Figure 15.2A has more in common with the trackball than with the joysticks shown in Figure 15.1!

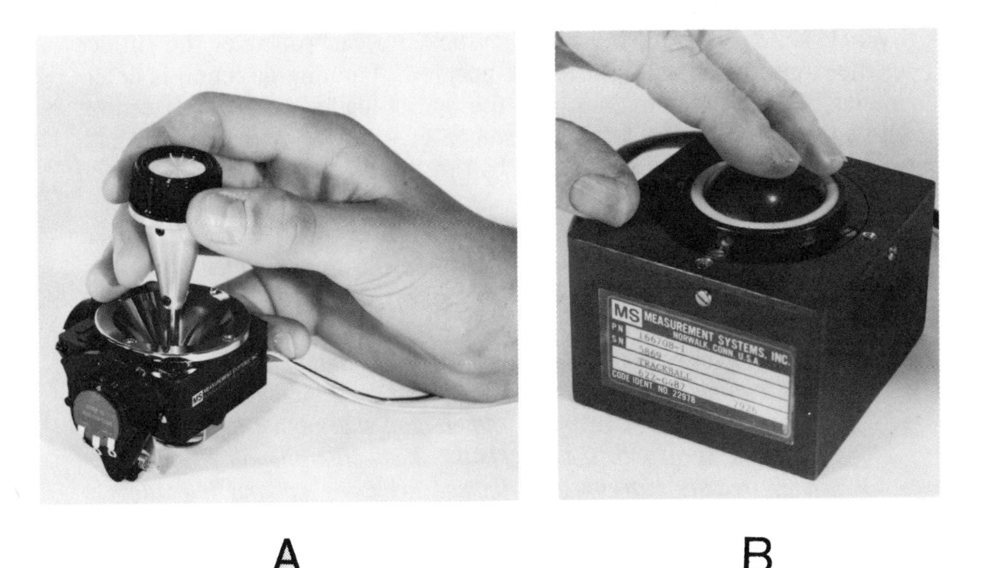

FIGURE 15.2. Comparison of joystick (A) and trackball (B). (Measurement Systems, Inc.)

If you are starting to wonder about the appropriateness of always characterizing input devices by names such as "joystick" or "mouse," then the point of this section is getting across. It is starting to seem that we should lump devices together according to some "dimension of maximum significance," rather than by some (perhaps irrelevant) similarity in their mechanical construction (such as being a mouse or joystick). The prime issue arising from this recognition is the problem of determining which dimension is of maximum significance in a given context. Another is the weakness of our current vocabulary to express such dimensions.

Despite their similarities, these two devices differ in a very subtle, but significant, way. Namely, it is much easier to simultaneously control all three dimensions when using the joystick than when using the trackball. In some applications this will make no difference. But for the moment, we care about instances where it does. We look at two scenarios.

Scenario 1: We are working on a graphics program for doing VLSI layout. The chip on which we are working is quite complex. The only way that the entire mask can be viewed at one time is at a very small scale. To examine a specific area in detail, therefore, we must "pan" over it, and "zoom in." With the joystick, we can pan over the surface of the circuit by adjusting the stick position. Panning direction is determined by the direction in which the spring-loaded stick is off-center, and speed is determined by its distance off-center. With the trackball, we exercise control by rolling the ball in the direction and at the speed that we want to pan.

Panning is easier with trackball than the spring-loaded joystick. This is because of the strong correlation (or compatibility) between stimulus (direction, speed, and amount of roll) and response (direction, speed, and amount of panning) in this example. With the spring-loaded joystick, there was a position-to-motion mapping rather than the motion-to-motion mapping seen with the trackball. Such cross-modality mappings require learning and impede achieving optimal human performance. These issues address the properties of an interface that Hutchins, Hollan, and Norman (Chapter 5) call "formal directions."

If our application demands that we be able to zoom and pan simultaneously, then we have to reconsider our evaluation. With the joystick, it is easy to zoom in and out of regions of interest while panning. One need only twist the shaft-mounted pot while moving the stick. However, with the trackball, it is nearly impossible to twist the ball at the same time that it is being rolled. The 3-D trackball is, in fact, better described as a 2+1D device.

Scenario 2: I am using the computer to control an oil refinery. The pipes and valves of a complex part of the system are shown graphically on the displays, along with critical status information. My job is to monitor the status information and, when conditions dictate, modify the system by adjusting the settings of specific valves. I do this by means of *direct manipulation.* That is, valves are adjusted by manipulating their graphical representation on the screen. Using the joystick, this is accomplished by pointing at the desired valve, then twisting the pot mounted on the stick. However, it is difficult to twist the joystick-pot without also causing some change in the X and Y values. This causes problems, since graphics pots may be in close proximity on the display. Using the trackball, however, the problem does not occur. In order to twist the trackball, it can be (and is best) gripped so that the finger tips rest against the bezel of the housing. The finger tips thus prevent any rolling of the ball. Hence, twisting is orthogonal to motion in X and Y. The trackball is the better transducer in this example *precisely because of its idiosyncratic 2+1D property.*

> *Thus, we have seen how the very properties that gave the joystick the advantage in the first scenario were a liability in the second. Conversely, with the trackball, we have seen how the liability became an advantage. What is to be learned here is that if such cases exist between these two devices, then it is most likely that comparable (but different) cases exist among all devices. What we are most lacking is some reasonable methodology for exploiting such characteristics via an appropriate matching of device idiosyncrasies with structures of the dialogue.*

APPROPRIATE DEVICES CAN SIMPLIFY SYNTAX

In the previous example we saw how the idiosyncratic properties of an input device could have a strong affect on its appropriateness for a specific task. It would be nice if the world was simple, and we could consequently figure out what a system was for, find the optimal device for the task to be performed on it, and be done. But such is seldom the case. Computer systems are more often used by a number of people for a number of tasks, each with their own demands and characteristics. One approach to dealing with the resulting diversity of demands is to supply a number of input devices, one optimized for each type of transaction. However, the benefits of the approach would generally break down as the number of devices increased. Usually, a more realistic solution is to attempt to get as much generality as possible from a smaller number of devices. Devices, then, are chosen for their range of applicability. This is, for example, a major attraction of graphics tablets. They can emulate the behavior of a mouse. But unlike the mouse, they can also be used for tracing artwork to digitize it into the machine.

Having raised the issue, I continue to discuss devices in such a way as to focus on their idiosyncratic properties. Why? Because by doing so, I hope to identify the type of properties that one might try to emulate, should emulation be required.

It is often useful to consider the user interface of a system as being made up of a number of horizontal layers. Most commonly, syntax is considered separately from semantics, and lexical issues independent from syntax. Much of this way of analysis is an outgrowth of the theories practiced in the design and parsing of artificial languages, such as in the design of compilers for computer languages. Thinking of the world in this way has many benefits, not the least of which is helping to avoid "apples-and-bananas" type comparisons. There is a problem, however, in that it makes it too easy to fall into the belief that each of these layers is independent. A major objective of this section is to point out how wrong an assumption this is. In particular, I illustrate how decisions at the lowest level, the choice of input devices, can have a pronounced effect on the complexity of the system and on the user's model.

Example 2: Two children's toys. The *Etch-a-Sketch* (shown in Figure 15.3A) is a children's drawing toy that has had a remarkably long life in the marketplace. One draws by manipulating the controls so as to cause a stylus on the back of the drawing surface to trace out the desired image. There are only two controls: Both are rotary pots. One controls left-right motion of the stylus and the other controls its up-down motion.

The *Skedoodle* (shown in Figure 15.3B) is another toy based on very similar principles. In *computerese,* we could even say that the two toys are semantically identical. They draw using a similar stylus mechanism and even have the same "erase" operator (turn the toy upside down and shake it). However, there is one big difference. Whereas the Etch-a-

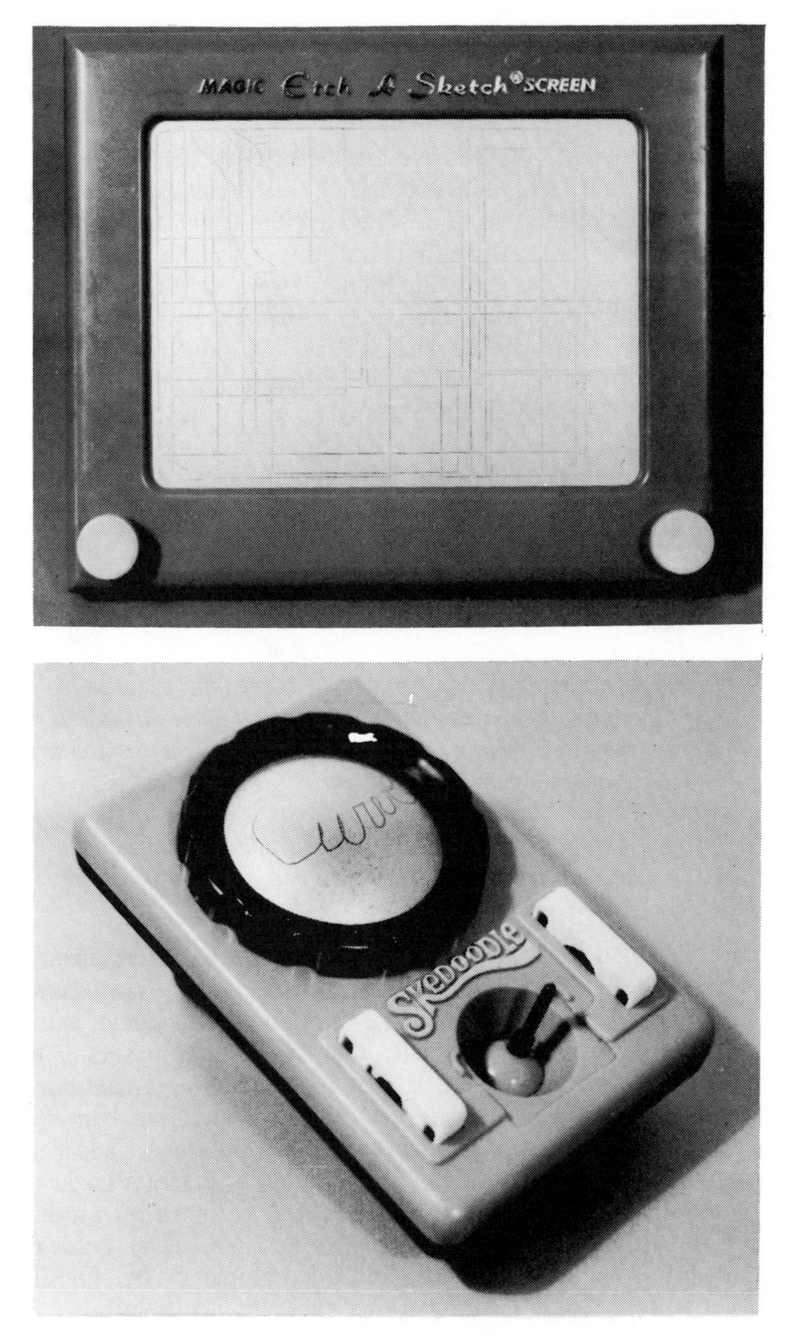

FIGURE 15.3. Two "semantically identical" drawing toys.

Sketch has a separate control for each of the two dimensions of control, the Skedoodle has integrated both dimensions into a single transducer: a joystick.

Since both toys are inexpensive and widely available, they offer an excellent opportunity to conduct some field research. Find a friend and demonstrate each of the two toys. Then ask the friend to select the toy felt to be the best for drawing. What all this is leading to is a drawing competition between you and your friend. However, this is a competition that you will always win. The catch is that since your friend got to choose toys, you get to choose what is drawn. If your friend chose the Skedoodle (as do the majority of people), then make the required drawing be of a horizontally-aligned rectangle. If they chose the Etch-a-Sketch, then have the task be to write your first name. This test has two benefits. First, if you make the competition a bet, you can win back the money that you spent on the toys (an unusual opportunity in research). Second, you can do so while raising the world's enlightenment about the sensitivity of the quality of input devices to the task to which they are applied.

> *If you understand the importance of the points being made here, you are hereby requested to go out and apply this test on every person that you know who is prone to making unilateral and dogmatic statements of the variety "mice (tablets, joysticks, trackballs, etc.) are best." What is true with these two toys (as illustrated by the example) is equally true for any and all computer input devices: They all shine for some task.*

We can build upon what we have seen thus far. What if we asked how we can make the Skedoodle do well at the same class of drawings as the Etch-a-Sketch? An approximation to a solution actually comes with the toy in the form of a set of templates that fit over the joystick (Figure 15.4). The point to make here is that if we have a general-purpose input device (analogous to the joystick of the Skedoodle), then we can provide tools to fit on top of it to customize it for a specific application. (An example would be the use of "sticky" grids in graphics layout programs.) However, this additional level *generally comes at the expense of increased cost in the complexity of the control structure.* If we don't need the generality, then we can often avoid this complexity by choosing a transducer whose operational characteristics implicitly channel user behavior in the desired way (in a way analogous to how the Etch-a-Sketch controls place constraints on what can be easily drawn).

FIGURE 15.4. Adding constraints to an input device. Templates on a Skedoodle joystick.

Example 4: The nulling problem. One of the most important characteristics of input devices is whether they supply *absolute* or *relative* values to the program with which they are interacting. Mice and trackballs, for example, provide relative values. Other devices, such as tablets, touch screens, and potentiometers return absolute values (determined by their measured position). Earlier, I mentioned the importance of the concept of the "dimension of maximum importance." In this example, the choice between absolute *versus* relative mode defines that dimension.

The example comes from process control. There are (at least) two philosophies of design that can be followed in such applications. In the first, space multiplexing, there is a dedicated physical transducer for every parameter that is to be controlled. In the second, time multiplexing, there are fewer transducers than parameters. Such systems are designed so that a single device can be used to control different parameters at different stages of an operation.

Let us assume that we are implementing a system based on time multiplexing. There are two parameters, A and B, and a single sliding potentiometer to control them, P. The potentiometer P outputs an absolute value proportional to the position of its handle. To begin with, the control potentiometer is set to control parameter A. The initial settings of A, B, and P are all illustrated in Figure 15.5A. First we want to raise the value of A to its maximum. This we do simply by sliding up the controller, P. This leaves us in the state illustrated in Figure 15.5B. We now want to raise parameter B to its maximum value. But how can we raise the value of B if the controller is already in its highest position? Before we can do anything we must adjust the handle of the controller relative to the current value of B. This is illustrated in Figure 15.5C. Once this is done, parameter B can be reset by

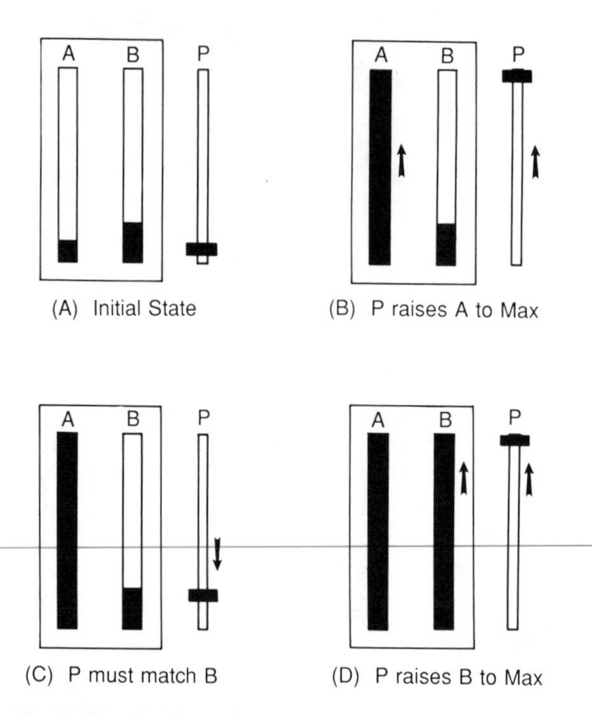

FIGURE 15.5. The Nulling Problem. Potentiometer P controls two parameters, A and B. The initial settings are shown in Panel A. The position of P, after raising parameter A to its maximum value, is shown in Panel B. In order for P to be used to adjust parameter B, it must first be moved to match the value of B (i.e., "null" their difference), as shown in Panels C and D.

adjusting P. The job is done and we are in the state shown in Figure 15.5D.

From an operator's perspective, the most annoying part of the above transaction is having to reset the controller before the second parameter can be adjusted. This is called the *nulling problem.* It is common, takes time to carry out, time to learn, and is a common source of error. Most importantly, it can be totally eliminated if we simply choose a different transducer.

The problems in the last example resulted from the fact that we chose a transducer that returned an absolute value based on a physical handle's position. As an alternative, we could replace it with a touch-sensitive strip of the same size. We will use this strip like a one-dimensional mouse. Instead of moving a handle, the strip is "stroked" up or down using a motion similar to that which adjusted the sliding potentiometer. The output in this case, however, is a value whose magnitude is proportional to the amount and direction of the stroke. In short, we get a relative value which determines the amount of change in the parameter. We simply push values up, or pull them down. The action is totally independent of the current value of the parameter being controlled. There is no handle to get stuck at the top or bottom. The device is like a treadmill, having infinite travel in either direction. In this example, we could have "rolled" the value up and down using one dimension of a trackball and gotten much the same benefit (since it too is a relative device).

An important point in this example is *where* the reduction in complexity occurred: in the syntax of the control language. Here we have a compelling and relevant example of where a simple change in input device has resulted in a significant change in the syntactic complexity of a user interface. The lesson to be learned is that in designing systems in a layered manner—first the semantics, then the syntax, then the lexical component, and the devices—we must take into account interactions among the various strata. *All* components of the system interlink and have a potential effect on the user interface. Systems *must* begin to be designed in an integrated and holistic way.

PHRASING GESTURAL INPUT

Phrasing is a crucial component of speech and music. It determines the ebb and flow of tension in a dialogue. It lets us know when a concept is beginning, and when it ends. It tells us when to be attentive, and when to relax. Why might this be of importance in our discussion of "body language" in human–computer dialogue? Well, for all the same reasons that it is important in all other forms of communication.

Phrases "chunk" related things together. They reinforce their connection. In this section I attempt to demonstrate how we can exploit the benefits of phrasing by building dialogues that enable connected concepts to be expressed by connected physical gestures.

If you look at the literature, you will find that there has been a great deal of study on how quickly humans can push buttons, point at text, and type commands. What the bulk of these studies focus on is the smallest grain of the human–computer dialogue, the *atomic task.* These are the "words" of the the dialogue. The problem is, we don't speak in words. We speak in sentences. Much of the problem in applying the results of such studies is that they don't provide much help in understanding how to handle *compound* tasks. My thesis is, if you can say it in words in a single phrase, you should be able to express it to the computer in a single gesture. This binding of concepts and gestures thereby becomes the means of articulating the *unit tasks* of an application.

> *Most of the tasks which we perform in interacting with computers are compound. In indicating a point on the display with a mouse we think of what we are doing as a single task: picking a point. But what would you have to specify if you had to indicate the same point by typing? Your single-pick operation actually consists of two sub-tasks: specifying an X coordinate and specifying a Y coordinate. You were able to think of the aggregate as a single task because of the appropriate match among transducer, gesture, and context. The desired one-to-one mapping between concept and action has been maintained. My claim is that what we have seen in this simple example can be applied to even higher-level transactions.*

Two useful concepts from music that aid in thinking about phrasing are *tension* and *closure.* During a phrase there is a state of tension associated with heightened attention. This is delimited by periods of relaxation that close the thought and state implicitly that another phrase can be introduced by either party in the dialogue. It is my belief that we can reap significant benefits when we carefully design our computer dialogues around such sequences of tension and closure. In manual input, I will want tension to imply muscular tension.

> *Think about how you interact with pop-up menus with a mouse. Normally you push down the select button, indicate your choice by moving the mouse, and then release the select button to confirm the choice. You are in a state of muscular*

tension throughout the dialogue: a state that corresponds exactly with the temporary state of the system. Because of the gesture used, it is impossible to make an error in syntax, and you have a continual active reminder that you are in an uninterruptable temporary state. Because of the gesture used, there is none of the trauma normally associated with being in a mode. That you are in a mode is ironic, since it is precisely the designers of "modeless" systems that make the heaviest use of this technique. The lesson here is that it is not modes per se that cause problems.

In well-structured manual input there is a *kinesthetic* connectivity to reinforce the *conceptual* connectivity of the task. We can start to use such gestures to help develop the role of *muscle memory* as a means through which to provide *mnemonic* aids for performing different tasks. And we can start to develop the notion of *gestural self-consistency* across an interface.

What do graphical potentiometers, pop-up menus, scroll-bars, rubber-band lines, and dragging all have in common? Answer: the potential to be implemented with a uniform form of interaction. Work it out using the pop-up menu protocol given above.

WE HAVE TWO HANDS!

It is interesting that the manufacturers of arcade video games seem to recognize something that the majority of main-stream computer systems ignore: that users are capable of manipulating more than one device at a time in the course of achieving a particular goal. Now this should come as no surprise to anyone who is familiar with driving an automobile. But it would be news to the hypothetical anthropologist that we introduced at the start of the chapter. There are two questions here: "Is anything gained by using two hands?" and "If there is, why aren't we doing it?"

The second question is the easier of the two. With a few exceptions, (the Xerox Star, for example), most systems don't encourage two-handed multiple-device input. First, most of our theories about parsing languages (such as the language of our human–computer dialogue) are only capable of dealing with single-threaded dialogues. Second, there are hardware problems due partially to wanting to do parallel things on a serial machine. Neither of these is unsolvable. But we do need some convincing examples that demonstrate that the time, effort, and expense is worthwhile. So that is what I will attempt to do in the rest of this section.

Example 5: Graphics Design Layout

I am designing a screen to be used in a graphics menu-based system. To be effective, care must be taken in the screen layout. I have to determine the size and placement of a figure and its caption among some other graphical items. I want to use the tablet to preview the figure in different locations and at different sizes in order to determine where it should finally appear. The way that this would be accomplished with most current systems is to go through a cycle of position-scale-position-. . . actions. That is, in order to scale, I have to stop positioning, and vice versa.

> *This is akin to having to turn off your shower in order to adjust the water temperature.*

An alternative design offering more fluid interaction is to position it with one hand and scale it with the other. By using two separate devices I am able to perform both tasks simultaneously and thereby achieve a far more fluid dialogue.

Example 6: Scrolling

A common activity in working with many classes of program is scrolling through data, looking for specific items. Consider scrolling through the text of a document that is being edited. I want to scroll till I find what I'm looking for, then mark it up in some way. With most window systems, this is accomplished by using a mouse to interact with some (usually arcane) scroll bar tool. Scrolling speed is often difficult to control and the mouse spends a large proportion of its time moving between the scroll bar and the text. Furthermore, since the mouse is involved in the scrolling task, any ability to *mouse ahead* (i.e., start moving the mouse towards something before it appears on the display) is eliminated. If a mechanism were provided to enable us to control scrolling with the nonmouse hand, the whole transaction would be simplified.

> *There is some symmetry here. It is obvious that the same device used to scale the figure in the previous example could be used to scroll the window in this one. Thus, we ourselves would be time-multiplexing the device between the scaling of*

examples and the scrolling of this example. An example of space-multiplexing would be the simultaneous use of the scrolling device and the mouse. Thus, we actually have a hybrid type of interface.

Example 7: Financial Modeling

I am using a spread-sheet for financial planning. The method used to change the value in a cell is to point at it with a mouse and type the new entry. For numeric values, this can be done using the numeric keypad or the typewriter keyboard. In most such systems, doing so requires that the hand originally on the mouse moves to the keyboard for typing. Generally, this requires that the eyes be diverted from the screen to the keyboard. Thus, in order to check the result, the user must then visually relocate the cell on a potentially complicated display.

An alternative approach is to use the pointing device in one hand and the numeric keypad in the other. The keypad hand can then remain in home position, and if the user can touch-type on the keypad, the eyes need never leave the screen during the transaction.

Note that in this example the tasks assigned to the two hands are not even being done in parallel. Furthermore, a large population of users—those who have to take notes while making calculations—have developed keypad touch-typing facility in their nonmouse hand (assuming that the same hand is used for writing as for the mouse). So if this technique is viable and presents no serious technical problems, then why is it not in common use? One arguable explanation is that on most systems the numeric keypad is mounted on the same side as the mouse. Thus, physical ergonomics prejudice against the approach.

WHAT ABOUT TRAINING?

Some things are hard to do; they take time and effort before they can be performed at a skilled level. Whenever the issue of two-handed input come up, so does some facsimile of the challenge, "But two-handed actions are hard to coordinate." Well, the point is true. But it is also false! Learning to change gears is hard. So is playing the piano. But on the other hand, we have no trouble turning pages with one hand while writing with the other.

Just because two-handed input is not always suitable is no reason to reject it. The scrolling example described above requires trivial skills, and it can actually reduce errors and learning time. Multiple-handed input should be one of the techniques considered in design. Only its appropriateness for a given situation can determine if it should be used. In that, it is no different than any other technique in our repertoire.

Example 8: Financial Modeling Revisited

Assume that we have implemented the two-handed version of the spreadsheet program described in Example 7. In order to get the benefits that I suggested, the user would have to be a touch-typist on the numeric keypad. This is a skilled task that is difficult to develop. There is a temptation, then, to say "don't use it." If the program was for school children, then perhaps that would be right. But consider who uses such programs: accountants, for example. Thus, it is reasonable to assume that a significant proportion of the user population *comes to the system with the skill already developed.* By our implementation, we have provided a convenience for those with the skill, without imposing any penalty on those without it—they are no worse off than they would be in the one-handed implementation. *Know your user* is just another (and important) consideration that can be exploited in order to tailor a better user interface.

CONCLUSIONS

I began this chapter by pointing out that there are major shortcomings in our ability to manually enter information into a computer. To this point, input has lagged far behind graphical output. And yet, as some of our examples illustrate, input is of critical importance. If we are to improve the quality of human-computer interfaces we must begin to approach input from two different views. First, we must look inward to the devices and technologies at the finest grain of their detail. One of the main points that I have made is that some of the most potent and useful characteristics of input devices only surface when they are analyzed at a far lower level of detail than has commonly been the case.

Second, we must look outward from the devices themselves to how they fit into a more global, or holistic, view of the user interface. All aspects of the system affect the user interface. Often problems at one level of the system can be easily solved by making a change at some other level. This was shown for example, in the discussion of the nulling problem.

That the work needs to be done is clear. Now that we've made up our minds about that, all that we have to do is assemble good tools and get down to it. What could be simpler?

SUGGESTED READINGS

The literature on most of the issues that are dealt with in this chapter is pretty sparse. One good source that complements many of the ideas discussed is Foley, Wallace, and Chan (1984). A presentation on the notion of virtual devices can be found in Foley and Wallace (1974). A critique of their use can be found in Baecker (1980). This paper by Baecker is actually part of an important and informative collection of papers on interaction (Guedj, ten Hagen, Hopgood, Tucker, & Duce, 1980).

Some of the notions of "chunking" and phrasing discussed are expanded upon in Buxton (1982) and Buxton, Fiume, Hill, Lee, and Woo (1983). The chapter by Miyata and Norman in this book gives a lot of background on performing multiple tasks, such as in two-handed input. Buxton (1983) presents an attempt to begin to formulate a taxonomy of input devices. This is done with respect to the properties of devices that are relevant to the styles of interaction that they will support. Evans, Tanner, and Wein (1981) do a good job of demonstrating the extent to which one device can emulate properties of another. Their study uses the tablet to emulate a large number of other devices.

A classic study that can be used as a model for experiments to compare input devices can be found in Card, English, and, Burr (1978). Another classic study which can serve as the basis for modeling some aspects of performance of a given user performing a given task using given transducers is Card, Moran, and Newell (1980). My discussion in this chapter illustrates how, in some cases, the only way that we can determine answers is by testing. This means prototyping that is often expensive. Buxton, Lamb, Sherman, and Smith (1983) present one example of a tool that can help this process. Olsen, Buxton, Ehrich, Kasik, Rhyne, and Sibert (1984) discusses the environment in which such tools are used. Tanner and Buxton (1985) present a general model of User Interface Management Systems. Finally, Thomas and Hamlin (1983) present an overview of "User Interface Management Tools." Theirs is a good summary of many user interface issues, and has a fairly comprehensive bibliography.

ACKNOWLEDGMENTS

This chapter was written during a work period at Xerox Palo Alto Research Center. During its preparation I had some very helpful input from a number of people, especially Stu Card, Jerry Farrell, Lissa Monty, and Peter Tanner. The other authors of this book also made some insightful and useful comments. To them all, I offer my thanks.

CHAPTER 16

Display Managers as the Basis for User-Machine Communication

STEPHEN W. DRAPER

PART I: INTER-REFERENTIAL I/O

An important feature of modern interfaces is the ability for one part of the user-machine dialogue to refer to another. The clearest familar example is when a pointing device (e.g., mouse) is used to specify an object displayed on the screen. This is an instance of the general concept of *inter-referential I/O*: The user input is a reference to an object in the system's output. The concept is not restricted to pointing. In natural language, references to objects in earlier utterances are typically achieved by phrases such as "the green one" or "the last one you mentioned": There is no asymmetry between partners marked by distinguishing one half of the dialogue as "input" and the other as "output."

The Four Classes of Inter-Referential I/O

Input to input. This kind of reference occurs when a new input is specified by reference to an earlier input. Editing a record of previous commands is the most common example, although this may be implemented via a uniform input-to-output referencing mechanism where the output referred to is a displayed record of previous inputs. A generalization of this, that would require a more profound implementation, would be to allow references equivalent to "go back to the context of that command" or "get me the file version used by that command." These would require the system to store not just a character-level record of past interchanges, but a record of referents as well. "Undo" and "redo" commands (although usually motivated by the need to support trial-and-error strategies) are applications of input to input reference.

Input to output. Currently, the most common example of input-to-output reference occurs when a pointing device is used to refer to some word or object displayed by the system. (Pointing devices can of course also be used to originate input, as when a line is drawn directly with one.) This is probably becoming the single most common form of inter-referential I/O, perhaps because, with present technology, output display rates are much greater than user input rates, and letting the user point to items rather than type them in is, among other things, a way of remedying the imbalance.

Output to input. A common example of output-to-input reference is compiler error messages, where their output includes a line number to refer to a specific item in the compiler's input. This technique is improved by variations that display the error message in one window and the relevant source code line in an adjacent window, or that automatically edit the error message into the source code file in a comment on an adjacent line. A more significant advance in using output-to-input references, however, would be to instantiate the references as lines drawn from the error message to the input it refers to. This would not only be faster to comprehend, it would allow some problems in the compiler's understanding to be sidestepped. For instance consider the error message "undeclared identifier on line 88." If, in fact, the word is a misspelled syntax word, then a message that said "unrecognized word" plus a drawn pointer would be easier for the compiler to generate. Moreover, it would save users from having to work out that "identifier" is being used to describe an item that they think of as a syntax word.

Output to output. This form of reference occurs when one part of a display is linked to another. A good example of this is the Smalltalk browsers (Goldberg, 1984), where a highlighted item in one panel (a window pane) is expanded into a list of components in the next. In current "integrated" browsers, the sequence of panels is fixed in advance both in number and in the type of expansion of the selected item at each stage. In a suitably general design, the visual linking would be managed centrally, and the processes behind the different panels could be entirely independent, receiving only a pointer to the display item as input. This would make it easy to implement more general browsers in which the user could dynamically select not only the item but the way in which it was to be expanded, while retaining the output-to-output references linking the selected items to their expansion in the next panel.

The Hierarchy of Abstract Objects

Pointing at an object does not, by itself, specify a unique object. If you point out of your living-room window and say "look," people will not usually know what you are pointing at. The direction has to be disambiguated by a specification of the type of object referred to, whether by context or by explicit description: Taken together, type and position denote a unique object. Thus you might say "look at that car," or "look at that car door," or "look at that car door handle" or, moving up the scale, "look at that traffic jam," or "look at that rush hour scene." In computer interfaces the same issue applies. For instance the *Star* text editor (Smith, Irby, Kimball, Verplank, & Harslem, 1982) takes one mouse click to refer to the character, two to the surrounding word, three to the sentence, four to the paragraph. In general, the user also might be referring to the screen pixel, or, on the other hand, to the surrounding window.

> *In many systems, the rule is that to indicate that the pointed-at object is the window itself, one must point at the border of the window. This convention violates intuitive pointing semantics in an attempt to avoid explicit object type specifications. Another tactic in some systems is not to support pointing at paragraphs (say), but to allow (i.e., force) the user to achieve the effect by pointing separately to the start and end of the paragraph. This is only consistent with a policy of discouraging the user's use of text structure, and promoting the view of a text as a set of characters.*

In the context of any particular task domain, there is a hierarchy of objects defined by that domain. The problem is to determine how the system can tell which level is being referred to. In general, to support inter-referential I/O, an interface must maintain information such that, for each screen position, it knows the hierarchy of displayed objects whose visible representations contain that point. A system might, for instance, handle this by using a mouse to point to a screen position, and a mouse button to summon a menu of the currently applicable hierarchy of displayed objects; selection of one of these completes the reference to an object. Unfortunately, the relationships among objects is not a simple hierarchy, but the more general multirooted, directed acyclic graph. For instance, a word in a text is both part of a sentence and part of a line, yet a line isn't always part of just one sentence: The word has multiple parentage in the object domain. Similarly, the distinction between display objects and the domain objects they represent gives rise to multiple parentage. For instance a pixel may form part of a character (and so part of a word, a paragraph, etc.), but it also forms part of a window display (screenful etc.). If you want to print out a snapshot of all or part of the display, these distinctions are important. A fully general implementation, then, will manage records that define this complex graph of relationships and so support pointing by the user to all kinds of object, leaving it to specific applications to present a subset of the possibilities to the user (constrained usually by the context of the task or the command to be executed).

Some Uses of References to Output

The concept of inter-referential I/O suggests an implementation that would support all four subclasses equally and generally. As we have seen, however, this would involve a considerable overhead in bookkeeping. As additional motivation for such a sweeping attempt, this section discusses three types of command that would become possible by using references to output displays as arguments. One important distinction is between computational objects and display objects. This is brought out by the first type of command.

Change Appearance

In general, a computational object will have more than one possible appearance, and "change appearance" commands allow the user to select an alternative to the current presentation. There are three basic reasons underlying this situation and militating against the general adoption of a simple one-object one-appearance policy.

1. Limited display space. Many systems use "icons" as a small representation of an object that can be "opened" into its normal, large, window representation. A screen can show many icons, but only a few opened windows. Overlapped windows are just another way of representing a large object by a small screen area (e.g., by the corner that is still visible), and "expose window" commands are a type of "change appearance" command.

2. Ask for a summary. Even if a full display is possible, a summary is sometimes more useful. A table of contents isn't a second rate substitute for the book itself—it serves a different function: You want both and the freedom to alternate between them. Another type of summary is Furnas' "fish-eye" representation which combines high local detail with an overview of more "distant" parts of a large object (Furnas, 1983).

3. The need for alternative perspectives. Changes of appearance may be requested not only to alter the total size and amount of the information displayed, but also its nature—that is, to present an alternative perspective on the same object. The prototype for this, viewed as a universal command, is the Xerox *Star* system's "properties" command, which essentially causes normally hidden properties of the selected object to be displayed. For instance, text on the *Star* is normally presented in its printed form, but the properties command shows you the name and size of the font used.

Error Messages as Displays of System Objects

One kind of display object is the error message. Often the message is not sufficient and the user would like to ask it to expand itself. This is actually a form of "help" request—give me more help information on this message. Some error messages refer to user input (e.g., "command not understood") and would require any further explanations to be made in terms of its relationship to the alternatives allowed by the interpreter. Many, however, are simple assertions of fact (e.g, "this file is protected") offered as implicit explanations of some failure to carry out a command. Asking for further help on these can be analyzed as asking for more information about the system entity they refer to (e.g., the protection bits associated with that file).

Because requests for help have the potential to be endlessly recursive (one can always ask for more information about the new entities mentioned in the answer to the previous request), the responses should be generated dynamically and interactively. Doing this properly would allow a reference to a message to be related to the system objects referred to in the message. Of course, such requests could be handled by a special help system that knew how to carry on a dialogue about the particular class of errors generated by the given program, but since it is reasonable to ask for help about any system object at any time, a more general approach is indicated. In particular, error messages could be quite brief as long as the user could follow them up. This could be done by a general help system if, when called with a pointer to an error message, it could easily retrieve the system object referred to by the message.

As an example of what such a facility could be like, consider possible dialogues after a user command failed with an error message about not being able to access a particular file.

"Can't access file /csl/drper/HCIbook/DisplayManager" might be the first message. If the file was supplied by the user who in fact had made a typing slip (perhaps typing "drper" where "draper" was intended) then this message may be enough. If, however, the user meant what was typed, then a more detailed analysis would be required. This might perhaps give: "/csl/drper does not exist" or perhaps if the directory does exist "directory drper: permission denied." Then, following the latter, the user might want to know either "Who can change the permission?"—answer "superuser"— and then "Who are the real people associated with the user name "superuser?" or, "Why is permission set that way?" (look for a stored reason, or tell the user who to ask) or, "When was it last changed?"

Note that some system designs try to eliminate many of these problems from the outset. For instance a Direct Manipulation system might attempt to cover the function of explaining system workings by displays integral to the rest of the interface (e.g., pictorial displays of the file system). Another tack might be to avoid "file does not exist" errors by only allowing the user to specify a file directly by pointing to a list that only shows legitimate files. Both of these approaches will prevent user slips leading to error messages. However, in modern networked systems, programs frequently contain wired-in references to files that must be fetched over a local network from another machine, or perhaps need material from specific floppy disks which must be loaded. File specification by pointers cannot guard against the need for textual messages in these cases.

In any case, if the system aims to give explanations of any kind, pictorial or textual, then it should support dialogue to accommodate the range of users and circumstances. I am arguing that Lewis and Norman's "Let's talk about it" category of system response (Chapter 20) is always needed, at least as a back-up, and that system responses of "doing nothing" or simple actions will not always be adequate as explanations. It is a variation on O'Malley's argument (Chapter 18) that users often want to follow up a previous question, but that one can't generally predict how.

Error messages can lead to protracted dialogues to satisfy the user's desire for more information. Although a designer might decide that users should at each stage be offered more information than they explicitly requested (chosen perhaps by a "semismart" algorithm), there is no general possibility of offering it all because the path of questions branches, with each branch potentially going on forever. This is why attempts to improve error messages by library routines will not suffice in general. Given, then, that error messages must be regarded as paving the way for a subsequent interactive query session by the user, they should leave a trace behind to allow a general help system to pick up the thread. A pointer to the system object is the appropriate trace. Hence, error messages that are not direct commentaries on user input can usefully be regarded as displays of system objects, even though they may have the form of English text, and may play a speech act role of explanation. They can thus be assimilated to a view of I/O consisting of object representations, rather than as messages.

What Generated This Object?

A desirable facility would be to allow the user to point to a display object and ask where it came from. In a sense, this is a kind of "help" request, but aimed at a display object rather than a system object, and needing an explanation, not of system concepts, but of why the user's instructions caused this effect. (Compare this distinction to O'Malley's "why did that happen" versus "what did I do wrong" in Chapter 18.) It could thus be a valuable debugging aid. Just as it can be useful while debugging to add print statements to a program, a program with full inter-referential I/O capabilities would tag all its output actions so that on inquiry about the generated object, the user could be shown the source code line responsible. This is based on a crude notion of a display "object" originating at a statement. In addition, code written to manipulate appearances of objects (for instance, code that calculates tradeoffs or responds to requests for changes in appearance) should "label" objects that it handles so that such requests can retrieve the passage of an object through the generators and managers on its way to the screen. This suggestion shows that keeping the records needed to allow general reference to displayed objects is closely connected with keeping a full history.

Programmers are sometimes asked to make all error messages begin with the name of the program or utility issuing them, so that the user will always know where it came from. Such a rule is clumsy, since in many cases the user will be in no doubt or will not care: A system that supports post hoc questions about the origin of output will solve this more neatly and without relying on enforcing conventions on a crew of programmers.

Summary: Relating Input and Output

A thorough implementation and use of inter-referential I/O, then, offers many useful consequences, above all a sense of supporting dialogue with the user as opposed to issuing isolated messages and commands. The implementation of this consists essentially of maintaining "back-pointer" records of the origins of each display. Like any compiler, the display specification language interpreter must not only perform the forward mapping from high-level language to machine actions, but also record sufficient information to allow the reverse mapping to be constructed (much as must be done to support a source-code debugger).

This concept applies equally to both graphical output and textual displays. In graphics, the reversibility requirement has been familiar from the earliest days. In a graphics editor, the user wants to specify the appearance of the current diagram, but in order to allow the user to move, expand, or delete not just pixels, but complex objects such as lines and boxes, the editor must maintain the representation of those objects and their mapping from the screen display.

The same idea can be applied to user input: Input is processed through successive layers of interpreters, and these transformations should be recorded as well as performed, so as to allow references to input to be addressed to any of these levels including the keystrokes (for correcting slips), the resulting command (for "redo"), and the final effect (both for "undo," and to allow references to the actual objects or actions resulting from that command in the context that pertained at the time it was executed). Thus, the request to undo a "delete all messages" command requires reference to objects that were deleted (and which, therefore, no longer exist), and so records must be kept of which objects were referred to by the command. As already noted, providing general support for references to previous I/O entails keeping a full history.

PART II. INPUT AND OUTPUT AS A UNIFIED ACTIVITY

In this second part of the chapter I argue that, although the proposals made in the first part are only a generalization of existing techniques, they add up to a change in our view of user–machine communication that is more important than its original components. Instead of distinguishing sharply between input and output (and treating them as two independent channels), it is more appropriate to note that both user and machine typically communicate by writing on a shared display, or by referring to something already on the display. This common medium is managed by software which does not originate communication, but simply translates actions by user and system into changes to the display, and maintains the necessary records for interpreting later references. This software might collectively be called a display manager.

The Shift in Our Conception of I/O

Inter-referential I/O provides one of the important features of human conversation: the capability of referring to previous parts of the conversation, and so of abbreviating some references. This is the key feature that allows us to cross the gap from monologue to dialogue, from the feeling of being lectured by the system in the way a manual page does (or even worse, by a pandemonium of unidentified fragments of a system), to a feeling of conversing with it. Another important technique is that of autodisplay—of displaying output automatically without a specific user request. Its most important application is in the form of state displays, such as those of "display editors" (text editors that display a screenful of text at all times) in which a representation of an internal state is constantly maintained and updated, and those of large simulation-based systems like Steamer (Hollan, Hutchins, & Weitzman, 1984). In this technique, output is used to mirror an internal state, rather than to signal events. It is one method of implementing the principle of Visible Programming (an important support for Direct Manipulation), which holds that computation should not be a black box, but instead be made comprehensible by having its components made visible to the user.

> *A suggestion compatible with the ideas of this chapter is that when the experience of Direct Engagement (see Chapter 5) fails, you should be able to fall back on conversing with the system, rather than merely being the target of messages (or silence) that do not allow for discussion.*

Ciccarelli's work (1984), which is close to the topic of this chapter, is wholly concerned with providing presentations of domains— i.e., with interfaces that use state displays. I have suggested, however, that even when an application program was not itself designed like this, a lot of benefits could be salvaged if the system it was embedded in had general display manager facilities. In that case, although autodisplay might be missing, an ability to point at monologue output and ask where it came from, or for more or different information about its components (e.g., the objects mentioned) would alleviate the problems considerably. This would, in effect, be covering the absence of an application-specific direct manipulation design through the use of a general-purpose dialogue facility.

Both inter-referential I/O and state displays have the effect of bringing system output into the domain of discourse between user and system: The system must keep track of its output in order to know how to update it and how to understand the user's references to it. In effect, I/O has become an object in itself. It can be referred to and operated on. It has been moved inside the computational world instead of being the stuff which leaves and enters it. The implied model of I/O has thus shifted from viewing output as messages that are signals of events dispatched into the void, to viewing output as an object shared between user and system. Communication takes place by user and system alternately modifying and referring to the objects on the screen. Thus, whereas the implicit model of the overall user-plus-machine system used to be one in which there were two main modules (user and machine) communicating by messages (I/O), the new model consists of three modules—user, machine, and shared I/O.

This shift in conception amounts to a sharp break with tradition. Hitherto, I/O has always been the ugly part of computation, treated as a "side-effect," and sometimes omitted from the definitions of programming languages as if computation could somehow be usefully designed without it. As a "side-effect," it is a problem for state-transitional treatments as well as for "functional" (i.e., applicative) ones, since its "effects" are not taken to be effects on the state of the machine. Indeed, the applicative approach in some ways deals with it more cleanly by conceiving of a whole computation as a mapping from user input to output in the same way that a function takes parameters and returns a value. This approach is most overt in UNIX pipes (Ritchie & Thompson, 1974), which employ a class of programs with a single input and a single output channel whose actual natures are not known by the program itself, but which are linked-up externally to files, printers, or other programs as required by the user. This leads to the implementation of terminals as special devices which, although they can be read from (to get user input), and written to (to display output), have the peculiar property that what you read from them is never what you write to them (unlike normal files).

The new concept of inter-referential I/O changes this: The terminal is now viewed as a display with a finite size and known contents. Like a disk, the contents can be read and changed; it is not a sink absorbing a one-way traffic, but a piece of memory shared by user and machine. Implementing this requires a different viewpoint of terminals than is usually provided by computer systems. The system must now maintain a program-accessible model of what is displayed at any time, allowing programs to specify changes to it directly. Whether the implementation of this is direct or built on top of the old model is not important here: The point is the character of the virtual machine presented to most programmers. The considerations explored in this chapter imply that this model should be presented to programmers and users at all levels. No longer would programmers be encouraged to print lines without being conscious that every new line displayed might disrupt some other line. This viewpoint applies even to voice output or to printers (with infinite rolls of paper), where each new output makes previous ones successively harder to recall or find. The shared memory represented by a fixed size display can make explicit what constitutes the current context for the objects being referred to.

The Relationship Between Inter-Referential I/O and Object-Oriented Programming

Some of the terminology of the chapter may remind the reader of the concepts of object-oriented programming, and it would indeed be especially easy to implement these proposals in a general way in a language such as Smalltalk (Goldberg, 1984). It is not true, however, that such languages already embody the conceptualization presented here, despite their importance as antecedents to it. The key shift is to treat I/O itself as an object distinct from the computational object it represents to the user, and to date I am not aware of any language that does this systematically.

The historical development of Smalltalk is particularly interesting in this connection. Smalltalk was born of twin dreams: to reconceive computation wholly in terms of objects and message passing, and to make the computation visible. These are connected by a single desire to make computation more natural (by adopting the structure of the phenomenal world of directly perceptible objects with persisting individual properties), yet the two themes have never been fully integrated. Smalltalk's major contribution to programming languages has been to demonstrate that the whole of computation can be reformulated in terms of objects and message-passing, in contrast to its precursor *Simula* and to the more recent *flavors* packages in Lisp, in which objects are grafted on to an underlying procedural language. Parallel to this was Smalltalk's pioneering exploitation of a mouse and a large display. The nearest approach to integration is represented by Borning's Thinglab system, written in Smalltalk (Borning, 1979), in which the user dealt only with visible graphical objects. This approach can only apply to systems consisting of a small set of objects, all of whose properties can be simultaneously visible in a single canonical visual representation. Even sophisticated text processing must depart from this to allow the user access to fonts and paragraph formats, concepts that are abstract even though they have visible consequences.

To integrate the twin dreams of computation by message-passing between objects and of visible computation, we must accept I/O as consisting of objects distinct from the computational objects that are ultimately of concern. The user must be conscious of and able to express this distinction between concept and representation in order to control the display, ask for alternative representations, conceal or expand information, and so on. Shared I/O is needed in order to go beyond the "directness" possible for simple systems and to support ideas of representation and symbolism in which the user knowingly treats display objects as a partial representation of an abstract (and therefore necessarily "hidden" or "invisible") object. Thus, shared I/O is needed for a true integration of Smalltalk's (and other systems') twin aims to be a general programming language and a self-revealing and, therefore, straightforward computational environment.

CONCLUSION

Display managers establish a different view of communication with the user: one consisting of shared use of the same display medium. When the user is engrossed in the presented domain and communication is flowing without a hitch, this can amount to an experience of direct engagement for the user: The user moves or changes a display item, and the program makes the corresponding internal change to the domain, reflecting any further consequences back to changes to the display. But when communication becomes self-conscious or self-referential in any way, as natural language communication can, then the shared medium supports this too. If users wish to change or modify the subject, they can point to an item and say "open this," "show me the inside of this," or "show me another view of this object." If communication falters because some semismart (or stupid) program displays a message that mystifies the user, the user can ask "where did *that* come from?" and summon an explanation of its origin. This provides a basis for conversational repair analogous to that in everyday discourse when a person asks for clarification or elaboration, and it permits useful communication between partners that share a medium, even if they do not share a complete understanding of each other's minds.

In one sense, implementing a display manager to support general inter-referential I/O could be classified as a low-level proposal for better I/O routines. I have suggested, however, that from the viewpoint of human–computer interaction, if not of computer science, it bears on the highest level concerns: whether breakdowns in the interaction can be smoothly repaired or not, and whether users experience the interaction as dialogue with a responsive partner, or as supplication to an alien entity who alternates silence and oracular monologues.

The re-conception of I/O as communication via a shared medium may, in the end, have its most important effect on designers as well as users at a "higher" level. When one designs a program and its user interface, one is also programming the user in the sense that the inputs (commands) that the program recognizes define the actions being permitted to the user, and the displays being created define the information the user works from. It is not constructive to regard I/O as sending messages into a void, nor even as interacting with some target user whose nature will be unearthed by human factors research or learned by fifth-generation intelligent software. Rather, the designer must view input/output as the delicate task of designing a co-routine. There is complete control of the internals of the machine side, but not of the user side. The two partners communicate: via the shared medium of the display. This puts demands on the designer to ensure that the shared communication can proceed.

The concept of a display manager continues the contributions of existing User Interface Management Systems: rejecting the attitude that communicating with the user is a matter of the unconnected side-effects of scattered "print" and "read" statements—the only model offered by most programming languages—and promoting the view that communication as a central activity that should be designed in a coherent and unified way.

SECTION VI

INFORMATION FLOW

Information Flow is meant to be a suggestive term, representing the flow of information that takes place among the systems, documents, and users of a system. At the least, the term "information flow" is used to emphasize the importance of a unified approach that embraces information sources other than manuals, including other people, system displays, all system messages, and prompts. In line with the emphasis of this book on new perspectives, the chapters in this section emphasize relatively neglected aspects of information flow.

Many issues in the field of Human–Computer Interaction revolve around the flow of information within the system comprising user(s), machine, and the relevant surroundings (which generally include printed manuals, phone lines to other users or experts, and colleagues near or far). On a fine scale, information flow analysis concentrates on all the items of information that must reach users during the execution of their given goals in order for them to be achieved. On a broader scale, information flow analysis concentrates on all the flows of information that affect a user over time.

It is obvious that information flow analysis includes for all users attention to feedback, such as the echoing of characters and the movement of cursors, and, for novices, how they pick up all the necessary facts. With large and complex systems, even very experienced users know only a minority of the large set of commands: Their expertise lies, not in their having learned enough to solve any problem immediately, but in their having become skilled in gathering information and supplementing what they know by an active use of external sources of information (Draper, 1985). Thus, information flow remains important for all users, although a shift is to be expected in the relative importance of different sources of information as a user gains experience: from a dependence on other people to a facility for interrogating manuals and source code, and for learning from experimentation.

There are two main topics for study here: identifying the information *content* and identifying information *sources* or *delivery systems.*

Examples of kinds of information content:

File names
Little recipes for the key strokes to get some neat or useful effect
Getting into a system or subsystem (e.g., how to invoke an editor)
A minimum set of commands
Advanced commands
Where to get information
Efficiency—neater ways to do certain things
How things work
Strategies for combining system programs (functions) for user tasks
Recognizing information when it's there
How to use information you have
How to recover from errors
What is commonly done on this system (i.e., standard user tasks and standard plans for accomplishing them)

Examples of information delivery systems:

Printed manual entries
On-line manual entries
"Cheat Sheets" (summaries of commands)
Menus
Owen's "DYK" System ("Did You Know": See Chapter 17)
People—replying to specific queries
People—fortuitous pickup (looking over someone else's shoulder)
Monitors that examine your input and periodically make suggestions about new commands, etc.

There are an unlimited number of possible delivery systems, since they can vary in who supplies the information (a public bulletin board versus system-designer documentation), in whether the information is specifically requested or spontaneously generated (error messages, monitors), in the access structures (i.e., by command name, English description, etc.), in how much information about the user is consulted in selecting the presentation (none, or records of past requests, of commands vocabulary displayed in practice, etc.).

The primary value of adopting an information-flow perspective is that it causes us to step back from any prior assumptions about the pre-eminent importance of learning, of conceptual knowledge, or of the role of certain "official" channels such as manuals and tutorials. Instead, the perspective encourages us to look at the amount of information vital to users that is picked up, used, and forgotten as well as at the importance of sources such as colleagues that do so much of the work supposedly done by manuals. Hence, the information flow perspective suggests that a designer's task in this area is first to calculate *what* information must be conveyed to users (the information content needed) and then to orchestrate the possible information sources to achieve this, perhaps designing special delivery systems. Some people promote slogans or codes of practice such as "good systems must have on-line help" and "always write a manual entry." Others have proposed rules for screen design, technical writing styles, and so on. Before we are ready to consider these assumptions and approaches to the effectiveness of particular information delivery systems, we should look at the broader perspective of the information flows actually important in current systems—both their content and the sources now supplying them.

A second aspect of the information flow perspective is to consider opportunities for replacing a requirement for users to know something by timely delivery of the needed information. In the simplest case, a menu might list available commands. However, there is also scope for aiding not just simple knowledge of items like names, but also understanding. By providing examples ("recipes") an action can be described to the user in such a way that it can be performed even if the user understands little about it. This is important whenever completion of an interaction is more important than education of the user. Thus, timely information delivery can stand in for memory, saving planning and understanding. This view can be developed by regarding the design of help systems as an attempt to design an extension of the user's memory.

Help Systems as an Extension of Memory

A simple but interesting viewpoint for considering help systems is to regard them as memory aids, for conserving the user's memory resources. Thus, we could consider all help systems as forming an extended external memory for the user—one kind of mind amplifier. From the cognitive point of view there is a spectrum of retrieval types. The best situation is when users can remember the item needed effortlessly. The next best is when they find the needed item already displayed. Third best is when they must ask. It is only third best because this counts as an interrupting activity, causing suspension of their current activity (see the discussions in Section IV of this book). This is the distinction made in the literature between active and passive help—between whether the user has to prompt delivery or not. Fourth best (i.e., worst) is when the user doesn't know how to ask. The activity requiring the information will have to be suspended indefinitely, and either a random search will ensue or the user must just wait to stumble on the answer. This is analogous to when you know you should know something—someone's name perhaps—but you just can't recall it. The information is there, but inaccessible.

The ideal, in fact, is not just to reduce the effort of seeking information but (subjectively) to abolish it—to achieve where possible the smooth flow of information. A familiar instance of this occurs in every display editor which revolves around the manipulation of the cursor's position. Users seldom note the subtask of retrieving the cursor position. Instead, they simply edit; moving the cursor to wherever it should go. Adequate information flows from the display at every moment. We should aim for a similar subjective effect in delivering all information. Currently not enough is understood to tackle this systematically.

The cursor example depends on the display being constantly updated without explicit commands from the user—the technique of *autodisplay.*

Autodisplay poses two major problems: what to display, and whether the user will pick up the information displayed. The design of compiler error messages illustrates the first problem. Since error codes are useful only as pointers to other, better descriptions of the problem, it is a safe bet to print the latter automatically. On the other hand, automatically displaying a piece of source code or a possible diagnosis is only as good as the reliability of the compiler at knowing which line of the code needs to be changed or knowing that a set of symptoms has only one possible cause. The second problem may be called "receptivity"—whether the user will pick up relevant information if it is displayed. One of us once watched the other discover that half his files seemed to be missing. It took several minutes before someone else pointed out that he was logged on to a different machine than he thought. Yet all the time the name of the machine was being displayed in every prompt, only three character widths to the left of the fixation point as he frantically typed commands. The machine name was not noticed, even though the user had himself designed that prompt with exactly this sort of problem in mind. This is a common kind of problem: failing to see a solution that is literally staring us in the face, even when we are actively searching for a solution. When the information presented does not fit the current hypothesis it may be ignored. Until we understand what governs receptivity a lot better, designing information flow will be a patchy business at best, and it will be hard to design information delivery systems that make users feel that the system's help facilities are an extension of their own memories.

The Chapters

Owen, in Chapter 17, reminds us that we often learn by first discovering information, then determining what question it answers—the inverse of the what is thought to be the standard method of seeking information. Owen calls this approach "*answers first, then questions.*" The point is that on many occasions we come across an idea or piece of information more or less by chance, and then recognize it as interesting or relevant in some way: i.e., an answer to a dormant question. He illustrates this mode of information flow in everyday life and presents a fantasy of one style of advanced graphics-intensive user interface. He describes the exploration of this method in an experimental facility he calls DYK ("Did you Know?"). DYK offers assorted factual tidbits on demand, providing a novel, but quite effective, method of conveying some kinds of information to the user population. (Furthermore his DYK facility is fun to use, making it an attractive displacement activity as well as a valuable source of information about computer usage.)

Another reason for the importance of "answers first, then questions" as an information delivery mode emerges from considering a tacit assumption behind most conventional help systems. The most common way to store and present information follows a database model: all the stored items have the same structure but different details. An example is the telephone directory: All entries (items) show a name, an address, and a phone number. Computer databases follow this rule, with standardized formats for its records. Computer help systems and manuals also usually have this form, with one entry per command and various standard parts to each entry.

To ask a question of such systems—to formulate a query—inquirers must know how to transform their original need for information into queries the system can respond to. This requires a prior knowledge of the query language syntax, of the structure of the database, and of the indexing system. For instance, to find a phone number you must know the person's name: in the case of standard telephone directories, knowing only the address will not work because they are indexed by name. A first improvement of retrieval systems is to provide more and different indices to the same database—for instance, computer system manuals can to be indexed by function as well as by command name. These improvements, however, do not overcome the problem that the inquirer must know the standard structure of the information and the legal queries, in order to use the index or phone directory. (We should remember also that many people prefer to ask another person rather than to use a directory themselves: phone companies mount active campaigns to move their customers away from asking information operators for help.)

O'Malley (Chapter 18) concentrates on the more usual case when the user initiates a query to the system: "*questions first, then answers.*" She asks why users so often prefer to ask other people for help rather than to use manuals. She suggests that a large part of the answer stems from the fact that to formulate a question for such systems, you already need to know many things.[1] This means that asking a question is in principle a multistep process. Failure in any step means failure of the whole. The apparent superiority of human help is often due to its robustness—the people one consults can aid in filling in the missing steps and in reformulating the question in an appropriate fashion. Thus, in both measuring the success of a delivery system, and in designing improved ones, we need to identify all the steps and ensure

[1] As Miyake and Norman (1979) once put it in the title of their paper: "To Ask a Question, One Must Know Enough to Know What Is Not Known."

that they are all supported by one means or another. It is not enough to provide answers, we must also support the formulation of questions.

Bannon focuses upon a relatively neglected aspect of information flow in human-machine interaction—the role of social interactions. In his first chapter in this section, Chapter 19, Bannon focuses on user-user information flows, arguing that they will always be an important part of the total pattern of information flow within the larger system of computers plus user community, despite present and future advances in system-to-user delivery methods. He suggests that facilities could and should be installed on computers to facilitate (not to replace) user-to-user flows—on-line human help, as it were. This chapter is complementary to the one by O'Malley, for it argues that human assistance is such a pervasive aspect of information finding and computer usage that systems ought to be designed with this in mind, to supplement and to extend the ability to get help from others.

Lewis and Norman (Chapter 20) discuss the possible responses of a system when it detects a problem: a class of system-to-user information flow, initiated by the system. This chapter is a product of the intensive workshop we held to review draft chapters of the book. There, Lewis and Norman discovered their common interests in the manner that contemporary systems handle cases where the system cannot interpret the information given to it. "Error," says the system: "naughty user." In fact, the error is just as much on the part of the system for failure to understand as it is on the part of the user. The main point is that there is a breakdown in the pattern of information flow. In normal conversation, as Lewis and Norman point out, listener and speaker do not jump to assign blame for a failure to understand. Instead, the listener simply does the best job possible at forming an appropriate interpretation and seeks help when that is not possible. Lewis and Norman discuss the range of responses open to the system and its designers to errors by the user. (These are points raised again in Chapter 22 by Brown.)

In the last chapter of this section (Chapter 21), Bannon again focuses on user-user information flows, but in a different sense than in his earlier chapter: not as part of information flow *about* the system, supplementing the documentation, but as an end in itself, for which the computer merely acts as a medium. He reviews some of the most common forms of computer-mediated communication (e.g., electronic mail and computer conferencing), and then some ways of distinguishing the effects the media have on the human interactions they support. He also shows how one might design an office by deliberately selecting some set of communication tools to fulfill the needs of the office members. This chapter adopts the largest scale of any in this section: that of augmenting and shaping human interactions and environments by the selection of technology. Bannon reminds us of the social component and consequences of human–computer interaction and discusses ways that the computer can be used as a tool to improve and enhance social interaction. All too often today the computer is thought of as an individual tool, used by a person in relative isolation from all others. But just the reverse is really the case. First, we need other people in order to learn how to use computers most effectively (the point of almost all the chapters in this section). Second, computers provide enhanced communication abilities, both synchronous and asynchronous. And third, the computer is a new tool that can enhance participatory work, allowing for shared workspaces, and shared communication channels, so that our ideas can immediately be made available to the others with whom we work, allowing for increased interaction and creativity.

CHAPTER 17

Answers First, Then Questions

DAVID OWEN

Within a day or two of arriving in California from England, Tom sees a Sparkletts truck pull up and deliver several large clear containers full of a clear liquid to a McDonalds. In England similar looking trucks carry similar looking bottles full of sulphuric acid, but only occasionally do you see them and they are normally delivering to factories. So they use a lot of acid in California? Tom then sees the advertising on the back showing smiling faces with gleaming teeth drinking the stuff by the glass. Tom fleetingly ponders on the quality of water he has been drinking from the tap.

Later that day, on his way home he sees a billboard advertisement telling him he can have a water purifier fitted to his home water supply.

Fundamental to the notion of information flow is the question "why do people need information?" It is important because the answer should influence both the kind of information delivery systems that are designed and the nature of the information they deliver. Implicit in delivery systems most commonly available in computing environments is the assumption that people need information because they do not know how a particular tool works. They are therefore based on the "Question-Answer" paradigm: "Ask me how 'foo' works and I will tell you."

In the world at large many delivery systems work in the same way. Telephone "Yellow Pages" and travel reservation services are examples. But there are others which seem to be based on a paradigm better characterized as "Answers First, Then Questions." Billboards and TV advertisements deliberately volunteer information to whomever will take notice. There are other examples which are not as contrived although the effect is much the same. For example, people who have a set of skis on the roof-rack of their car may not be intending to volunteer the information that there are ski-slopes available, but to an observer, the effect may be just that. In the story at the beginning of this chapter, Tom benefited from just that kind of delivery system. This second paradigm, the central topic of this paper, results from a second answer to the same question. "Why do people need information?"—To be able to decide what to do.

I begin with a simple model of how people conduct themselves which indicates the role that a delivery system based on the "Answer-Question" paradigm could fulfill. Then, by reference to this model and examples from the world, I examine the way the "Answer-Question" paradigm works in practice. Finally, I discuss how this paradigm may be exploited in the computing environment. In particular I describe DYK, a program that illustrates these ideas.

WHERE THE ANSWER-QUESTION PARADIGM FITS

An approximate model (one that serializes what is no doubt a complex parallel process) of how people further their objectives begins with an assessment of the currently perceived state of affairs with respect to their objectives. Based on this assessment, tentative goals of varying degrees of specificity are generated and decisions made as to which should be pursued to their conclusion. Crucial judgments concerning resource allocation are implicit throughout. In how much detail should the current state be analyzed? Should the generation of possible goals be exhaustive? How much effort should go into ensuring the selection of the "best" of the candidate goals? Thus, in planning what to do we face an optimization problem. Moreover, the effort required to solve the problem also figures in the calculations. On the one hand one may meticulously plan a vacation taking great care to find out and assess all possibilities. On the other hand one may wander into a cinema in the vague hope that whatever is showing will provide some relaxation and enjoyment.

Generating Tentative Goals

There are three obvious sources for the generation of tentative goals. First, one may have already experienced a state of affairs that would be more satisfactory than the current one. Not being in pain, and avoiding misspellings, are examples. Second, some new more favorable state may be creatively fantasized. The humble sandwich is a famous piece of creativity on the part of John Montagu, the fourth Earl of Sandwich. In both these cases, the information necessary to formulate the goal is by definition, already known. A third alternative is to perceive some situation or event which essentially provides the opportunity to take advantage of the experience or creative fantasizing of others. For example, one may see someone with pink hair, i.e., the specification of a novel state. On other occasions the situation may only reveal an action sequence, leaving us to deduce the goal. Why do people in southern California put a cover on their car when it clearly is not going to rain? This third alternative relies on information delivery systems which are implicitly "Answer-Question" based.

Selecting a Goal

To decide which of the tentative goals to pursue, some kind of cost-effectiveness analysis has to be made. It involves many questions. Is it theoretically possible to achieve the goal? Are states Y and Z to be achieved first? Do the facilities necessary to achieve Y and Z already exist? How can I find out if they exist? How long will it take to find out? Will the "answers" be reliable? To what extent will this further my objectives? If answers to these questions are not already known, then the resource requirements of evaluating the goal, and hence the cost of pursuing it, increase. Another issue which has relevance here is captured by the title of a paper by Miyake and Norman (1979): "To Ask a Question, One Must Know Enough to Know What Is Not Known." The recognition of one form of this problem is implicit in the fact that people regard all new courses of action as involving some risk: "Is there anything I have not thought of?" The overall result is an understandable bias towards doing what one already knows how to do, a phenomenon which has been noted many times before (Luchins & Luchins, 1959; Zipf, 1965). The game played by an "Answer-Question" based delivery system is one of tipping the balance in favor of a new strategy by reducing the perceived cost of adopting it. If some of the questions are answered "for free" then that safe but suboptimal strategy may finally be replaced.

Taking a different perspective on the same point, the questions that will typically be posed vary enormously in levels of specificity. For example, at one extreme one may ask "How can I improve the appearance of this document?" and at the other extreme "How can I append a section of the file I am currently editing to the end of another file in my directory." This poses formidable problems for an information delivery system based on the "Question-Answer" paradigm for two reasons. First, it requires that available information be classified in many overlapping ways and second, it has to deal with the infinite variety of ways in which essentially the same question may be posed by the seeker. In Chapter 18, O'Malley discusses this point in some detail.

Pursuing a Goal

Eventually some action has to be performed: I wish to delete a file from my directory, or I wish to book a flight from Los Angeles to London. On the one hand this often leads to specific questions such as how does the "remove" command work? This is where the Question-Answer" paradigm is probably most appropriate. On the other hand if the removal of the file is marked by the absence of its iconic representation, then the follow-up question, "Did my file get removed?" can be avoided (see Draper's discussion of inter-referential I/O in Chapter 16).

In summary, although the "Answer-Question" paradigm can only supplement the more conventional one, it has at least three potential roles. First, it can be effective in providing information that people would not think to ask for. (See Chapter 9 for the significance of this to user understanding.) Second, it may be effective in reducing the perceived cost of pursuing some goal that has already been formulated. And third, it may release people from the effort of actually having to ask a particular question.

THE ANSWER-QUESTION PARADIGM IN ACTION

The essence of the paradigm is the presentation of unsolicited information, but to be successful it must be done in a way which at worst is not fruitlessly distracting or confusing to the perceiver and at best is of some value. Three key factors which influence success or failure are *when* the information is presented, *what* is presented and *how* it is presented. These questions are not independent, and have to be asked from the point of view of the perceiver rather than the purveyor of the information. The trick in designing an "Answer-Question" based delivery system is to juggle the three parameters to match the state of an individual and some examples will show how this can be done.

What and When

Few contrived systems that are successful instantiate the purest form of this paradigm in which there is no solicitation of information. Most rely on implicit or explicit indications from individuals, to identify not only when they will be receptive to volunteered information, but also the kind of information which would be received favorably. People are likely to be least flexible in this regard when they are intent on some specific task. Exceptions occur when the activity is so well automated as to require little attention. Thus, walking to the shops may be regarded as an indication that attention is not in heavy demand. Billboards exploit this cue explicitly, and the informative aspect of the the actions of others (e.g., the skis on the roof rack) do so implicitly. Reading a magazine indicates receptiveness and the choice of the magazine indicates a broad, acceptable topic. Using a personal appointment book and leaving reminder notes in conspicuous positions are particularly interesting examples, since both are set up by the individuals themselves, a characteristic that will be referred to again later. They both have a reminding function (see Cypher, Chapter 12, and Miyata and Norman, Chapter 13) and are relatively specific in the information they will convey. However, they are quite different in terms of the behavior on which they rely. In particular, the appointment book example exposes the fact that the two paradigms belong on a continuum. The appointment book could also be classed as a "Question-Answer" based delivery system, since it is generally consulted deliberately in search of an answer to a specific question. Reminder notes rely on their profile being high at just the right time, with no special action on the part of the individual beyond the original creation of the note.

What and How

Most of these systems are conservative in that they minimize the risk of being intrusive by maintaining a relatively low profile in comparison to the indication of receptiveness given by the user. Run-time advice and tutoring systems walk a thin line between being disruptive and usefully informative. They are, in general, operating in a situation in which people are least flexible in this respect. In general the fact that people are using a particular facility is not a reliable indication that they will be receptive to general information about that system. The problems associated with volunteering information are exacerbated by the fact that it is not always possible to predict what information people will construct from what is presented to them. The significant information content in the billboard advertisement for a domestic water purifier, for example, was that the local water supply was of suspect quality. The name of the product may not have been noticed. This, coupled with the propensity people exhibit for construing what is perceivable as also being relevant can lead to convoluted confusions in the computing domain. In their work on "abduction," Lewis and Mack (1982) give some examples that arise when apparently innocuous pieces of information find their way to a terminal screen uninvited (see Chapter 8 by Lewis). However, such *selective construction* on the part of the perceiver can be exploited by delivery systems especially when the information delivered is so low in profile as to be hardly ever distracting. For example, the page numbering on a book volunteers information on how far through the book the reader has reached. It is in general not perceived as distracting, but a voice which automatically whispered "page 42" when the page was turned might prove to be intolerable. This aspect of information delivery is worth noting because it is difficult to exploit in the computing domain; it is discussed in more detail in the next section.

THE ANSWER-QUESTION PARADIGM IN THE COMPUTING ENVIRONMENT

How can this paradigm be used in a computing environment? One way is simply by the observation of others using novel approaches or special techniques. There are two factors that make this less effective than it could be. First, personal workstations and terminals are often situated in individual offices or homes, minimizing the likelihood of accidentally observing the computing activities of others (see Bannon, Chapter 19). In general, computing is a solitary pursuit, with lamentably few facilities for cooperative problem solving (see Brown, Chapter 15). Second, the kinds of activities that people engage in and their observable manifestations are not very self-revealing. In general, only terse textual statements appear on the screen, and action sequences are rapidly executed keystrokes. It may be that one of the spinoffs of the richer screen displays that result from the direct manipulation systems described by Hutchins, Hollan, and Norman in Chapter 5 will be that the interface will reveal more to the casual observer.

Selective Reconstruction

If the way in which unsolicited information is presented is low in profile, then it is not so important to avoid redundancy and ensure relevance. People can selectively construct what they need from what is presented. Information inherent in the form of an object is an example. One can estimate how far through a book one has reached from a glance at the thickness of the pages that remain. The soft feel of a sponge and the hardness of a nail are strong clues to their appropriateness for some task. We can tell roughly how much beer is left in a can by shaking it. Of course the value of the information in these last two examples is crucially dependent upon touching the objects. Other examples rely more on vision. From an apparently unorganized room full of folders, the location of one next to the telephone or on top of the filing cabinet may be the clue which identifies it as the one needed at the moment. The batch of filecards that were written in purple ink, in a hurried scrawl just as the library was closing, may identify themselves as just the ones containing the relevant facts, or the ones that can be skipped over when they are reviewed 6 months later. What happens to this kind of information in the computing domain? In general it gets filtered. One of the powers of the computing medium is that it compresses a very large amount of information into a small space. The price paid is the loss of much "unnecessary" information, and the loss of multidimensional accessibility, inherent in the bulky three dimensional form that the same information took on in the world.

The Computer as an Information Filter

The computer acts as the perceiver of information presented to it, and then a source when it re-presents it to the user. The filtering is done during a range of translation, representation, and reconstruction stages (the most obvious of which are shown in Figure 17.1). Each stage provides its own capacities for losing information.

If the information cannot be expressed conveniently through the input device, the translation suffers (e.g., it is hard to express graphical concepts through a keyboard). The system may not be able to represent some aspect of what the input device provides; for example the velocity with which the mouse cursor is moved across the screen. (Although see Buxton, Chapter 15, and Minsky, 1984.) If a user wishes to annotate some information on the machine, those annotations must be structured to match the structures acceptable by the machine. There are only limited facilities for users to exploit for themselves the "Answer-Question" paradigm by placing the equivalent of reminder

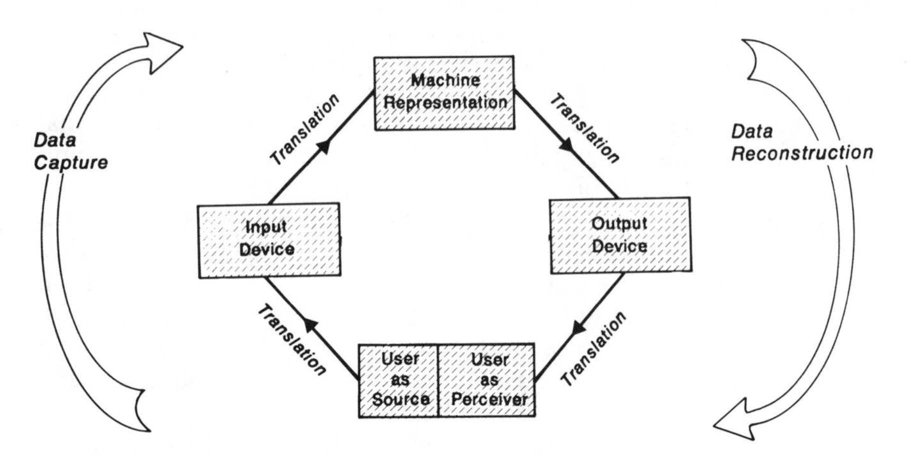

FIGURE 17.1. Information filtering. When data in one form are translated into another form, significant information may be lost. The typical data-capture/data-reconstruction cycle of a computing environment involves several such transactions.

notes as temporary marks on documents, or pages within documents, for example.

The next compromise comes with the reconstruction of the data into a form which becomes a "source" for the user. Here is where the loss of information becomes manifest, either because it never made it as far as the machine's representation, or because the way it is reconstructed makes it hard or impossible for the perceiver to access. In summary, when data in one form are converted to another form decisions are made as to what constituted significant information in the original. Other information, particularly that which implicitly exploits the "Answer-Question" paradigm via its form, may or may not survive the translation.

Some Fanciful Examples[1]

Suppose I am browsing through a document on a computer and want to go back to a page on which something caught my attention. I can try

[1] Some of the ideas in these examples have been explored elsewhere in a much more elaborate way, and in a different context. See for example the Spatial Data Management project of the Machine Architecture Group at MIT (Bolt, 1985). The examples given here serve to clarify one aspect of the theme of this paper. Also see Hooper's Chapter 1.

to think of a key word and then search for that, or if I know how many pages there are in the whole document, and if I know roughly how far into the document the page was, I can do the arithmetic to find out which page it is close to and request a move to that page. Compare this with thumbing through the document directly. On the one hand I lose the ability to specify a search string, and that is a substantial loss. On the other hand I can feel how thick the paper is and maybe estimate the position of the relevant section that way. I can flick through the pages and possibly catch a glimpse of a diagram, a textual pattern, or possibly my own scribbled annotations.

It is not obvious how the advantages of both representational forms can be captured. A fast scroll facility may allow me to "flick" through the document, noticing gross features to be used as reference points. But here is where the argument about information loss in translation bites back. Something of the feel of flicking through a paper may have been lost by translating it into its apparently significant characteristics: i.e., speed and flexibility of browsing. Would the mechanism depicted in Figure 17.2 feel qualitatively different? This is a 2-D fantasy simulation of a 3-D environment, which includes a 2½-D depiction of a sheaf of papers. There I can see how many pages there are. The screen is pressure sensitive. I put my finger on the top page and drag some pages over, the number depending on how hard I press. I then put my thumb and finger on the exposed page and drag it to the left, it lifts above the top page. So I end up after a while with a completely unordered set of papers. Instead of spending several minutes reordering them by hand, a tedious task, I press a button and they are ordered for me.

> *Is this an example of direct manipulation? It passes the tests posed in Chapter 5.*

It's not obvious how it would feel, but there is no question that the data have been reconstructed for me in a very different way from either a book or current computer conventions. I can both "thumb" through the pages and search for a string. This kind of simulation follows the "Answer-Question" paradigm in the sense that it presents information that wasn't requested, and probably wouldn't be consciously wished for (let alone asked for), and yet may be very useful.

Let me continue with the example of document browsing. I typically make pencil marks to indicate corrections or to draw my attention to sections as I read. Suppose it were possible to write by hand on the visible pages of Figure 17.2, and that the annotations remain with the page as it is subsequently hidden, revealed, stored, or recalled. Screen

FIGURE 17.2. A screen simulation. The screen mockup, together with its description in "Some fanciful examples," suggests ways in which the filtering of the information flow between user and computer might be minimized.

manifestations of file names might profit from the same facility, essentially allowing the owner to associate arbitrary graphical annotations with its iconic appearance (some examples are just visible in Figure 17.2). In writing directly on the screen, there is no separation of the designation of position and the act of making the marks, a separation which is forced on the user in most computing environments. It restores the "direct engagement" present in the use many other writing methods from the stone tablet and chisel to pencil and paper and is mentioned throughout this book (especially Chapters 5 by Hutchins, Hollan, and Norman, and 4 by Laurel).

There is a world of locations, objects, and relations among them, locked inside the machine. It can only be glimpsed in fragments making it hard to construct or retain a map of the space and where things of interest can be found. But it should be possible to construct that space in a way that makes the map more accessible, that allows users to construct their own maps and to have the spaces and objects take on three dimensional form that can be manipulated directly, and taken in as a whole. Think of a chess-board. Instead of representing fighting units let the pieces represent files or utilities. The board could be partitioned into project areas equivalent to screen layouts. I pick up a piece representing the source code for some program and place it in a project area. When I come to display the screen associated with that workspace the icon for the file is there, in the position that I placed it on the board.

But can all this not be done with icon manipulation and multiple screens? This may be the case, but there is an assumption lying behind the interfaces just described. It is that the information (and accessibility) delivered by the form and spatial relations between objects in the world implicitly exploits the "Answer-Question" paradigm, and should be restored in the computing environment beyond the use of simple icons. The defense of the assumption hinges mainly on the quality of the interaction—how it feels to the user—and only indirectly on quantitative criteria. Rather than making it possible to undertake new tasks, the increased dimensionality distributes both the mental and physical effort involved in interaction, across a wider range of people's inherent abilities.

THE DYK PROGRAM

In contrast to the fantasies just described let me now describe an existing program called DYK (for "Did You Know" and pronounced "Dee Why Kay"). DYK provides an information delivery system based on the "Answer-Question" paradigm.

Taking a Cue From How People Already Behave

The goal is to exploit the "Answer-Question" paradigm to disseminate general information about the local computing environment and I wanted the program to play a role similar in spirit to a billboard, or of noticing something over someone's shoulder. The first problem was to determine an ecological niche for such a system and the informally observed phenomenon of "displacement activity" seemed to be an appropriate candidate. As light relief from some tedious task, to delay starting it at all, or just to unwind at the end of the day, users often engage in some peripheral displacement activity. Local favorites include finding out who else is on the system, checking for new mail, listing system statistics, and playing games. These distractions seem to attract users into repeated and often redundant use, just the kind of behavior an "Answer-Question" delivery system could exploit.

Activities informally discernible as fulfilling a "displacement" role are either brief, or can be abandoned rapidly without cost. They often foster some kind of informal contact with the rest of the community. Checking for new mail, and seeing who else is logged in both have this property. Many of the games offer interestingly different screen layouts from the conventional forms.

These observations guided the system design in several ways. First, it would be user-invoked. Second, the information presented would be limited in length to small, suggestive fragments rather than longer, more complete diatribes. Third, the presentation style would differ from the usual form and in some way be engaging.

So What Does DYK Look Like?

The program is briefly described in the caption to Figure 17.3. There are facilities for users to add their own DYK items introducing the implicit contact with others in the community evident in some other displacement activities. Also, over a period of time, it will provide data on the usefulness to DYK users of information formulated by other system users. After reading the DYK item, a user may indicate that it is inaccurate whereupon the author is informed via the mail system, and asked to check it. A similar mechanism will probably be used to ensure that each DYK item is checked by its author at regular intervals after its creation. This will go some way towards distributing the task of maintaining the integrity of the database across the user community. A prototype version queried each user to find whether they found the information interesting. It was clear that some DYK items were found to be very useful and others fared very badly. It is too early to be able to characterize the differences reliably between the two groups, and thereby give some guidelines to authors, but the data may be used to remove persistently unpopular DYK items from the database automatically. It is not obvious that this is a good policy. It is equivalent in some ways to removing infrequently used words from a dictionary. The usage statistics also indicate that the prototype has been successful as a displacement activity. During the first year the database was limited to 30 fixed entries, but many users continued to call the system although they must have seen each entry at least three or four times.

Shell Tool 1.1f-> edit
q-quit h-help r-repeat-topic p-pick-for-me c-censor-topic u-uncensor-topic
*dyk cshell unix *C #suns
other
hmi-reviews pdp-reviews admin *vi
Using 'x' and 'p' in combination in "vi"
The x command deletes the character under the cursor,
the p command reinserts the last thing deleted... so
If you transpose letters in typing, which often happens,
position the cursor over the first offending letter
and just type xp This will delete the first character
and stick it back in the right place.
Please answer these questions (y, n, q to quit or h for help)
Is this information of interest?:
Censor this DYK in future?:
Is this DYK inaccurate?:

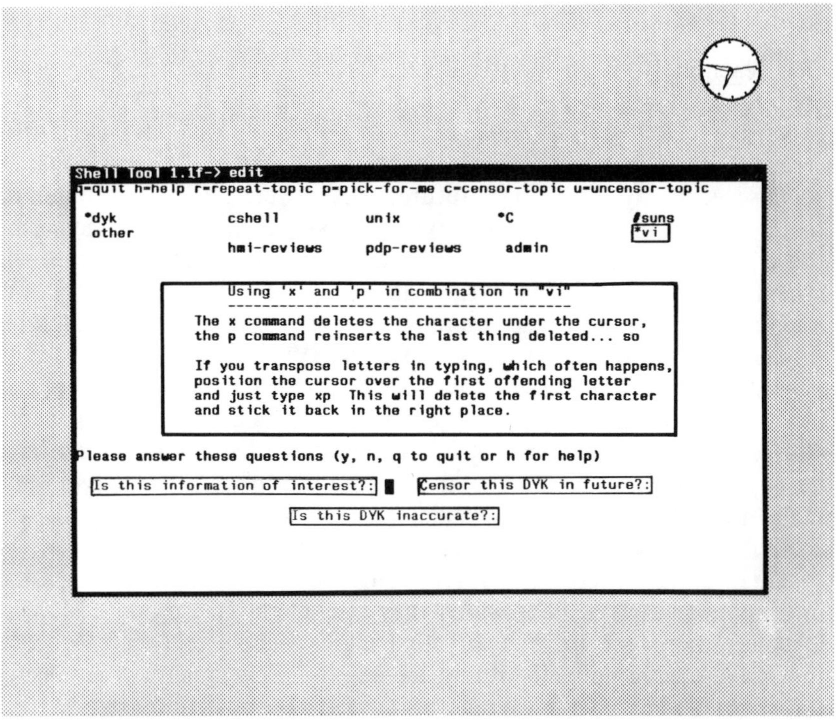

FIGURE 17.3. The topic groups are highlighted in turn and pressing <return> will select the currently highlighted one. Alternatively, pressing 'r' (repeat) will force the reselection of the previously selected topic group. In this example the "vi" topic has been selected.

The topics marked with a # ("sun" in this example) contain at least one DYK which this particular user has not yet seen. If one of these groups is selected, the new additions will be shown first.

Topics marked with a * contain only DYKs the user has seen before and did not censor. If the user has censored all the DYKs in a topic group, it remains unmarked.

Users are asked to indicate whether the DYK is of interest and if it should be censored for them in future. The final question concerns the accuracy of the information. If they find it to be inaccurate, then they may choose to correct it themselves; otherwise the author will be informed.

But Our DP Manager Would Go Crazy About the Unproductive Cycles

One reaction from an industrial member of the audience at a recent conference at which this system was described was that DYK was too expensive: "*But our DP manager would go crazy about the unproductive cycles.*" It may be a reasonable response today, but will almost certainly be false economy in the near future. Several people have commented that systems will become more comprehensive in the future, increasing the chances of getting lost and making it harder to find out what the system can do (Joyce, 1984; Kay, 1984). There is also evidence that manuals and conventional training sessions meet with considerable resistance from users (Scharer, 1983). A DYK-like system provides an alternative way of addressing these problems. However, it does rely on users having access to the computer on a regular and relatively unconstrained basis. For example, it is unlikely that a bank teller would have the opportunity to engage in the kind of displacement activity that DYK represents, unless management can be convinced that it is a cost-effective way of providing employee training.

DYK currently conveys several different kinds of information. It informs people of facilities which they might not have thought of. For instance, there is a local facility that allows people to check to see if they have electronic mail without logging in. This is particularly useful for people who have terminals at home. There is information that extends people's knowledge of systems with which they are already familiar. The topic covering the most heavily used editor is particularly popular in DYK. This sometimes takes the form of a straightforward statement about the availability of some command in the system; alternatively, strategies for usefully combining commands are given.

> *In "vi" there is a command 'x' to delete a character and a command 'p' to insert the last thing deleted. It turns out that a common typing mistake is to transpose adjacent letters in a word. By positioning the cursor over the first offending character and typing 'xp' the letters are re-transposed.*

Relating what particular commands do in a strategy that tackles a particular situation is not always obvious until it is pointed out. Relevant strategies may also depend on local combinations of common tasks and tools which system documentors cannot predict. The advantage of presenting this kind of information to users in a DYK like manner is that they can select for themselves which pieces of information fit in with what they need and their level of sophistication.[2] They can safely ignore those they do not yet understand. It is also possible that simply creating a DYK item will enhance the author's understanding of some aspect of the system. There is some evidence that putting specific pieces information into one's own words encourages the recognition of its wider implications and connections with existing knowledge (Mayer, 1981). It will take time and experimentation to find ways of disseminating information well. The system outlined here is not proposed as the solution but is one attempt that serves to raise some of the issues.

CONCLUDING REMARKS

I have tried to draw attention to the fact that people seem well disposed towards acquiring information by chance encounters with situations, as well as specifically seeking answers to questions. The approximate model presented in the early pages offers some reasons as to why this might be, and examples from the world show how it can work in practice. The tacit collusion between the world and our sensory and mental capabilities takes on a variety of subtle guises. It is by no means obvious that the same communicative devices will profitably transfer to the computing medium, but it seems likely that some will. Two possible applications have been presented here. The first was concerned with reconstructing the data held in the machine in a way that allows people to exercise their own powers of selective reconstruction, discrimination, and contextual memory. The second, the DYK program, is an attempt to produce a synthetic alternative to chance encounters from which users will both learn and derive some satisfaction.

ACKNOWLEDGMENTS

I am most grateful to several people for their help in producing this paper. Don Norman's detailed comments and criticism improved it immeasurably. The other contributers to this book, and especially Sondra Buffett and Steve Draper, gave valuable feedback on earlier drafts and Phil Zakhour helped build DYK.

---

[2] Vygotsky uses the term Zone of Proximal Development to describe those embryonic levels of understanding that can be encouraged to maturity by the timely presentation of information which perhaps confirms, clarifies, or relates them to what the individual already understands. (See Vygotsky, 1978).

CHAPTER **18**

Helping Users Help Themselves

CLAIRE E. O'MALLEY

This chapter is motivated by two observations, or rather by two views of one observation: First, that users prefer to consult other people rather than to use the manuals and other types of help provided for them; second, that the people that users consult may be characterized not simply by how much they know, but also by their ability to get access to the relevant information.

The introduction to this section points out that the "information flow" perspective means paying attention to a *variety* of types of information that must reach users during the execution of their tasks—from such low-level feedback as the echoing of characters on the screen to the higher-level information necessary to plan the next activity. This perspective involves considering not just what information users actively seek, but also what they pick up by chance (Owen, Chapter 17); it involves considering not just the information that can be obtained from the computer system and the "official" documentation, but also what help can be obtained from other people (Bannon, Chapter 19).

In this chapter I focus on the information that users *actively seek*, rather than what they might be taught, or what they might pick up by chance. The situation is one in which a question has arisen, and an answer is sought: Questions first, answers later. The emphasis, however, is not so much on the issue of how to help users find information so much as on how to help users *ask the question* in the first place.

USERS HELPING EACH OTHER

Despite the variety of forms of computer help and documentation that are available, it is widely observed that users tend not to use them but prefer to consult the "local expert" or other users.[1] There are several possible reasons why users give up on manuals or help systems. Common complaints include:

- "I can't find the information"
- "There's too much information to wade through "
- "I don't have the time to spend searching"
- "I just want to check up on some detail"

> *Users want to be able to describe the "symptoms" of their problem and be told what to do, but they aren't able to because they often don't know how to describe the symptoms in the terms that can be found in "the book." In the case of computer manuals, users have the task of inferring the nature of the "illness" from a description of the functioning of a healthy "organism." Just as they would go to a doctor who can perform a diagnosis and make a prescription rather than looking up their symptoms in a medical handbook, users go to other people because they are better able to "diagnose" the problem and prescribe a solution. Taking this analogy further: Just as the doctor often doesn't cure people's illnesses directly, but tells them where to get the medication that will cure them, the computer consultant doesn't necessarily solve users' problems for them but instead tells them where to get the information that will help them solve the problem. The "expert," in both these cases, acts as an interpreter between "users" and "the system."*

When users get help from other people, it is not always the case that the other person knows the solution. Often it is more the case that the other person knows *where to find the information* that will answer the

[1] Cf. Lang, Auld, and Lang, 1982; Lang, Lang, and Auld, 1981; Scharer, 1983.

user's question. One of the characteristics of the "local expert" in a user community is an ability to use the documentation and manuals. This is just as likely to be the reason why such people are frequently consulted, rather than the fact that they "know more" than other users.

This simple observation bears closer analysis: Users may go to the "expert" because they are unfamiliar with the way the available information is organized, and it is less effort to ask someone who is familiar with it. But what does this really mean? Knowing something about the way information is organized means being able to ask the right question to get at the required information. So another way of articulating this observation is that "experts" are able to point users towards the information they need because they are able to formulate the question better than users can.

Users have trouble finding the answer themselves because there is usually a huge gap between the initial internal (mental) form of the query and the information they need, as expressed by the system. Other people are often better at closing parts of that gap, by helping users translate their intentions into specific questions that can then be mapped onto the form in which the information is presented and organized in the system or the documentation.

HELPING USERS ASK QUESTIONS

How could you help users find the answers when they don't even know the questions to ask?

1. You could present some likely answers so that they could recognize what they want (i.e., you can give them the answers so that they can find the questions: see Owen, Chapter 17). For example if you ask travel agents to suggest a "good" holiday, they may try asking a few questions but will often end up dumping some brochures in front of you, waiting for you to recognize a "good holiday."

2. Rather than presenting the information in an unorganized way, you could help users select what they need by giving them an overview of the organization (structure) of the information. Figure 18.1 shows some examples of ways of structuring information.

3. You could adopt a middle path and present examples, then allow users to refine the examples in some way (e.g., by altering certain properties, or by criticizing certain aspects of the

A

P100 CEEFAX 100 Fri 30 Jan 16 09/57
BBC CEEFAX

NEWS HEADLINES 101
NEWS IN DETAIL 102-119
NEWS FLASH 150
NEWS INDEX 190
NEWSREEL 199
FINANCE HEADLINES 120
FINANCIAL NEWS 121-139
SPORT HEADLINES 140
SPORT PAGES 141-159
FOOD GUIDE 161
FARMING 160

FOR THE DEAF 169
SUBTITLES 170
TV/RADIO 171-4
TV CHOICE 177
TOMORROW 178
WEATHER/TRAVEL INDEX 180
GENERAL INDEX 193-5

VALENTINE MESSAGES 201 (BBC 2)

Index to the BBC's Ceefax Teletext System

B

```
HELP PRINT
PRINT
  Queues one or more files for printing, either on a default system printer or on a specified device.
   Format:
     PRINT file-spec[, . . .]
  Additional information available:
  Parameters Qualifiers
  /AFTER=absolute-time      /CHARACTERISTICS=(c[, . . .])
  /DEVICE=device-name[:]                   /FORMS=type               /HOLD
  /NOHOLD (D)               /IDENTIFY (D)                /NOIDENTIFY
  /JOB_COUNT=n (D = 1)      /LOWERCASE /NOLOWERCASE (D)             /NAME=job-name
  /PRIORITY=n               /QUEUE=queue-name[:]         /BURST      /NOBURST
  /COPIES=n (D=1)           /DELETE     /NODELETE (D)                /FEED (D)      /NOFEED
  /FLAG_PAGE /NOFLAG_PAGE               /HEADER          /NOHEADER   /PAGE_COUNT=n
  /SPACE[=n]
```

Print Help on the VAX/VMS

C

PERMUTED INDEX

	pr: print file.	pr(1)
print:	pr to the line printer.	print(1)
monitor, monstartup, moncontrol:	prepare execution profile.	moncontrol(3)
monitor, monstartup, moncontrol:	prepare execution profile.	monitor(3)
monitor, monstartup, moncontrol:	prepare execution profile.	monstartup(3)
colcrt: filter nroff output for CRT	previewing.	colcrt(1)
types:	primitive system data types.	types(5)
cat: catenate and	print.	cat(1)
lpr: off line	print.	lpr(1)
fortune:	print a random, hopefully interesting, adage.	fortune(6)
date:	print and set the date.	date(1)
cal:	print calendar.	cal(1)
hashstat:	print command hashing statistics.	csh(1)
jobs:	print current job list.	csh(1)
whoami:	print effective current user id.	whoami(1)
pr:	print file.	pr(1)
fpr:	print Fortran file.	fpr(1)
history:	print history event list.	csh(1)

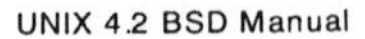

UNIX 4.2 BSD Manual

FIGURE 18.1. Examples of ways of structuring information. **(A)** The British Broadcasting Corporation's *Ceefax* ("see facts") teletext system enables the viewer to choose certain pages (screenfuls) of information to display, by entering a page number via a remote control unit. **(B)** On the VAX/VMS system, a user wanting information on the print command types the command HELP followed by the argument PRINT. The system presents a brief explanation of the print command, followed by a list of possible parameters that can be specified along with it. More detailed assistance is available on each of these parameters by typing HELP PRINT/ followed by the parameter qualifier, e.g., HELP PRINT/IDENTIFY. **(C)** The permuted index to the UNIX manual allows the user to find manual entries for commands in several ways.

example that they don't like). This is the approach taken by the Rabbit system,[2] for example, and combines aspects of the first solution (present the answers) with aspects of the second (structure the information). Figure 18.2 illustrates how Rabbit does this.

However, the problem with solution 1 (present the answers) is that users are faced with the very problems they have with most existing help systems and manuals—there is too much information to wade through. The problem with solution 2, and to some extent, solution 3, is that one kind of organization of information will not necessarily work for all types of problem. Users' tasks more often than not cut across the boundaries forming the organization of the information.

For example, on our UNIX system, if you wanted to print a document with tables and equations on the laser printer, you might have to type the command line:

tbl filename | eqn -s8 -fI | ltroff -mlcsl

This single command-line contains three separate programs with their various associated flags or options. In order to get help for this task you would have to access information via the online manual from three different entries for the programs tbl, eqn, *and* ltroff*—and you would have to spend some time on each query to find the relevant information, assuming you knew the names of the appropriate commands.*

The point is that you often have to break down a single task into several components, and if help is required in subsequently constructing the command line, it has to be obtained for each of the components separately. The documentation is structured at the level of the system modules (commands) and not at the level of the task, nor at the level of a legitimate executable command line. This is not a gripe about one particular system: It is a problem for any system that offers users the power of combining units.

---

2 Tou, Williams, Fikes, Henderson, and Malone, 1982.

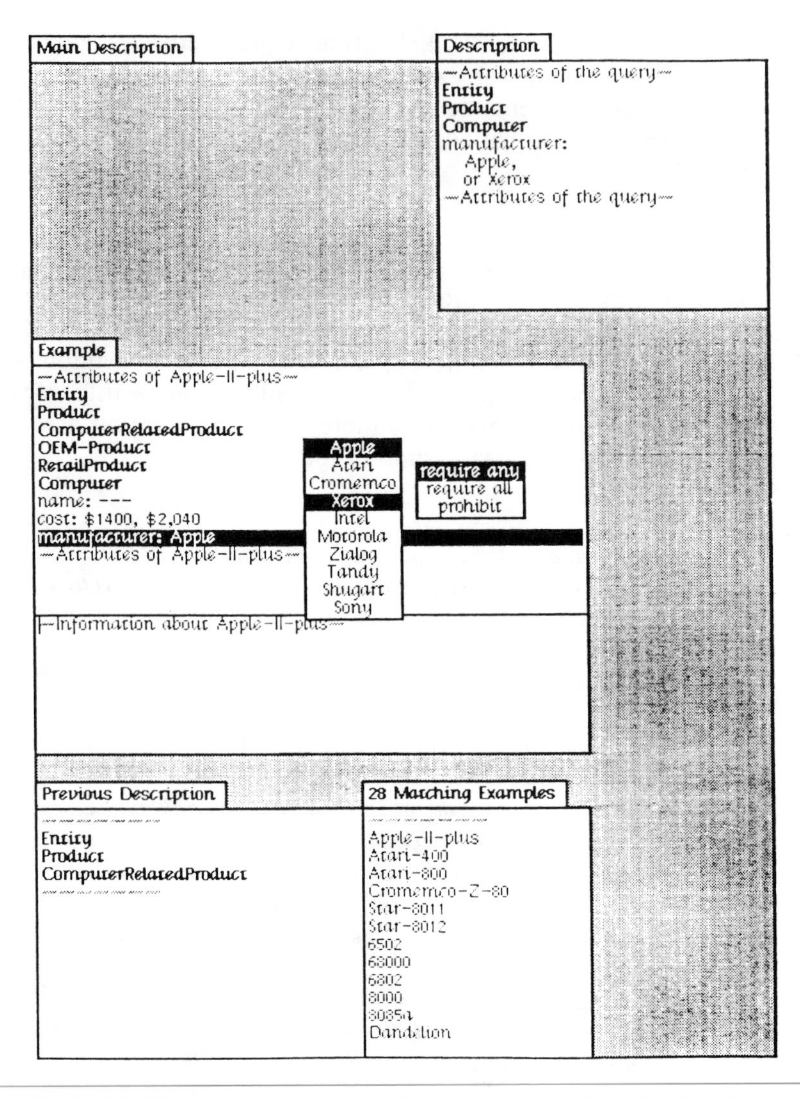

FIGURE 18.2. The Rabbit system allows the user to access information by reformulation of examples. In this particular example the user wants to find a computer with a bit-map display and a certain type of processor. An image of an example in the database matching the current partial description (*entity product*) appears in the example pane. The user then specifies that the manufacturer be either Xerox or Apple. The process of "critiquing" the example instance continues until the user is satisfied with the current partial description. Then another example is retrieved based on the new description, and so on until the desired item is found. (Example provided by M. Williams, personal communication.)

Formulating Questions Is a Multistep Process

There are two common types of problems that arise in formulating questions, both of which are often specifically addressed in existing help systems: First, knowing the name of the command to type in order to get help; second, knowing the name of the arguments to give.

For example, in UNIX, in order to ask the question "*how do I search for a line in a file?*" I have to formulate the question as "*man grep.*" Even if I know that the command for getting help is *man*, I still may not know which command I should ask about. A simple way of handling this problem is with keyword search facilities, such as the program on UNIX called *apropos,* (assuming I know its name and, indeed, of its existence). For instance, with our local version I could type "*apropos search for line in file*" and I would be given a list of possible commands, including the one that is needed:

grep, egrep, fgrep (1) — search a file for a pattern.

This is an example of the "bootstrapping" problem discussed in the section introduction. In order for users to ask for help they must already know about the help commands.

There are other kinds of attempts to take care of both of these problems. For example, the use of menus obviates the need to know how to ask for help by showing the user what information is available. This is a small-scale application of the "*answers first, then questions*," approach within the framework of the "*questions first*" paradigm which "patches" the problem that users don't know how to ask questions. Of course, menus (and their pictorial counterparts icons) have their own set of difficulties, e.g., a complex hierarchical structure.

However, users often go to other people rather than consult the "official" sources of help because other people can help them ask the right question. This is more complicated than the simple case of the other person telling the user what command to type to find information. In other words, "asking the right question" goes beyond the problem of using the right language or terminology. The user often needs help in actually *formulating* the question, that is, in specifying it to a detailed enough level. Moreover, this process often involves several steps, not just one. It can be multistep in at least two senses: going from the general to the specific (internally generated questions) and diagnosing problems (externally generated questions).

Going from the general to the specific: Internally generated questions. There may be several stages to go through from the first general question to the kind of specific question that can be asked of the system or documentation. Take an example where users have some general intention, e.g., they are trying to figure out how to format a letter for printing. They have some idea in mind of the eventual product, and they know how to create the actual text (i.e., how to edit a file), but not how to specify how they want it formatted. In order to get help in realizing their intention they have to translate it into specific questions so that they can make use of the help facilities to find the information they need. The translation process may involve a one-to-one or a one-to-many mapping—that is, one question may map straight onto another (e.g., "*how do I search for a line in a file?*" translates into "*man grep*"), or several questions may be involved (e.g., "*how do I print a file on the laser printer*" translates into "*man tbl*," "*man eqn*," "*man ltroff*," and so on).

Norman points out in Chapter 3 that there are several stages in going from very general intentions to the level of actually executing commands on the system. The complexity involved in the planning process has also been highlighted by several other studies (e.g., Moran's "external-to-internal-task mapping" model, 1983; "planning nets," Riley & O'Malley, 1985). An implicit corollary of these analyses is that users also need to be supported in formulating and asking questions of the help system that correspond to each of these stages. In other words, each step potentially involves one or more questions to be formulated and asked.[3] Such support is needed particularly where questions are generated "internally"—that is, where they are generated from some plan that the user has not yet made explicit.

For example, Figure 18.3A illustrates what a typical unformatted letter might look like just after the user has typed it in. Figure 18.3B shows what might be desired as the final formatted product. Figure 18.3C shows the formatting commands that would actually have to be inserted in the file on our system. Finally, the user would have to type the command line *lnr file*, in order to have the letter printed on the laserprinter. In this case it is clear that it would be rather unhelpful if the person who was consulted simply told the user to type *lnr*. For

[3] In fact, you could look at this point in another way: Observation of the questions generated naturally at each stage might serve to validate these analyses.

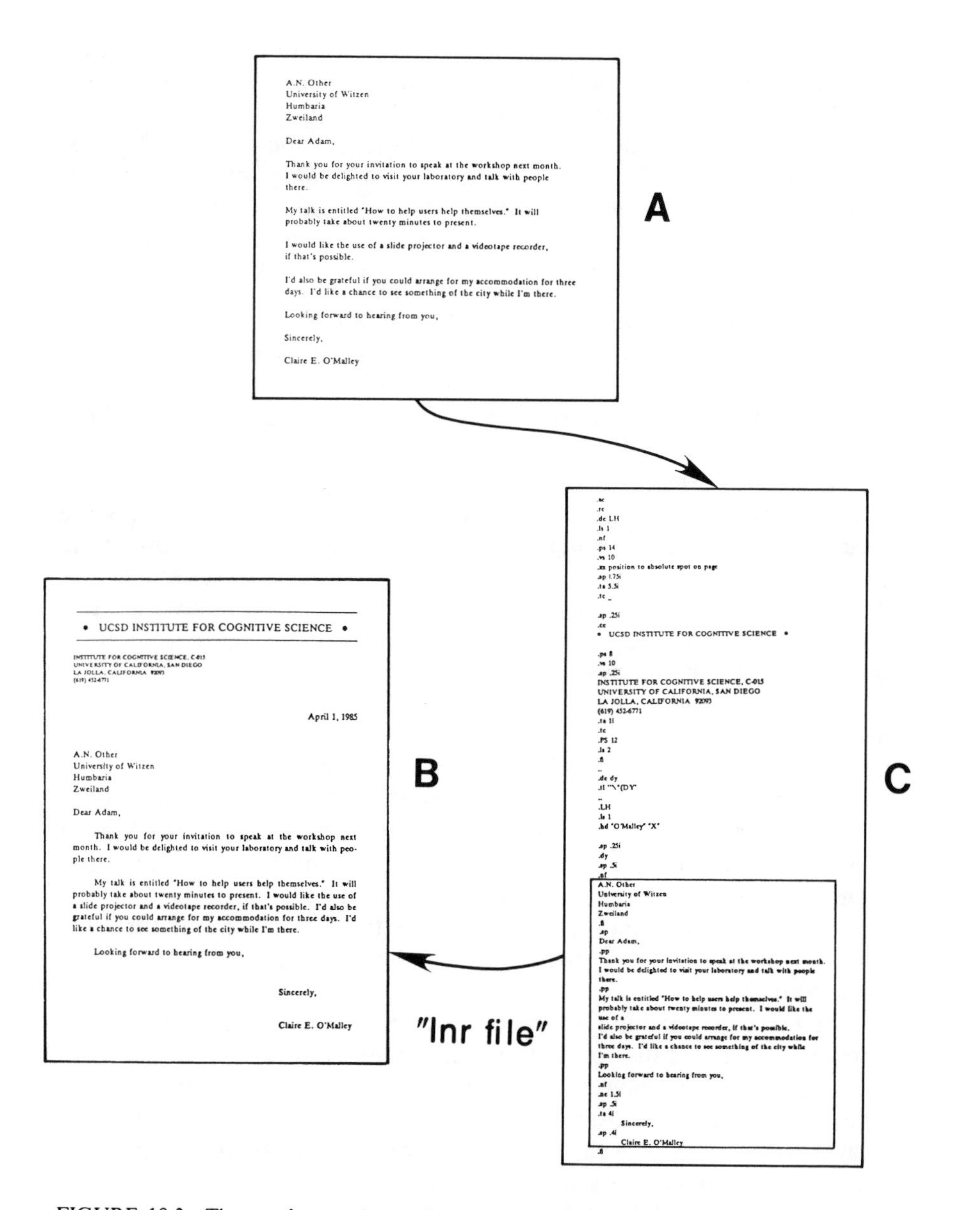

FIGURE 18.3. The user's question is "*how do I get a formatted printout of a letter?*" **(A)** shows what a typical unformatted letter might look like after the user has typed it in. **(B)** shows what is desired as the final formatted product. **(C)** shows the formatting commands that the user would have to insert in the file on our system. Finally, the command "*lnr file*" has to be typed in order to get a printout.

example, let's say the user's most general level question is: "*how do I get a formatted printout of a letter?*" Assuming the user actually knows that they have to put formatting commands in the file in order to get a formatted version printed, the next specified question might be "*how do I get a letter heading?*" or "*how do I get the date printed on the right hand side?*" "*how do I indent the paragraphs?*" and so forth. What is needed is to go through a whole series of questions and answers in order to determine what is required for accomplishing the whole task.

Diagnosing problems: Externally driven questions. Let's now look at a slightly different example concerning the problem of formulating questions. Figure 18.4 is adapted from the example given by Norman (see Chapter 3). This figure illustrates that, although there may be a one-to-one mapping between the initial diagnosis of the problem (e.g., "*I need to make this paper look better*") and the intention ("*I have to change .pp to .sp*"), several steps need to be taken to determine the specific problem. For example, the user has used the right formatting commands in the file, at least as far as the system is concerned, but then finds that the formatted version of the letter is not really what was wanted. However, there may be problems in figuring out exactly what is wrong. All that's known is that the letter "doesn't look right."

Since there was no error as far as the system is concerned (the formatting commands are "correct"), the user somehow has to convey the fact that it wasn't what was wanted—which is difficult to do on most systems. For example, if I went to another person for help, I could point to a printout of the letter and say "*this doesn't look right,*" or "*I don't want so much space here on the left side.*" Someone who knows that there are several alternative formatting commands for obtaining different forms of paragraphs will recognize from the printout that, for example, blocking them rather than indenting them would make it "look better."

There are two points I want to draw from this example: (a) There is a concrete result in the form of a visual representation to focus on; and (b) There are some objective or external criteria for solving the problem. In this case the user doesn't have to specify in any great detail what is wanted, because there are some objective standards for "neat" as opposed to "messy" letters.

I am contrasting the case of going to a travel agent and saying "I want a good holiday" (cf. "*how do I format a letter?*") with "I want to go somewhere less touristy than Mallorca" (cf. "*how do I make this look better?*") In the second case, both the travel agent and the person consulted about the letter start out knowing primarily what is not wanted. This is precisely the approach taken by Rabbit (see Figure 18.2), where

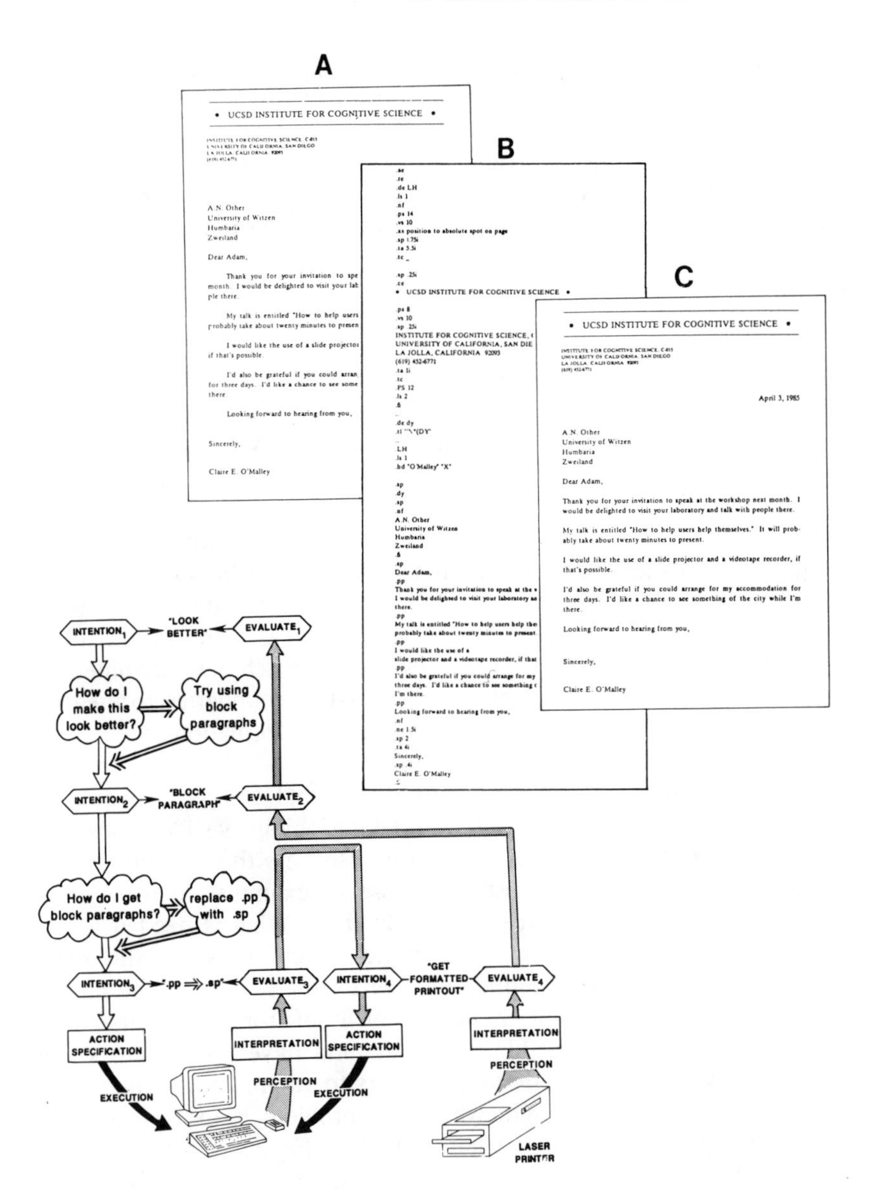

FIGURE 18.4. Example adapted from Norman, Chapter 3. **(A)** shows the starting point. The letter doesn't "look right," so the initial intention is "improve the appearance of the letter." **(C)** shows what is desired as a final formatted product, but the user has to try to figure out how to get to **(C)** by changing the formatting commands in the file **(B)**. Thus, at each stage the user needs to be supported in asking the question that will give the information that is needed in order to form the next intention.

the interaction is arranged around presenting examples, which may then be criticized—i.e., you can say what you do or don't like about it. In other words, if the user's question is generated from some external representation, the designer can use "*answers first, then questions*" as a technique, within the basic question–answering paradigm, in order to circumvent the need for the user to know how to formulate questions.

Summary:

1. Questions may be driven by "internal" factors, such as goals and plans. There is a direct parallel here with the kind of analysis given by Moran's "external-to-internal task mapping" model (Moran, 1983)—the formulation of specific intentions from general goals and plans is a complex process, involving several stages. Support is needed for questions that arise at each of these stages. One approach to providing help for "internally-generated" questions, especially when users don't know what questions to ask, is to present an overview of the information, but as I have argued earlier, such an approach is undermined by the fact that users' tasks cut across the boundaries formed by any particular method of organizing the information.

2. Questions may also be driven by external objects or events in interacting with the system. Making things visible helps the user identify problems, but it can also help the system identify how it can help the user. The use of examples in the Rabbit system, Figure 18.2 capitalizes on this. What is needed is the ability for both the user and the system to share the *same* representation of the problem. For example, it would be nice if the user could point to an area of a soft-formatted version of their file, say something about it, and have the system know that what they are pointing to refers to an area in the input file, so that it can highlight where the problem is located, a point discussed by Draper in Chapter 16 (see Figure 18.5).

DIFFERENT TYPES OF QUESTIONS REQUIRE DIFFERENT TYPES OF DELIVERY SYSTEMS

In the course of trying to go from an initial vague question to the final answer—that is, in the process of carrying out planning supported by questions—various different types of questions can crop up, and should

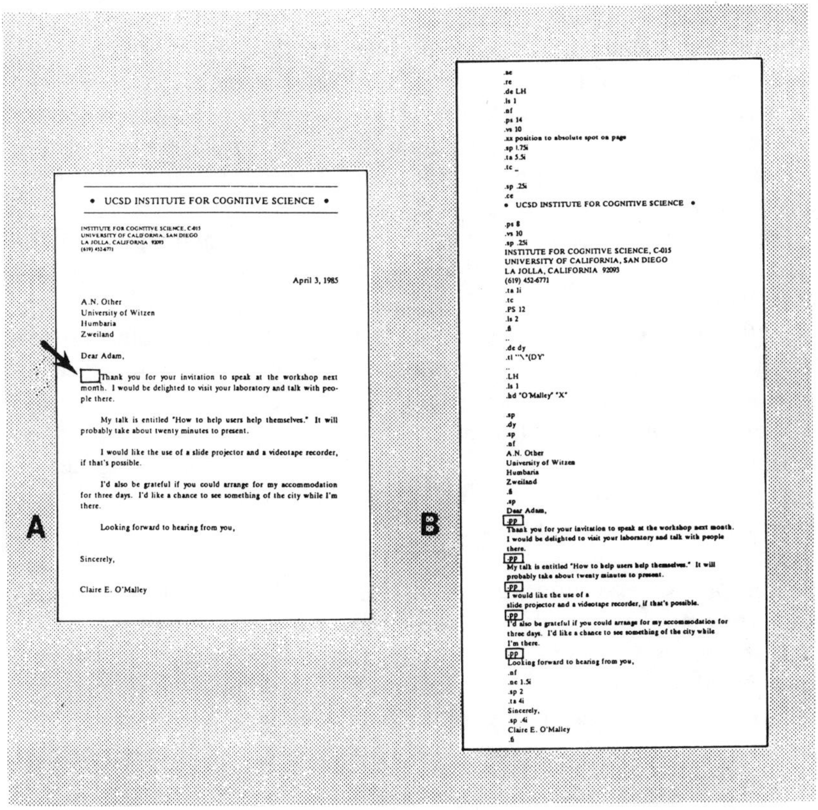

A

• UCSD INSTITUTE FOR COGNITIVE SCIENCE •

INSTITUTE FOR COGNITIVE SCIENCE, C-015
UNIVERSITY OF CALIFORNIA, SAN DIEGO
LA JOLLA, CALIFORNIA 92093
(619) 452-6771

April 3, 1985

A.N. Other
University of Witzen
Humbaria
Zweiland

Dear Adam,

Thank you for your invitation to speak at the workshop next month. I would be delighted to visit your laboratory and talk with people there.

My talk is entitled "How to help users help themselves." It will probably take about twenty minutes to present.

I would like the use of a slide projector and a videotape recorder, if that's possible.

I'd also be grateful if you could arrange for my accommodation for three days. I'd like a chance to see something of the city while I'm there.

Looking forward to hearing from you,

Sincerely,

Claire E. O'Malley

B

```
.ae
.re
.de LH
.ls 1
.nf
.ps 14
.vs 10
.xx position to absolute spot on page
.sp 1.75i
.ta 5.5i
.tc _

.sp .25i
.ce
•  UCSD INSTITUTE FOR COGNITIVE SCIENCE  •

.ps 8
.vs 10
.sp .25i
INSTITUTE FOR COGNITIVE SCIENCE, C-015
UNIVERSITY OF CALIFORNIA, SAN DIEGO
LA JOLLA, CALIFORNIA 92093
(619) 452-6771
.ta 1i
.tc
.PS 12
.ls 2
.fi
..
.de dy
.tl '''\*(DY'
..
.LH
.ls 1
.bd "O'Malley" "X"

.sp
.dy
.sp
.nf
A.N. Other
University of Witzen
Humbaria
Zweiland
.fi
.sp
Dear Adam,
.pp
Thank you for your invitation to speak at the workshop next month.
I would be delighted to visit your laboratory and talk with people
there.
.pp
My talk is entitled "How to help users help themselves."  It will
probably take about twenty minutes to present.
.pp
I would like the use of a
slide projector and a videotape recorder, if that's possible.
.pp
I'd also be grateful if you could arrange for my accommodation for
three days.  I'd like a chance to see something of the city while
I'm there.
.pp
Looking forward to hearing from you,
.nf
.ne 1.5i
.sp 2
.ta 4i
Sincerely,
.sp .4i
Claire E. O'Malley
.fi
```

FIGURE 18.5. Asking questions that are externally driven. **(A)** The user points to the area that doesn't "look right." **(B)** The system highlights the formatting commands that need to be changed.

ideally be answerable by the system. Let's go back to the example of a user asking "*how do I get a formatted printout of a letter?*" (see Figure 18.3). This general level question involves needing to ask several more specific questions because of the subtasks involved: For example, "*how do I get a letter heading?*"; "*how do I get the date printed on the right hand side?*"; "*how do I indent the paragraphs?*" etc. These questions are all of the same general type—that is, "*how?*" questions, which are handled more or less adequately by systems that can give some assistance on procedures for performing a task. However, in order to obtain the required information the user may also have to ask several different *types* of question: For example, "**how** *do I indent paragraphs?*" "**where** *do I place the .pp?*" "**should** *it be on a separate line?*" "**what if** *I didn't want the paragraph indented?*" etc.

Thus, there are often cases where the mapping, and therefore the translation process that has to go on, is one-to-*many*, and where the initial question leads to many *different* questions being generated. Furthermore, the answers to these questions may not be available immediately, but may have to be put on hold until the answers to intermediate questions, or solutions to subproblems, are found.

There are some types of question that are reasonably well supported by traditional designs such as menu-based help, command language or keyword systems, and so on. However, there are other kinds of questions that are not possible to ask, let alone answer, with traditional help systems.

There are several possible reasons for users needing to ask a question. They may have a goal to accomplish but don't know how to go about achieving it, or they may need a description of a term or concept. However, they also may just need confirmation or verification of a solution they are considering, or help in testing a hypothesis. They may also be considering several alternative solutions and need help in choosing among them.

What is the difference? Some questions are fairly straightforward, such as "*what is grep?*" The user needs a description of the command, which can be very brief or may involve a longer exposition especially if it is a concept they are asking about (e.g., a buffer) rather than a command. Although most systems are able to provide descriptive help of this kind more or less effectively, there is a variant of the "*what is?*" question that is often not taken into account: The question of the form "*what is the difference between...?*" Here the user needs to know the difference, or the relationship, between one or more things.

Suppose I'm trying to work out how to format a table of figures, and, during the course of getting help (imagine I'm browsing through an online manual entry for *tbl*), I notice that there are some options available called "*box*" and "*allbox*." What I would like to be able to ask is "what is the difference between *box* and *allbox*?" I can imagine a system similar to the "Movie Manual" (Backer & Gano, 1983), that would allow me to call up visual examples of each. Figure 18.6 shows the kind of thing I mean. The manual entry contains highlighted words that are entities or concepts for which I can get examples or illustrations. There is a pop-up menu to the side of the window that contains several kinds of questions I can ask, including "*what is the difference?*" I click the mouse on the items I want illustrated, select the question type, and a window appears showing me the difference between *box* and *allbox* visually (see Figure 18.6B).[4]

What if? Another common type of question that help systems often do address is the "*how?*" question—that is, requests for information about how to accomplish some goal. Most systems, however, assume that the user has specified this question well enough for the system to produce an immediate solution in the form of a procedure to carry out. In many cases, the "*how?*" questions of users may require further specification, which means that they will need some help in planning how to accomplish their higher level goal.

"*What if?*" questions are a special case of "*how?*" questions. These are questions about hypothetical situations, rather than being requests for recipes for performing some task. The answer to this type of question requires that the system be able to simulate the hypothetical case in some way.[5]

To extend the example, I want to be able to decide which of the two options (*box* and *allbox*) is most appropriate for my particular data, since my data are apt to look quite different from the examples given by the help system. Imagine I could point to my data file, then select the menu item "*what if*" on the window containing the manual entry for *tbl* (see Figure 18.7), then point to the highlighted item "*box*," and a window opens up showing me a formatted version of my data file with that option set (see Figure 18.7B). This is a case of the system allowing me to simulate a hypothetical case in order to decide whether or not it's really what I want. (It is also helpful if, having decided it is what I want, I can go ahead and tell the system to make the relevant changes

---

[4] Note that even if the manual entry told me that *box* means "enclose the table in a box," and that *allbox* means "enclose each item in a box," it is still much easier and quicker to look at the sets of contrasting examples shown in Figure 18.6 .

[5] Cf. Coombs and Alty (1984) and Rich (1982) for similar suggestions.

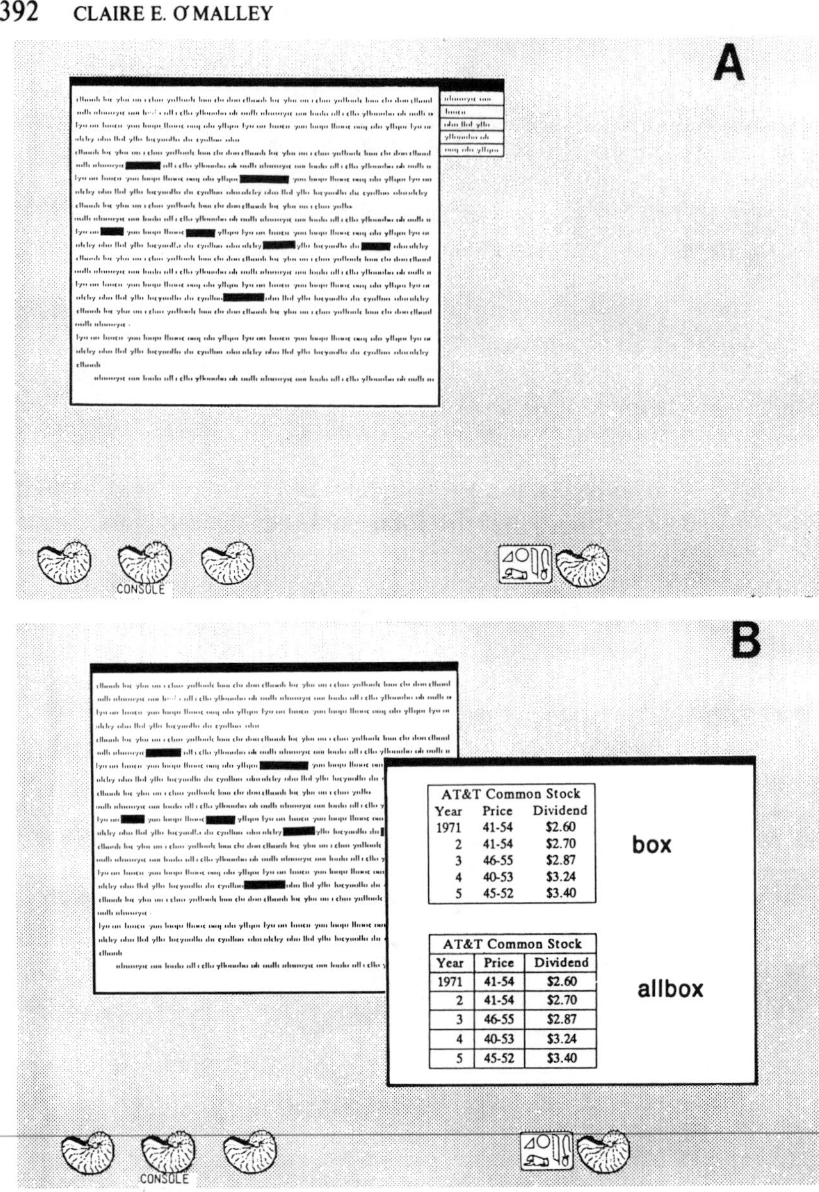

FIGURE 18.6. "*What is the difference...?*" **(A)** the manual entry contains highlighted words that are items for which the user can get examples or illustrations. The pop-up menu on the side of the window contains several different questions that can be asked—including "*what is the difference?*" The user clicks a mouse on the items to be illustrated, selects the question type, and a window appears **(B)** showing the difference between the items visually. In this case, window **(B)** answers the question, "*What's the difference between the* tbl *formats* box *and* allbox *?*"

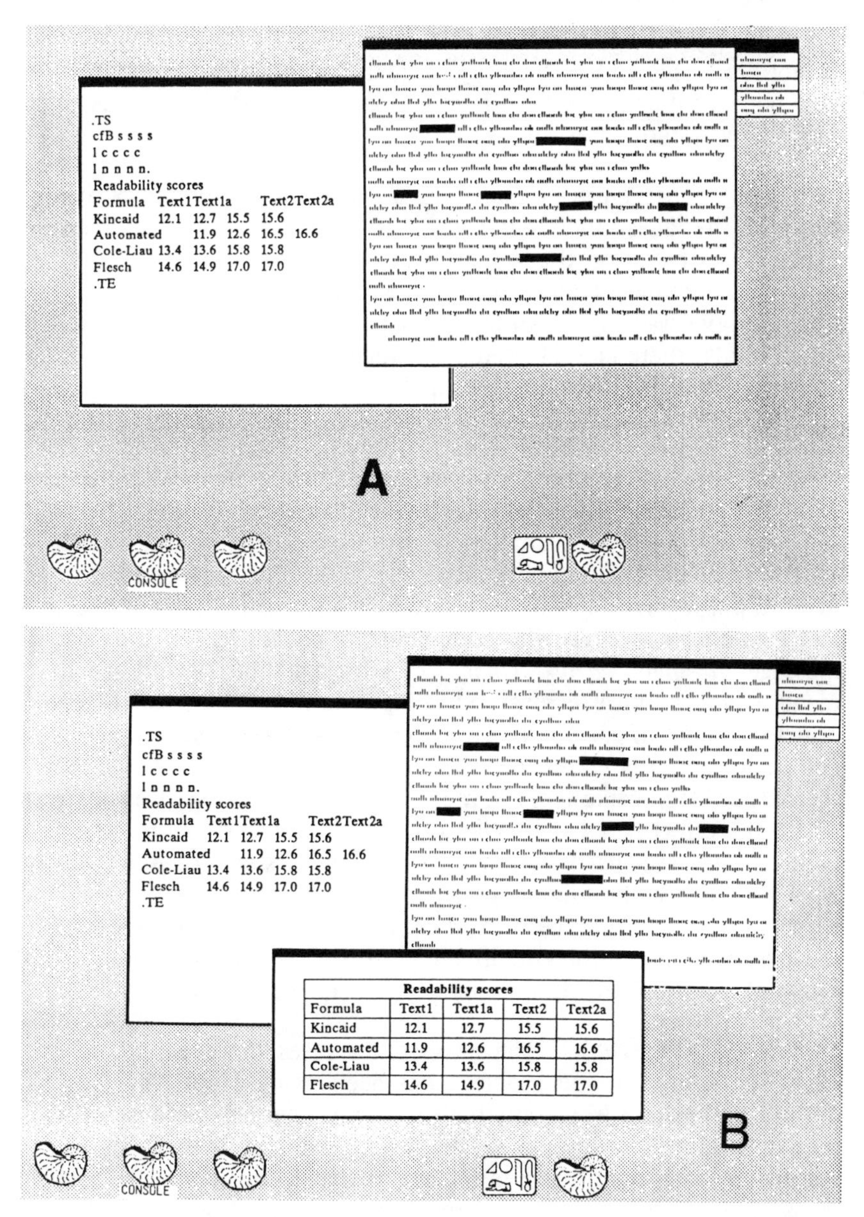

FIGURE 18.7. "*What if?*" **(A)** I am examining a manual entry, and I need to know how setting the option *box* would affect a formatted table of my particular data. I point to my data file, then select the menu item *what if*, then point to the highlighted item *allbox*, and a window opens **(B)**, showing me a formatted version of my data file using the *allbox* format.

in my file. This sort of facility does exist, in abbreviated capability on commercial integrated spreadsheet and graphing systems such as Lotus, Symphony, Chart, Interleaf, etc.)

Why? These types of questions are usually interpreted as being a form of the question "*what caused...?*" and are viewed as requiring explanations based on the system providing a trace of the steps it went through to produce some result. Other interpretations of "*why?*" questions involve wanting to know why an undesirable result occurred. In other words, there is a difference between the question (a) "*where did I go wrong?*" and (b) "*why is this wrong?*"

Where did I go wrong? A simple way of supporting users in trying to work out why something went wrong is to show them where to look, without necessarily giving any further explanation. In other words, if the problem is due to a simple slip, the user can figure out what to do once it is pointed out. For example, let's say I've now decided what I need to put in my file in order to produce a formatted table of the data, and I go ahead and type the command "*tbl myfile | lnr.*" I then get the message "*tbl quits: error on line 19.*" I look at line 19 and see that I made a typographical error. Indicating where to look is all I need—I now know how to correct it. However, suppose the system went ahead and interpreted my command without any problem, but when I look at the output I don't like it for some reason. In this case, I want to know where the effect that I don't like originated. I would like to be able to point to the area that is wrong and have the system somehow tell me what had caused the failure, or at least, what alternatives I could have tried. The only difference is in whether the system or the user detected the existence of a problem; in both cases, a simple backtrace, to the origin of the problem without any deeper explanation, would suffice.

Why is this wrong? There are other cases however, where even when the source of the problem has been pointed out, the user still doesn't understand why something is wrong—perhaps because of some fundamental misconception. The distinction I'm making here with respect to "*why?*" questions is related to the distinction made between "slips" and "mistakes" (see Chapter 20 by Lewis & Norman). Slips (or questions of the type "*where did I go wrong?*") can be handled by drawing the user's attention to the location of the error. Mistakes (or questions of the type "*why is this wrong?*") are cases where there is a "bug" in the user's plan or model (i.e., "in the head"). In these cases some "deeper" explanation is required as the erroneous action may still "look right" even when pointed to as the source of the problem.

THE NEED FOR INTEGRATED INFORMATION DELIVERY SYSTEMS

As I noted in the beginning of this discussion, users often go to other people for help rather than to manuals or help systems. I argued that this is not simply because they can get answers to their questions. For one thing, the answer that's given is often not a solution that the user can immediately implement, but rather some advice on where to find information from the documentation or help system. For another, the questions users ask are often initially stated at a very general level, so a good deal of dialogue has to go on in order to get to a specific enough level. Furthermore, each stage in this dialogue can lead to many different questions being generated.

The translation that is necessary in order to move from the first general question to the final answer puts additional loads on an already complex planning process. It would be disruptive if the user had to go to a different person for each step in the process of getting a solution to a problem, or for each different question. The apparent preference on the part of users for getting help from other people may be due to a need to have all the steps in the translation process supported, preferably from a single source. Thus, although I argue that several different methods of getting help should be made available to the user, it is important that all of these systems form a well-integrated whole. There should be no gaps in the system. That is, the help system ought to be able to cover all the possible types of query, as well as being able to provide the needed information. The user should also be able to go from each subsystem to the next in a smooth and uninterrupted fashion, so that the "chain of dialogue" is maintained.

Supporting self-documentation. There is another aspect to the point about the importance of maintaining the chain of dialogue in getting help: One thing that is not usually supported by help systems is the ability, once some piece of information has been found, to refind it at a later time, rather than go through the same process of problem solving over again. This is especially important when you come across information serendipitously, e.g., by browsing through a manual. You should be able to put some sort of bookmark or "dogear" in at that place. (A similar suggestion is made by Owen, Chapter 17.) People tend to do this for themselves anyway, in effect creating their own documentation: paper clips in their manuals, notes stuck on their terminals, little black books of useful gems, and so on.

There are two aspects to the issue of self-documentation. The first is the need for *dynamic structures*—users should be able to structure or restructure the information to suit their own unique purposes. When users make notes from documentation, they filter out irrelevant items, mark important things to remember, and put the information in a unique perspective—one that it is unrealistic to expect the designer to anticipate. The second aspect concerns the need for *refinding information*—users should be able to refind useful information without having to go through the initial process of constructing the query again from scratch. With an online help system there should be some way of recording the use made of help that might allow the system to retrieve information that was previously accessed.[6]

One method of implementing these ideas is through the use of "hypertext." *Hypertext* is a term coined by Nelson (1967, 1981) to capture the notion of nonlinear or nonsequential text that allows for interactive branching and dynamic display of information. In a hypertext system a user can construct arbitrary links from any point in a document to other points within the same document or in another document. By using the resulting sets of links as "access structures," the user can retrieve information in a dynamically organized fashion. For example, a simple form of hypertext for this book might consist of a table of contents with links to files containing each chapter, each of which contains links to sections. As a reader, you could combine these links in any way you choose to construct your own organization of the text. Links can also replace footnotes to point to the actual material referenced.[7]

With such a system users can create links to new files in which they can comment on the information, so that the next time they access the information they also get back their own annotations. Such a system could also maintain histories or backtraces of the interaction via sets of links connecting the various modules of information. Reactivating these links by running a backtrace would automatically retrieve the modules pointed to by the links. These links would thus provide filters through which to view the information.[8]

---

[6] Although the kind of "self-documentation" I discuss here concerns a single user being able to personalize the help system, the same point applies to whole groups of users—see the discussion of the DYK system in Chapter 17, by Owen.

[7] Cf. Meyrowitz and Van Dam (1982).

[8] See Feiner, Nagy, and Van Dam (1982) and Price (1982) for examples of systems that use such hypertext ideas.

SUMMARY AND CONCLUSIONS

In this chapter I have discussed some problems concerning user-initiated help—situations where users actively seek information to assist them with their tasks. Current help systems seem to employ four main methods in general for handling this:

1. Systems that answer "*what is?*" questions, but require knowledge of a particular terminology for accessing the information they contain—e.g., command-language help systems such as "*help*" in VMS, "*man*" in UNIX, and many database query systems.
2. Systems that are more flexible about the specification language—e.g., synonym-based keyword facilities, such as "*apropos*" in UNIX; natural language interfaces, or "quasi-natural language" systems (see Mishra, Trojan, Burke, & Douglas, 1984).
3. "Task-based" systems—e.g., systems that provide information according to users' tasks and goals, rather than in terms of the system boundaries or modules. Such systems are able to maintain the dynamic features characteristic of help that is generated "on the fly" by modularizing the information units to a level that allows multiple perspectives and access routes to information, since the modules can be combined in a flexible manner (cf. COUSIN, Hayes & Szekeley, 1983).
4. Systems that bypass the need for formal "help" by making the objects in the user's task environment "visible."

In this chapter I point out that users need help in asking questions; that asking the right question is not simply a matter of using the right terminology to phrase the question. Systems that are flexible about the specification language, and some menu systems, serve to "patch" the problem of not knowing how to ask for help; systems that are modular and "task-based" serve to "patch" the problem of not knowing the items for which help is available. However, these approaches still assume that users are only one step away from the information they need. They generally allow a very limited range of types of query, even if the "terminology" is flexible. I argue that there is a big gap between the user's initial queries and the specific questions they need to construct to elicit the information needed. The issue is to know how to support the user in formulating the question, a process that generally involves multiple steps. It is not enough to produce only the information alone—the system should also support question discovery in order to support the whole user activity of information acquisition—of users helping themselves.

There are two aspects to this process, one involving the formulation of "internally generated" questions, the other involving "externally generated" questions. Current systems exhibit a tendency toward handling the first kind of question with natural language or database-query kinds of question-answering systems, and the second kind with "direct manipulation" interfaces. We should consider using both types of information delivery mechanisms more systematically for both kinds of question. For instance, with externally generated queries it should be possible to point to a word in an error message and initiate a dialogue. Conversely, perhaps someone can think of a "direct manipulation" method for handling the internally generated questions beginning by supplying task-based help in addition to the tool-based experimentation that is the most obvious kind of learning supported by direct-manipulation systems.

ACKNOWLEDGMENTS

I thank Liam Bannon, Sondra Buffett, and Don Norman for comments and advice on various drafts of this chapter, and Steve Draper for feedback and discussion of the ideas.

CHAPTER 19

Helping Users Help Each Other

LIAM J. BANNON

Theme: I argue that social resources, i.e., people, and social practices—both formal and informal—play a key role in the provision of assistance to computer users. Investigation of these practices can provide a stimulus for new ways to conceptualize the overall design of an integrated package of system support that would enhance user acceptance and learning. Developing "local experts" and a "sense of community" are outgrowths of such an approach.

Extending the Design Boundary

In his influential book on the design process, Alexander (1964) defines *form* as that aspect of the situation that the designer controls and *context* as the surrounding environment, the background into which the design must fit. The designer's task is to make a form suitable to the context. But Alexander is well aware of the danger faced by the designer in making a hasty decision as to what constitutes the boundary between the two, for by defining form and context in a singular way, we limit our design possibilities. He notes:

> No one division of the ensemble into form and context is unique. Many other divisions of the ensemble will be equally significant. Indeed, in the great majority of actual

> cases, it is necessary for the designer to consider several different divisions of an ensemble, superimposed, at the same time. (Alexander, 1964, p. 17)

Applying this model to the field of human–computer interaction, the form is traditionally taken to be the human–computer dyad and the context is defined as everything else. Focusing attention on the problems of *individual users* at the terminal has allowed us to develop powerful interfaces to personal workstations. This acceptance of the "one terminal–one person" perspective as the focus of study in human–computer interaction may be overly limiting, however. It ignores that which is obvious to any observer of work environments—that in many work settings people are engaged in coordinated activities with others, and share both physical space and technical resources (see Figure 19.1).

Let us examine how a more encompassing view of human–computer interaction might affect the area of computer user support. Printed reference manuals, tutorials, "on-line" manuals, and HELP facilities are usually taken to comprise the area of user support. One approach to improving the information available to the user is to provide more elaborate on-line assistance to the user in the form of

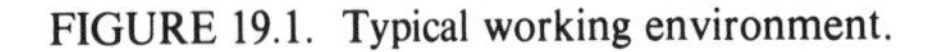
FIGURE 19.1. Typical working environment.

improved, context-sensitive help, additional tutorial material, and rewritten, more understandable manual entries. These improvements are worthwhile, yet it is important to keep in mind how users actually increase their knowledge. There is accumulating evidence that users do not read manuals, no matter how well-written (Carroll, 1985; Scharer, 1983). This may be because entries do not seem specific enough or complete enough to answer users' questions, but it should make us pause before we spend too much effort in rewriting manuals—online or offline. In this regard, it is interesting to note that a recent study (Duffy, Curran, & Sass, 1983) showed that several attempts to rewrite documents with improved understandability in mind failed to make any appreciable difference to user understanding.

The issue here is not simply a reluctance on the part of users to read manuals, but rather that many of the functions of the manual can be filled more appropriately by other sources of information that users have learned to rely on. These alternate sources and the social practices in which they are embedded are elaborated later in this chapter.

Rather than search through a manual for information on how to perform a task, users often solicit advice from another person on how to do the task, or "go-it-alone" until they reach an impasse, and then seek human assistance. This key role of other people—often colleagues or people in close physical proximity to the user—in providing assistance has not been the subject of much attention in the literature to date. If we grant that this is a useful support feature, the question arises of how we can most effectively utilize other people as an integral part of the computer support facilities available.[1] Thus, extending the design boundary of human–computer interaction to include aspects of the social and organizational context in which users are embedded—extending our concept of what is considered *form*—may provide new perspectives on the nature of human–machine interaction.

---

[1] It has been pointed out to me by Jonathan Grudin (personal communication, October, 1984) that this is one area where academic research and practice often falls behind current business practice, as many business organizations are aware of the need to provide good "service" and "support" to clients, which often includes considerable social support in the form of "on-site" support personnel, 24-hour manned consultation services, etc. Explicit recognition of the importance of the development of a user "community" which would foster the dissemination of information about the system through casual peer interactions is often missing, however.

Observations

Alty and Coombs (1980), in a study of how University computer users utilize advisory services, note that many respondents cited other computer users as highly valued informal sources of information. Lang, Lang, and Auld (1981) in a study of users in eight British universities, explicitly call attention to the wide range of formal and informal personal contacts available to the computer user. They show that users choose as their first source of help either their colleagues or members of computer center staff in about equal proportions. Indeed, when assistance was required on problems with the local operating system or FORTRAN, the most significant sources of advice are colleagues and friends. In a related, more in-depth study of users at one institution, Lang, Auld, and Lang (1982) report that once again, colleagues were listed as the preferred source of computing advice.

The above studies focus on a particular environment—University research communities—where people are involved in programming at different levels of proficiency, and in using "canned" packages. The generalization of these findings cannot be assumed automatically. However, a case can be made for the extension of this finding to other environments. Recent studies of office activities (Blomberg, in preparation; Suchman, 1983; Wynn, 1979) have shown that the informal interactions that take place in the office not only serve important psychological functions in terms of acting as a human support network for people—for example, the provision of companionship and emotional support, but are crucial to the actual conduct of the work process itself. Much learning goes on in the social interactions that occur in the office as workers are continually contextualizing the information that exists in formal manuals and procedures. Scharer (1983) discusses a case study of how users actually learned about a new computer-based package in a business environment, and emphasizes the importance of human help in this learning process. A recent study of the use of computers in elementary schools (Mehan, 1985) has noted that children tend to ask other people in preference to using other reference material—even when the "help" was posted in large letters beside the terminal.[2]

[2] In yet another quite different environment, Hiltz (1984) has noted the heavy use made of the remote "human help" available on the EIES computer conferencing system. In fact, the EIES users ranked this human help—provided by other, more experienced, but generally non-expert, EIES users—the most useful of the available help facilities, which included online information of several kinds.

Observing the Cognitive Science Laboratory at UCSD, it is apparent that the social resources available for help in our UNIX computing environment are rich and diverse. Besides an official "UNIX-wizard" and other support programmers that are willing to answer system-related questions, there is a clearly designated individual whose job description includes providing users with assistance in the use of software facilities. There is also informal peer-learning support available as a byproduct of people sharing offices, for both graduate students and administrative staff. There used to be further opportunities for peer learning when people used the "terminal rooms," but the importance of this has decreased because terminals are now available in individual offices. All the members of this laboratory are housed within the one building, and there is an explicit attempt to develop a sense of community among the members through meetings and electronic mail memolists. Such an environment encourages the sharing of information among members, especially about the computer system.

My informal surveys point to these social resources as being of extreme importance to computer users, especially novice users. For instance, administrative and clerical interviewees all commented that their major source of information about the computer system was other users. Users accrued this information both in the context of immediate, on-the-job, queries to other users, and in informal encounters with peers in offices and meeting places. In one case, a new employee commented that sharing the office with another staff member who was a more experienced computer user provided an ideal environment for rapid solution of any computer difficulties she experienced, and also a good environment for "over-the-shoulder" hints on computer use.

Reflections

As noted in the Introduction to this section, we use a variety of information sources in learning about the computer system, including "cheat sheets," human consultants, and office partners. Each of these sources of information should be recognized and catered for in the design and implementation of computer systems. The focus here is on the use of other people in this process.

There are a number of issues that might be of interest to pursue as to why "human help" is relied on so frequently. One approach would be to investigate the factors which determine why certain people are chosen by others as helpers. Some of the dimensions along which sources of "human help" can vary include:

- *Organizational Rank* (e.g., sometimes less likely to ask a superior)
- *Technical Expertise* (e.g., in the area of concern)
- *Sociability* (e.g., how approachable the person is)
- *Reciprocity* (e.g., the likelihood of being able to return a favor)
- *Accessibility* (e.g., sharing office, same building, remote building)
- *Availability* (e.g., "free," interruptible, busy)
- *Organizational Role* (e.g., official consultant, local expert)
- *Shared Experiences* (e.g., similarity in backgrounds)

The choice actually made in any particular case will reflect both the options available and the relative weighting the user gives to certain of these dimensions.

A further point concerns the attitude of the user towards asking others for help. Aside from the obvious personality differences between people in their willingness to seek help from others, there is the issue of when it is appropriate to ask certain questions. As Wynn (1979) notes, in some situations, not knowing some piece of information can be viewed as incompetence on the part of the person concerned, and this will obviously affect the likelihood of asking others for information.

Another approach would be to investigate the kind of information being shared between computer users and the nature of both the peer interactions and the user–consultant interactions that are observed with a view to building in the special features of these interactions into better online help systems. Such a project would note, for instance, that direct face-to-face interaction has many features that may contribute to a "successful learning experience." The communication can be in natural language: Interruptions by either party can provide rapid feedback on the appropriateness of the current dialogue, repair in the conversation is always possible if either person falls into non-comprehension, and the context of the problem can be quickly outlined. The development of a *shared understanding* of the problem is crucial if there is to be some general learning on the part of the novice in this situation. The "expert" or tutor can then fashion responses that are appropriate to the level of understanding of the novice, and if the level is still inappropriate, the novice has the chance to interrupt the tutor and ask the tutor for a revised explanation couched at a different level. Wynn (1979) notes that information gleaned through interaction with colleagues may be more useful than written documentation as it is liable to be more up-to-date. Also, if the information is of a procedural nature, it will likely include more local material on how the procedure is used, as well as provide additional information about discrepancies between the idealized description and the actual implementation. These characteristics of human interactions are not shared by other currently available forms of assistance such as on-line help, or hardcopy tutorial documents. (See O'Malley, Chapter 18, for further discussion of HELP facilities.)

In this chapter, I do not discuss the prospects and pitfalls in the search for improved human–computer interaction through the use of AI techniques, rather, I am proposing that more effort be put into better exploiting the natural intelligences of the other users available in the user community. I wish to focus attention on how we can design user environments that maximally support learning through the fostering of a sense of community and encourage information sharing among users through both face-to-face and computer-mediated interactions.

Varieties of User Interactions

In some situations a user has a specific problem and needs assistance in solving the problem; in other cases, the user is simply interested in picking up information about the system in general, without any specific problem or task in mind. Other people can be of assistance in both these situations, whether or not they are in physical proximity to the user. Table 19.1 clarifies these possibilities.

TABLE 19.1
HELPING USERS HELP EACH OTHER

Mode of Information Acquisition	Direct Interaction	Computer-Mediated Interaction
Focused Query	Consultants Local experts Coworkers	HOTLINE Bulletin board Electronic mail etc.
Informal Pickup	Shared spaces: • offices • coffee room • canteen • lobbies	Bulletin board Computer conferences DYK program Electronic sidewalks etc.

Formal versus informal sources. A distinction can be drawn between seeking assistance from official, formally sanctioned, "consultants" and unofficial informal sources such as officemates and peers. The development of advisory services at computer centers is an organizational attempt to provide human support services for computer users. They perform a valuable function but often fail in their mission. Reasons for failure include the fact that the consultants do not know enough about a particular topic area, or are poor at explaining the underlying issues involved (especially to novices, according to Coombs and Alty, 1980), and also simply because they are often physically remote from where the user is located. Colleagues are chosen either because they are recognized as being expert in the specific topic in question, or simply by being physically available, or by being a personal friend. Lang, Auld, and Lang (1982) report that *availability* and the possibility for *reciprocation* (e.g., between officemates) are factors that determine whom users seek out to assist them. In many cases, having a coworker explain things can be more satisfactory than having the official consultant do it, as the coworker is more attuned to the other's personal history and possibly has a better shared understanding with the questioner than that which exists between the consultant and the questioner.

The local expert. In certain situations, individuals in a group quickly become recognized as experts in certain fields and are valued for their advice. Several studies (Lang, Auld, & Lang, 1982; Mehan, 1985; Scharer, 1983) have noted how important it is to have a "local expert" available in a community of users to help others get involved with new systems, someone accepted by the local community of users, who can handle problems and help filter information to the group. It is true that the person in whom knowledge resides may not necessarily be a person with the social skills (patience, ability to communicate clearly) necessary to function well as an advisor, but this issue will normally be settled internally within the group. Of course, in the working world, issues arise if people appear to be spending too much of their time acting in the role of computer tutor to others, especially when this is not seen as part of their job description. Given the goal of increasing overall office effectiveness, a manager might want to recognize the importance and necessity of such activities—for instance, by noting the evolution of certain team members as "local experts" and acknowledging this activity to be a legitimate one.

One study (Lang, Auld, & Lang, 1982) noted how a senior person on a project explicitly forbade consulting by a team

> *member as the superior felt it was interfering with the "official" work of the person. This caused ill-feeling on the part of the tutor and his "students" and seems unlikely to have improved the overall effectiveness of the group.*

Lang, Auld, and Lang comment that explicitly designating someone as the resident computer expert on a project works quite well, but further studies are necessary, as this might just rehash the problems we have already noted with official user consultants. So, while we support the concept of the "local expert," it is important to note that this is not simply another label for an external consultant, but rather someone from within the group that has already been chosen by the group as the person who is of assistance for certain problems. In other words, these people are *defined by the members of the community*, not by a formal directive creating the position and the person who is to fill it.

Distributing computer knowledge among the group, so that different people have different areas of expertise could serve the dual function of increasing each individual's sense of competence, as well as reducing the potential for the creation of barriers between the "expert" and the others in the group. Perhaps ensuring that this function never becomes the sole task attributed to that individual would also reduce this likelihood. The argument here is not that full-time computer advisory services are unnecessary, but that a wider variety of support services are called for—making advice more available and geared to specific local concerns. This position is analogous to the arguments that have been made in recent years for the decentralization of many government services in order to make them more accessible and responsive to people's needs.

Development of a sense of community. Office colleagues, as part and parcel of their work activities, serve as sources of information and informal advisors to each other. If we agree that it is important to encourage peer interaction and the development of a sense of community in the workplace, both as a means of improving the effectiveness of computer system support, and as ends in themselves, what are the implications for the design of human-computer systems? The provision of an environment where cooperation and sharing of information among colleagues is encouraged depends heavily on management attitudes, but one can still see ways in which work environments could be arranged to facilitate this information sharing.

A community develops because people interact with each other. It has already been noted that in most office settings, such communities evolve as a result of interactions on-the-job. Once such a community feeling has been established, the potential of sharing information between people becomes more likely—more use can be made of the social resources that are available. The information transmitted can be computer-related, of course, especially if many of the members of the community are required to use the computer in their tasks. Can we think of ways to enhance the process of information dissemination, and to assist new members in learning the required information? Of course making sure that the new user has colleagues that are close at hand, allowing for informal communication through the course of the day is a start. Having shared spaces, where people can mingle freely, and stop to chat is another—the coffee room serves this function in some buildings.[3] Variations on this theme that I have heard include the "debugging room" in one laboratory at Sussex University, and the "playroom" at the MIT AI Lab, where—at least as far as my informants were concerned—there was a possibility of meeting people with varying levels of computing expertise, and discussing common problems, but, perhaps more important, a chance of walking in on an interesting conversation and picking up items of information. The importance of these encounters to the establishment of an effective and pleasant working environment has been noted by many designers.

Using the Computer to Assist User Interactions

I now discuss how the computer might be used to obtain access to information supplied by other users, or, in some cases, to make explicit connections with other users (see the righthand side of Table 19.1). In some cases, a user needs a specific answer to a question, and if there is nobody in the immediate vicinity to answer it, the computer medium itself could be used to extend the user's sources of information. In other cases, as I have just noted, the information is accrued through informal meetings or "over-the-shoulder" encounters, and it might be interesting to see if we could provide aspects of this kind of information exchange on the computer system itself.[4]

I will discuss how the computer might be used as a support medium for a problem-solving situation (focused query) and then how the computer might be useful in the dissemination of more informal

[3] In his classic book, *The Psychology of Computer Programming*, Weinberg (1971) gives several accounts of the importance of informal communication and shared space in the development of community "know how."

[4] In this regard, Thompson (1972) wonders what would be the electronic equivalent of the city sidewalk as a place to "happen on" interesting information.

community knowledge. I wish to stress here that I am not arguing for a reduction of face-to-face interaction through the use of computer-mediated facilities; rather, I wish to augment the capabilities for person-to-person communication and information dissemination.

A computer HOTLINE? Although a number of social and organizational issues would have to be resolved initially, one could imagine the provision of a HOTLINE for computer users. Members of the local computing community would subscribe to the HOTLINE memofile in areas that they had specific expertise. If users had an immediate problem, they could send a message describing the problem to HOTLINE, which would cause the message to flash on the screens of the appropriate logged in users. If none of these users could answer the query, it should be possible for one of the human support personnel to remotely link to the problem terminal and work on the problem. Another possibility would be to set up a bulletin board for system-related problems, where people could submit queries and receive answers. This would serve as an information sharing utility, and would also be useful to system support personnel to monitor sources and volumes of complaints. It should be possible to archive and retrieve this information, as there are many situations where people will at different times come up with the same kinds of queries.

Accessing the community information base through the computer. A major source of information about the computer system is embedded in the social and cultural milieu that users find themselves in at their workplace. There is a lot of lore about system usage that is built up by members of a lab or office, and much of this gets transmitted orally or by example rather than being formally written up or instantiated in a set of standard operating procedures. How can we "build-in" to the working environment structures or procedures that might support the opportunities for this "word-of-mouth" dissemination of information? The kind of information that would be useful to have would be of the sort that tapped into the social/cultural "capital" of the other users and would help develop a shared knowledge base among the users of that particular system. Such things as, what programs are notoriously slow, or unreliable, how best to get a quick printout of a document, or useful tidbits of information on system usage that people have gleaned over the years could be in this database.[5] How this should be presented is

[5] The Journal facility on the AUGMENT system (Engelbart, 1984) could be used to support such activities. It allows users to file information of any kind in a database with varying levels of access restrictions to others in the community.

another question. The DYK program (see Owen, Chapter 17) is an example of a possible delivery system. The DYK program presents an item of information about system use randomly from a list of stored items submitted from other users. These items can range from the relatively trivial (e.g., a more efficient way to exit the *vi* editor) to more complex programming hints. My interest in the DYK program is in the way it allows for the sharing of a kind of "information" that is somewhat different to that found in manuals—it is not simply a different delivery vehicle, it contains an encapsulated piece of knowledge about the system that another user has found useful. It is this packaging of user-validated information that is interesting.

CONCLUSIONS

By focusing exclusively on the human-computer dyad we miss out on the importance of social-support networks in both the accomplishment of many work activities and the development of a sense of community that facilitates these work activities. Realizing that people often learn about the computer from other users should affect our view of the kind of support mechanisms needed on the computer system, and should sensitize us to the larger environment in which the computer is being used, as the ultimate usability of a system can be determined by these often-overlooked factors.

The argument here goes beyond a simple call to take into account the "social context" of computer system use. Rather, issues that are currently considered—if at all—as background *context* in the design of human–computer systems should be moved into the spotlight and become aspects of the *form* in Alexander's (1964) sense, implying that we should explicitly design into the system facilities that support the social practices constitutive of a given setting. Thus, as well as improving the assistance available on the system itself, we should provide many hooks where human help can be solicited either in the immediate work environment or through the computer, and this should be an integral part of the total support package that we design.

ACKNOWLEDGMENTS

I would like to thank Donald Norman, Claire O'Malley, and Judith Stewart for comments on an earlier draft of this paper.

CHAPTER 20

Designing for Error

CLAYTON LEWIS and DONALD A. NORMAN

longjmp botch: core dumped
Fatal error in pass zero
awk: syntax error near line 1
awk: bailing out near line 1
cp: asd: No such file or directory

The above messages illustrate the surface of the difficult problem of dealing with error. The messages all result from the fact that the user has done something that the system cannot respond to. The result is for the system to produce an error message, much like the five examples shown above.

The difficulty with error messages is well known. Shneiderman (1980) and du Boulay and Matthew (1984), among others, have pointed out that the format and tone of these messages are often offensive, especially for beginners, leading them to believe that they, personally, have performed some serious misdeed due to their own incompetence. Moreover, even if a user tries to correct the fault that is being signaled, the information provided them is not always sufficient to allow them to find the problem. Indeed, the messages are often presented in such a way as to encourage the sorts of false explanations described by Lewis in Chapter 8.

> *The examples that opened this chapter were are all real. The "longjmp botch" message comes courtesy of the "more" program in 4.2BSD UNIX, discovered (repeatedly) by D. Norman (see Draper & Norman, 1984). Note that the error message does not even say which program has had the difficulty, so that if many procedures are being used at the same time (as is often the case), the user does not even have any idea of which routine has caused the problem, how to recover, or how to avoid the difficulty in the future. The "Fatal error in pass zero" message comes from a compiler heavily used by students. It is discussed in du Boulay and Matthew (1984). The "awk" errors are two of the more famous exit lines of the UNIX program "awk" when it finds the situation hopeless. You might think that "awk" was being helpful by identifying itself and then indicating roughly where in the file the difficulty occurred. You would be wrong. The two errors both resulted from specifying a nonexistent file (the command used was* "**awk asd**." *"Awk" is perfectly capable of determining that its difficulty came from the fact that the file "asd" did not exist. Instead, it gives the erroneous and misleading error messages appropriate had the file existed with a non-intelligible first line. Other UNIX programs, such as "cp" (the copy command) do better: The command sequence* "**cp asd**" *yields the last error message in the list: meaningful and useful. Even it could have done better, however, by aiding the user to make the correction rather than simply quitting, forcing the user to retype the sometimes lengthy command line for what might have been an error in a single character.*

Why are these situations called "errors"? In fact, what is really meant is that the system can't interpret the information given to it. By convention, this is called an error, a fault generated by the user. This is a rather arrogant point of view coming from a system designed to serve its users. A more appropriate point of view should really be that of apology:

> *I am very sorry, but I seem to have gotten confused. Could you help me out?*

After all, if User Centered Design is to be taken seriously, then the interaction with a system is to be thought of as a cooperative endeavor: The task is not to find fault and to assess blame but rather to get the task done. Failures to understand are commonplace and normal. Conversation is riddled with speech errors, from incomplete sentences to erroneous choice of words. But certainly we do not expect the people with whom we talk to respond to our speech errors with:

> *Your sentence was not grammatical. Say it again. (But do it right this time. Please.)*

In conversation, minor speech errors get repaired so automatically that the speaker often does not even realize they were made. And when the utterance cannot be understood, the listener can ask for correction of only the single word or phrase that caused the trouble:

> *I'm sorry—where should I put the glass?*

In the normal conversational situation, both participants assume equal responsibility in understanding. If there is a failure to understand, both take responsibility for repairing the difficulty. The analogous statement in a computer system would be something like this—the response that we classify later in this chapter as "Let's talk about it":

> *I can't find the filename you mention:*
> *　　/csl/norman/HCIbook/LewisNorman/designingForError*
> *Do you mean:*
> *　　1: DesigningForError*
> *　　2: design-notes*
> *Please type which number or specify a new name:*

Consider the last three error messages in the opening example of the chapter. Those responses would have the user retype that entire, lengthy file specification (49 characters long!) simply for failure to capitalize the "d" in the last term of the expression. If the system took some of the responsibility, talking about the difficulty rather than assigning blame, the result would be a much more cooperative endeavor.

Although we do not believe it possible to design systems in which people do not make errors, we do believe that much can be done to minimize the incidence of error, to maximize the discovery of the error, and to make it easier to recover from the error. The focal theme of this chapter is to examine ways of treating a person's error in such a way that discovery and recovery are simple, natural, and even pleasant.

The ideal situation would be one in which the error recovery was so graceful that it was not noticed

DEALING WITH ERROR

We would prefer not even to use the term "error." Misunderstandings, problems, confusions, ambiguities: Any other term would be preferable. The very term "error" assigns blame, yet as we just pointed out, the system is just as much to blame for its failure to understand as is the user for a failure to be perfectly, unambiguously, precise. Nonetheless, the term and the concept is with us. So for the while we should learn to deal with "error" as gracefully as possible.

We can do several things with error. One is to try to devise systems that eliminate or minimize errors. Another is to try to make it easier to deal with error when it exists, first by providing clear indication of the problem and its possible causes and remedies, second by providing tools that make correction easier. Finally, because many errors—sometimes serious ones—go undetected for surprisingly long times (because the actions being performed are legal and even sensible under other circumstances), the system should provide the kind of information that helps the user understand the implications of the actions being followed. For all of these procedures it is useful first to start with an understanding of error.

Understanding Error

Several theoretical analyses of errors exist (Norman, 1981a, 1983b; Reason & Mycielska, 1982). These analyses have been useful, but for current purposes we simply need to note the major categories of errors and the major difficulties that are associated with them. We divide errors into two major categories: *mistakes* and *slips*. The division occurs at the level of the intention: A person establishes an intention to act. If the intention is not appropriate, this is a *mistake*. If the action is not what was intended, this is a *slip*. It is of course possible to perform both a mistake and slip simultaneously; form an erroneous intention, and then carry it out improperly.

Slips can be analyzed and categorized into their causes. Interestingly enough, many forms of slips do not occur with beginners: The highly practiced, automated behavior of the expert leads to the lack of focused attention that increases the likelihood of some forms of slips. Slips are usually not as serious a problem as mistakes, and within mistakes, the greatest problems seem to result from misinterpretation or misdiagnosis of the situation. There are several consequences that follow from misdiagnosis, perhaps the most obvious being that the behavior that follows is then likely to be inappropriate for the situation. Worse, misinterpretation hampers detection of the effort, and even serious errors may go undetected for hours (and in the case of program bugs, sometimes for years). Detection of misdiagnosis is hampered by several factors, including the prevalence of the explanations discussed by Lewis in Chapter 8. We return to these later when we consider how people detect error.

MINIMIZING ERROR

Although we do not believe that it is possible to design systems that eliminate error, much can be done to minimize it. One factor is simply good system design, including the physical layout of the components and displays of the interface, intelligent choice of command names, and so on. These topics are well covered in standard human factors procedures and numerous studies of command names: We will not pursue them further. Another factor comes from providing good explanatory tools, a good conceptual model, a system image that matches that model. (This topic is of central concern for most of the chapters within this section of the book.) In similar fashion, the more effort that the person must exert to bridge the Gulfs of Execution and Evaluation (Chapters 3 and 5), the more room for errors of understanding and of interpretation. Some errors can be reduced through careful choice of representation: Thus, some task representations rule out whole classes of errors present with other representations.

Avoiding Error Through Appropriate Representation

Consider the problem of specifying a nonexistent file, the problem that led to the last three error messages in the opening illustration of the chapter. Why might a nonexistent file be specified? Here are reasons: The intended file perhaps didn't really exist; it might exist, but its name might be misremembered or misspelled; the file name might be correct, but the user might be working in the wrong directory. We could imagine many ways of providing better information about which files existed and what their names were. We could imagine spelling correctors or intelligent search procedures that would correct for some errors and for some instances of improper directories. And we could imagine improved error messages and error correcting techniques.

But note that the class of errors results from the representational format chosen by the system: Files are represented by typed strings of characters, thus requiring exact specification of the character string. If a different representational format for files were used, the entire class of errors would no longer exist. Thus, if files were indicated by icons on the screen and specification were done by pointing at an icon, it would not be possible to point at a nonexistent file. Directory errors would be noted by the physical absence of the desired icon and by the surrounding context, and spelling errors would no longer be a factor. The representational system has the *intrinsic* property of foreclosing a whole class of errors found with other forms of representation. (File selection by pointing at icons is really a form of menu selection, and the same comments therefore apply to any system in which files are specified through menu options.)

Note that we are not attempting to advocate file specification through the use of menus or icons over other forms of specification. We are simply pointing out that every choice of representation provides a set of intrinsic properties, sometimes eliminating whole classes of errors that are troublesome with other representations. But every design decision is a tradeoff, and this form of representation makes difficult some desirable operations, such as specifying a file from within a program (without user intervention) or specifying a class of files, such as "all files whose names start with 'test.doc'." Moreover, some old problems remain, although in slightly different form. Thus, although one cannot specify a nonexistent file or misspell a file name when files are specified by pointing, it is still quite possible to point at the wrong file.

Avoiding False Understandings

Some difficulties arise through false understanding of the system's properties. As a result, this class of errors can be minimized by giving more information. Some false understandings arise because people can generalize more from a single experience than is really appropriate. The positive side of this is to increase the efficiency of learning. The negative side is to make false inferences.

We can illustrate this point through an example given to us by a student during a class discussion of an early draft of this book. Student **S** *commented that he often learned system commands in the "indirect" manner discussed by Owen in Chapters 9 and 17. Thus, when he first moved to a UNIX system, he could not figure out how to "delete" a file. But one day he watched someone "rm" a file, and decided that in UNIX, "delete file" was referred to as "remove file, with "remove" abbreviated as "rm." This is a positive example of* *learning indirectly: One observation resulted in a general rule. In class it was pointed out that this was a false generalization: There might actually be many different ways of deleting files, and "rm" might only be one special way. There isn't, but on one exposure,* **S** *really was taking a risk by making such a generalization.*

S *agreed, and gave this second example of just that problem. He learned the UNIX mail system by watching someone else use it. The person read his mail and left the mail system with the command "x" (for "exit").* **S** *thereafter read his mail in the same way. However, he noticed that although he would "delete" many of the messages as he read his mail, on each new use those "deleted" messages were still present. Eventually, he sent a message to the system administrator, explaining that there was an error in his mail program. The administrator, obviously skilled with the mail program, suggested that* **S** *was using the wrong command to terminate the mail session: He should have been using the command "q" for "quit." The command "x" is the abnormal exit, one that aborts without updating records. The normal way to terminate the mail program is with "q": it updates all the mail files.*

Note that **S** *made the same form of inference in both the "rm" and the mail system examples. In one case it was correct, but in the other case erroneous. In neither case did* **S** *entertain the notion that there might be several versions of the command.*

These are examples of learning by induction, a procedure that is both powerful and problematic. When learning is accomplished in this manner, its success depends critically upon getting the right sequence of examples (see especially the work of VanLehn & Brown, 1980, and Winston, 1975). This implies that the problems faced by the learners won't be solved by such things as improvements in the system image. Rather, we need to contrive ways to get users to try some alternatives—and just the right alternatives at that.

Minimizing Errors That Result From Slips

Analysis of slips can yield suggestions for the design of systems that minimize the incidence of this class of errors. One such analysis was done by Norman (1983b): We summarize here the major findings of that study. There are several categories of slips, but the most relevant for system design are mode errors, capture errors, and description errors. Let us examine each briefly.

Mode errors suggest the need for better feedback. A large class of errors are classified as mode errors: doing the operation appropriate for one mode when in fact you are in another. Modes will be with us as long as we wish to be able to do more operations than it is sensible to have special keys for. Most complex devices end up having them, be they digital watches, aircraft automatic pilots, or text editors. Even "modeless systems" have modes, but disguised. *Smalltalk*, the text editor *Emacs,* the *Macintosh* operating systems: all have numerous states where the first operation performed changes the interpretation of subsequent keypresses or mouse clicks. Avoiding modes entirely is not practical, but one can minimize modes. Feedback is essential so that the user can tell in what mode the system is in: Make sure that the modes are distinctively marked.

Description errors suggest the need for better consistency. A description error occurs when there is insufficient specification of the action, and the resultant ambiguity leads to an erroneous act being performed. The name emphasizes the fact that the description of the action was correct, but insufficiently specified, so the act that got done in error matched many characteristics of the correct one. One class of description errors occurs when a person attempts to rederive an action sequence and does so improperly, forming a sequence appropriate for an action different from the one intended. This occurs primarily through a lack of consistency in command structure, so that the appropriate structure for one command is not the same for another, even though the commands appear to be related and share a common description of purpose, action, and even part of the command format. The basic concept involved here is that when people lack knowledge about the proper operation of some aspect of a machine, they are apt to derive the operation by analogy with other, similar aspects of the device. The "derivation" may be unconscious, and it can influence behavior without the person realizing that it is happening.

Capture errors suggest the need for better feedback. A capture error occurs when there is overlap in the sequence required for the performance of an infrequent and a frequently performed action, and in the course of attempting the infrequent one, the more common act gets done instead. One possible way of avoiding this class of error is to minimize overlapping sequences, but this may not be possible, especially when the infrequent action sequence is simply a modification of the frequent one. A second way of avoiding the error is to try to catch it where it occurs. The error occurs at the critical place where the sequences deviate, so it is here that the problem must be faced. If the system knows what the intention of the user is (perhaps by requiring the user to indicate the overall intention), it could be designed so that at the critical choice point the proper path was flagged or in some other way brought to the attention of the operator. In addition, sufficient feedback about the state of the system should be provided to offer reminders as to the deviation from the intention. A major issue here is simply to know the critical place at which the errors occur so that remedial action can be built into the system at that critical point.

DETECTING ERROR

Detecting an error is the first step toward recovery. Early detection of error is extremely important. Delayed detection obviously wastes time and effort as the user works through steps that will not be effective because of an earlier error. As steps pile up, the diagnosis of the error becomes harder: The more that has happened, the more places there are to look for something wrong. This is especially a problem for the new user, who cannot partition what has been done into the surely right and the possibly wrong—the "bootstrapping" problem discussed in the Introduction to this section of the book. And in most systems the repair aids, "undo"s or inverse operations, often can only be applied right after the error, or within a small number of steps of it. Making errors show up clearly and quickly is a key design goal.

Slips are easier to detect than mistakes. In a slip, the action that was performed differs from the action intended: Detection occurs by comparing the outcome with the intention. In a mistake, it is the intention that is wrong, so if the person compares the outcome with the intention, it matches, so the person is apt to say, "yup, things are going just fine." As a result, it can be remarkably difficult to discover an error. Slips, mind you, are not always easy to discover. A major problem in the discovery of slips is the problem of levels: The level at which actions take place in the world differ from the level at which the intention is formed. Detection of misdiagnosis is hampered by a *cognitive hysteresis*—the tendency to stick with a decision even after the evidence shows it to be wrong. [1] As we argue later, this cognitive hysteresis appears to result from a bias to search only for confirming evidence, the danger of partial explanations, and the similarity between the actual and the perceived event. Developing a system that can provide sufficient information to help in effective diagnosis and in getting out of the cognitive hysteresis should be a major design consideration.

How Should the System Respond?

Now let us get to the heart of the matter: How should the system respond when it can not interpret the information given to it by the user? There are two goals here: one is to figure out what is intended so that the system can proceed with its operations; the other is to warn when something inappropriate has taken place (or is about to take place). Let us first examine this latter goal, that of providing a signal to the user thus signifying that something has gone wrong.

One basic response that helps signal the occurrence of difficulty is to construct the system in such a way that difficulties prevent continued operation. We call this property a *forcing function*: something that prevents the behavior from continuing until the problem has been corrected. This guarantees self-detection. Suppose you get into your automobile and start to drive away, but without ever starting the engine (the error does actually occur). This error cannot go undetected—the car simply will not move: This is a forcing function and the error is self-correcting. The fact that something is wrong is brought forcefully to attention. Forcing functions are sometimes carefully constructed by the designer, but can often fall out of the system operation naturally—as in the example of the automobile keys.

Note that the forcing function is not to be thought of as an error message. Rather, the sequence of operation is so designed that one simply cannot proceed, but the observation occurs naturally, in the normal course of events. No error bells or messages need occur in the automobile example. Starting the engine is a natural prerequisite to driving the automobile: The lack of movement serves as the indication that something has gone wrong. It is not always possible to develop such natural sequences. Thus, removing the ignition keys from the

---

[1] We use the term *cognitive hysteresis* because it takes less information to reach a particular interpretation of the situation than it does to give up that interpretation. Indeed, it sometimes appears that people can form judgments of a situation rapidly, but that it takes an enormous amount of information, time, and energy to cause them to discard that initial judgment. The psychological terms "set" and "functional fixation" refer to related concepts.

automobile at the end of the trip is not on a natural sequence of prerequisite events. As a result it is not uncommon to leave the keys in the car. (Requiring the use of the keys to lock the auto doors may seem inconvenient to some, but it does provide a natural forcing function, making it impossible to lock one's only set of keys in the car: an event that happens with some frequency in cars that can be locked without the use of keys.)

Forcing functions guarantee detection of a problem, but not proper identification of it. Failure to load the proper floppy diskette into a computer has a forcing function: The program will not work properly. However, the user may be only aware of the fact that something wrong has happened, but unaware of the cause. Determining the cause requires more feedback from the system.

> *In the early days of the UNIX text editor* Ed *it was possible to spend hours on the terminal editing a file, then quit—failing to save the work in permanent storage (the computer disk). As a result, all the work would be lost. (The event happened more frequently than might be expected.) The solution was to add a forcing function: An attempt to quit without first saving the text will fail. In the UNIX* vi *text editor (a grandchild of* Ed*) the failure is accompanied by the message:* No write since last change (:quit! overrides). *This response is both a forcing function and an explanation, describing why the command has not been executed, and stating how the command can be done, if that is really what is desired. On many small business and personal computers, it is still possible to stop editing without saving the work by removing the diskette. The Macintosh computer attempts to prevent this with a forcing function: The diskette cannot normally be removed unless the application has been terminated properly. This still does not prevent the user from simply turning off the power, however.*

A forcing function is simply one of many responses the system can make when it senses difficulties. In fact, we identify six possible responses: *Gag, Warn, Do Nothing, Self Correct, Teach Me,* and *Let's Talk About It.* In the next sections we discuss these six kinds of responses.

Gag. The automobile that will not move unless the key is in the ignition provides a good illustration of the technique we call *gag.* "Gag" is a forcing function, one that deals with errors by preventing the user from continuing, thus preventing the user from expressing impossible intentions. Some systems deal with error by locking the keyboard, preventing further typing until the user has "reset" the terminal. This is an effective gag, but one that at times appears to use brute strength when a friendly reminding would do: If the goal is to antagonize the user, this method works well.

A more successful illustration of "gag" comes from a tutorial language system called FLOW, developed by Raskin (1974). In this system, commands are typed in and processed one character at a time. If the user attempts to key a character that does not have a continuation into a legal command, the character is not accepted. You just can't type anything that isn't a legal command in the given context.

> *FLOW was an elementary programming language, designed by Raskin for beginning computer users from the fine arts, students who normally cringed at the thought of mathematics or computers. Raskin used to demonstrate this system by closing his eyes and pounding upon the keyboard, gleefully pointing out that not only did no system error messages appear, but that at the end, there would be a legal, syntactically correct program. (Of course the program was meaningless.) The result, explained Raskin, was that his students were not afraid to experiment. The lack of error message associated with syntax and text entering increased their confidence. (An excellent tracing program that stepped through the program line-by-line, showing which statement was being executed and what result would occur, helped the understanding of the semantics of the language.)*

"Gag" transfers the users' concern from trying to do things to trying to say them: The front-line problem is typing something the system will accept. Rather than letting the user try out various intentions, which may not work, the system intervenes to block the expression of the intention, perhaps before it is fully formulated. This may interfere with learning, especially learning through experimentation. But, as in Raskin's FLOW system, it can also relieve the frustrations associated with low-level typing errors, allowing the user to concentrate upon the major themes to be learned.

Warn. Several types of warnings are possible. In many automobiles, a buzzer warns when the seatbelts are not fastened, but the auto can be operated even if the warning is ignored. In some computer systems, similar warnings tell of files that are not write-protected, or other conditions that are not set up in normal fashion. Here, the philosophy is to note the occurrence of a potentially dangerous situation, to tell the user, but then to let the user decide how to respond. "Gag" forces itself on the user: "Warn" is less officious.

The Macintosh interface uses a related approach. Here, the menus show all possible actions, including ones that are not legal at the current state. The user is warned of the illegal actions: The commands not available at the moment are signaled by being shaded in gray as a warning that they cannot be selected at the moment. The user is not forced: The user can still attempt to select the command. In some sense, "warn" is presenting the error message before the fact, and as such might also be considered as an error-preventing procedure rather than as an error message.

Do nothing. Some direct manipulation interfaces use the *do nothing* approach. If you attempt an illegal action it just doesn't work. Nothing special happens, and you are left to infer that you tried something forbidden. The lack of movement after the attempt to drive an automobile without starting the engine is an example of a "do nothing" response. In the *Pinball Construction Set* (Budge, 1983), if you try to change the color of an object whose color can not be changed, you can go through all the motions, but nothing happens. "The paint won't stick" is how one user characterizes it. Note that the users of neither the automobile nor the construction set consider the "do nothing" response to be an error message. It is simply the way the system works: In the real world, some operations don't work on some objects. The pinball world is similar: Not all operations have effects. One could argue the merits of the "do nothing" strategy, but in this particular type of situation it seems to be very effective, both in conveying the error and in avoiding any disruption or ill feeling on the part of the user.

The "do nothing" method relies on visibility of the effects of its operations to convey the gap between intentions and outcomes. "Do nothing" is mostly seen in direct manipulation systems. But it isn't linked to the direct manipulation concept in any fundamental way, except that if it is to be effective, the lack of action must be readily apparent: In the terms used in earlier chapters of this book, the Gulf of Evaluation must be small or nonexistent.

> *Not all system designers seem aware of the requirement that "do nothing" can be used only where there is good feedback. Thus, in the Berkeley 4.2BSD UNIX system, the "remote copy" command, "rcp," copies files between computers. It*

> *requires a minimum of two arguments. If only one argument is given, the command fails, but with no indication. If two arguments are given, but neither refers to a legal file, the command still appears to work. "rcp" follows the "do nothing" philosophy of error message, much to the frustration of the user. UNIX does not provide automatic display of the state of its file system, so in this case the "do nothing" philosophy is inappropriate. (The normal copy command, "cp," does not misbehave in this way: It responds to a non-existent file with a message, as shown by the last example at the start of the chapter.)*

"Do nothing" is the simplest error technique and, when used properly, seems to have some important advantages. The user stays focused on the domain of actions and their effects, rather than being drawn out to process error messages. The user may have to do some experimenting to discover why what was tried didn't work, but this has its good side: The user is kept in a mode in which trying things out seems reasonable.

"Do nothing" also seems superior to "warn" in situations as complex as the *Pinball Construction Set.* Thus, even though when the user invokes the paint brush, the system knows that the intention is to color some objects, but when the display is complex with a large number of objects, an attempt to "warn" which objects were colorable and which not might result in a display that was visually more confusing than revealing.

Self correct. Another approach is for the system to try to guess some legal action that might be what the user would like to do. Simple spelling correction is an example of this approach. The ultimate in such systems is probably *DWIM: Do What I Mean,* the corrector available on the InterLisp system (Teitelman & Masinter, 1981). Note that automatic correction facilities, such as DWIM are only acceptable when there is a good "undo" facility, so if DWIM does something inappropriate, it is easy to "undo" it. As the designer of the system (Teitelman) once put it: "If you have made an error, you are going to have to correct it anyway. So I might as well have DWIM try to correct it. In the best case, it gets it right. In the worst case it gets it wrong and you have to undo it: but you would have had to make a correction anyway, so DWIM can't make it worse."

DWIM becomes a way of life, a fundamental part of InterLisp that changes the way people think of the system. DWIM is powerful enough that it is used deliberately. Because DWIM remembers corrections given it, one can deliberately introduce a new term that will cause the system to call upon DWIM, then tell DWIM that it means whatever it is that was really wanted, and forever after type a simple term instead of a more complex one, letting DWIM "correct" the term, simplifying the typing. DWIM has even been used to introduce Algol-like notation to Lisp: When the user typed in Algol expressions, DWIM "corrected" it to Lisp syntax. DWIM also can be annoying, for it can sometimes get carried away, correcting things that ought not to be corrected. Some users swear by DWIM, some swear at it. As an error correcting facility, DWIM seems superb. Its negative features result from the overgeneralization of its facilities. Although DWIM can be asked to query the user before making changes (and it can be turned off entirely), the problem is that the user wants some of the features, but not all, and this is not easy to arrange.

DWIM—and presumably other "intelligent" error correction systems, can sometimes be too smart, frustrating the user. Here is one user's description of the problem:

One especially annoying thing that DWIM can do is this: You are composing Lisp code and in true Lisp tradition, you want understandable variable and function names. Suppose you define and call a function "(get-next)." DWIM sees the expression and helpfully decides that you have made a typing error: You clearly mean to subtract "next" from "get" but forgot both the spaces and also the proper Lisp notation that requires the action to be listed first. So, DWIM politely, elegantly transforms your name "(get-next)" into the Lisp code "(minus get next)." When you try to run the program, you get an error message: "get" was discovered to be an unbound variable, found while Lisp was executing IDIFF, the routine for taking differences between integers. The problem, of course, is the hyphen in the function name. Using hyphenated names is common practice in other Lisp environments, but is to be avoided in InterLisp. There the convention is to capitalize initial letters, so the function would be called "GetNext" instead of "get-next." One learns these conventions after being burned a few times, but the first few times it happens, it can be very difficult to debug. Remember you never thought you mentioned a variable "get" and perhaps you didn't do any subtraction. Worse yet, maybe you did do some subtraction, and so you waste much time trying to fix the place in your program where you were subtracting. Argghhhh!

Simpler instances of "self correct" can be found. Some operations are insensitive to the order of arguments, yet the command insists upon a particular ordering. This requirement could be weakened, or at the least, the system could try all orders and ask the user if it was OK. Calling the text editor upon a nonexistent file is usually treated through a "self correct" philosophy. Most editors interpret this as wanting that file to exist—a guess that is often correct, and when not, not costly to repair. "Self correct" can probably be used a lot more than it is. There are numerous simple cases where not much intelligence is required to determine what the user might really have meant, if only the designer would have thought to try.

Let's talk about it. Some systems respond to problems by initiating a dialog with the user: We call this strategy "Let's talk about it," and we consider it a major step forward toward true interaction. A good example is the way that many Lisp systems respond when they get into trouble: a message describing the problem to the best of the system's ability, but then, automatic entry into the "Lisp Debugger" allows the user to interact with the system and explore the various candidates for trouble. For the class of experienced users for whom such systems are designed, there is a high likelihood that the user can simply fix the problem and proceed. Here we have shared responsibility between user and system to explore the problem and come up with a solution.

Teach me. And finally, there is the response we call "Teach me," the system that queries the user to find out just what a particular phrase or command might have meant. In the best of these systems, for example, *Clout* (a natural language inquiry system for the *Microrim* relational database system), when the system finds a word it does not understand, it asks the user for a definition. Any words in the definition that are not understood are then themselves asked about. This continues until all the terms have been defined. The new terms are then retained by the system so that the same phrases can be used in the future.

The Problem of Level

One major problem associated with error messages is to convey just where the problem is. It is often the case that the system detects an error, the user recognizes that an error has been made, but there is still considerable difficulty in determining just what the error is. A major factor is the existence of numerous levels of intentions. The situation that led to the "longjmp botch" error with which we opened this chapter illustrates the problem. It was first found while formatting files on the terminal, using a program that internally invoked "more," the screen-display program. "More" got into trouble and collapsed most ungracefully, displaying the message *longjmp botch: core dumped.* What is a "longjmp botch?" Why is the user being told this? Of what use is this information?

One problem with this message is that it is presented at the lowest level of program execution whereas the user is thinking at a fairly high level of intention: The user is formatting a file, unaware that "more" is even involved. But the error message does not give any hint as to what level the trouble is at: The user assumed it was a problem either with the text file or with the formatter.

Suppose a task is performed, but the end result is not satisfactory. Why not? The reason could lie at any level. There are many places for error, many places where intentions could fail to be satisfied. If the operation were carried out manually, one step at a time, then it might be relatively easy to detect the place where the problem lies. But in many situations this is not possible: All we know is that the intention has not been satisfied. Many of us have experienced this problem, spending hours "fixing" the wrong part of a program or task because we did not have the information required to judge the level at which the problem had occurred. The question, however, for the system designer is: What information is most useful for the user?

The question is very difficult to answer. For the system programmer who is trying to debug the basic routines, the statement *longjmp botch* might be very useful—just the information that was needed. The problem is not that the error message is inappropriate; the problem is that sometimes it is appropriate, other times not.

One approach to the levels problem is to know the intention. If the program knew it was being used by a person who was intending to format a file, it could make one set of responses. If it knew it was being used by someone trying to track down a problem, it could make another set. However, although knowing user intentions and levels often helps, it does not guarantee success. In studies of human errors there are found numerous cases where knowledge of the intention would not help. Consider the following example:

> *X leaves work and goes to his car in the parking lot. X inserts the key in the door, but the door will not open. X tries the key a second time: It still doesn't work. Puzzled, X reverses the key, then examines all the keys on the key ring to see if the correct key is being used. X then tries once more, walks around to the other door of the car to try yet again. In walking around, X notes that this is the incorrect car. X then goes to his own car and unlocks the door without difficulty.*

It is easy to generate a large collection of examples like this, some involving cars, others apartments, offices, and homes. The common theme is that even though people may know their own intentions, they seem to tackle the problem at the lowest level, and then slowly, almost reluctantly, they pop up to higher levels of action and intention. If the door will not unlock, perhaps the key is not inserted properly. If it still won't work, perhaps it is the wrong key, and then, perhaps the door or the lock is stuck or broken. Determining that the attempt is being made at the wrong door seems difficult. Now perhaps the problem is the error messages are inappropriate: The door simply refuses to open. It would be better if the door could examine the key and respond "This key is for a different car." Can programs overcome this problem? (See Chapter 16 by Draper for one possible direction—a solution based on the notion of "responsibility.")

The "Let's Talk About It" strategy provides a possible solution to the problem of levels. Suppose the message indicating difficulty was always presented at the highest level, indicating that there has been a problem and its seriousness. Then, let the user be given tools to explore the problem to whatever depth is desired. Let the user be able to trace down the levels, to see where the original mismatch occurred, how that level was reached, and the state of the system at each level. The programmer can explore in detail: The experienced user can explore until the basic problem is clear, and the uninterested can remain uninterested. Such exploratory tools will require some skills to be constructed, but the multiwindow, run-time support for Lisp Machines (and even Macintosh Pascal and Basic) offer suggestions of how to proceed. But this kind of constructive interaction with the system has the possibility of solving the levels problem along with the error message problem.

The Failure to Detect Problems

Often actions will accomplish a legal effect, and so cannot be detected as errors by the system, but are still problems in that they neither advance the user towards the goal nor do what the user intended. As we saw in the chapter by Lewis (Chapter 8), the users' ability to detect such states is unreliable. Word-processor learners can explain away even disastrous errors: They adjust their idea of what should be happening to fit what appears to be happening. Our studies of error bring out the same phenomenon: Serious failures may not be detected even though the available evidence appears (in hindsight) to be fully adequate. Norman (1984b, 1986) proposes three hypotheses that may account for these difficulties in detecting errors: relevance bias, partial explanation, and overlap between model and world.

Relevance bias. People seem to have an apparent bias toward seeking confirmatory evidence when evaluating a hypotheses, though looking for disconfirming evidence is often more useful (Johnson-Laird, 1983). This bias may be forced by the need to select evidence to consider from a large pool, too large to permit full examination. In selecting evidence one has to make judgments of relevance: One cannot afford to waste time processing irrelevant evidence. But judgments of what is relevant must be based on what one thinks is happening: in short, based on one's hypotheses. If confirming evidence is more likely to look relevant to a hypothesis than disconfirming evidence, the bias toward confirmation can be reinterpreted as a bias toward relevant-seeming evidence.

An example discussed by Lewis (Chapter 8) seems to fit Norman's analysis. When the learner searched for a page number and found a line number instead, the learner was satisfied, even though the label on the number made no sense in this context. The message looked relevant because it contained a number. And because it looked relevant, its very presence lent support to the hypothesis that it was what was wanted.

Partial explanation. The second conjecture is that errors are not detected because people accept crude agreement between what they expect and what they see. The example just discussed shows this mechanism at work as well. The learner interpreted the number in a way that accounted for the number, but not for the rest of the message. As noted, the rest of the display message indicated that the number was a line number, not a page number, but the learner did not require a complete interpretation. Like relevance bias, acceptance of partial explanations can be seen as a necessary response to limitations in processing capacity. One simply could not function if everything in the environment had to be fully interpreted.

Overlap of model and world. The final suggestion is that error detection is impeded by the fact that one's model of the world is likely to agree to a great extent with the way the world really is, even if the model is wrong in part. If one's model were too far off it would have to be adjusted. As a result of this rough agreement, finding points of disagreement is hard. Most things that one does will produce the results predicted by one's model, which is thus supported.

CORRECTING ERROR

If an error is detected the user needs to recover from it. Some existing systems provide general (or intended to be general) "undo" facilities, so that one can back up to an earlier state after making an error. Other systems aim to structure their operations so that they have natural inverses: Erasing a line is the inverse of drawing, and vice versa. It seems clear that, at the very least, *undo* is a desirable element of a system. It allows users to experiment more freely, it reduces anxiety and tension, and it permits ready recovery of many simple errors. Implementation of *undo* is nontrivial, however. An "undo" command that recovers from the last command is relatively easy, but many operations that the user thinks of as a single event can launch several different actions, only the last of which will be recoverable from with a single-step undo. But as soon as we go beyond recovery of the last operation, numerous technical and conceptual issues arise that have not yet been solved. Even single-step undo routines can be expensive of resources if one wishes to be able to recover from any command, including writing and reading from files and accidental or erroneous exits from the program.

Nievergelt has argued strongly for the essential kind of information a user must have in order to recover from error (Nievergelt & Weydert, 1980). Essentially, the user needs to know what the current state of the system is (*sites*), where one came from (*trails*) and what the possible alternatives are (*modes*). He bases the design of his experimental system around these three concepts: sites, modes, and trails.

The "Let's Talk About It," Teach Me," and the "Self Correct" strategy of dealing with problems seem like important directions to pursue. Certainly human-to-human communication makes very heavy use of these techniques.

CONCLUSION

In this chapter we have examined some of the ways that a system designer can deal with error. We think that error is the natural result of a person attempting to do a task. An error results for several reasons, including accidental mishaps, slips of action, and sincere mistakes. In many cases, the erroneous action represents the person's best attempt at the desired action. It is this case that most interests us, for here is where the spirit should be one of genuine cooperation. The user tried, and got it wrong. Think of the input not as an error, but as the user's first iteration toward the goal. The system should do its best to help. If it can figure out what was intended, so much the better. If not, it should explain gracefully where its problem is, perhaps making suggestions. It should not require retyping a long sequence or redoing a long set of operations when only one detail was in error. Human conversation does not require such correction.

Sometimes the responses of the person are not easily detected as being in error. That is, they are legal operations, but not appropriate for the situation. This is a more difficult case, for only the most intelligent of systems can detect this situation, and probably not even then. The solution here is to provide sufficient feedback and exploratory aids that the person is aware of the actions and of the system state. It should be easy for the user to examine the system, to find the consequences of actions, and to modify those actions. Undo facilities are desirable, but undo with the extra information that allows for intelligent use of the facility, including knowledge of how the operation got to its current state and what alternatives exist.

At times the best strategy is for the system simply to provide warnings, or not to respond at all. Sometimes it is possible to correct the error without further input from the person, although because the system too can make errors, this should only be done with consultation and permission. More important, it is often possible to avoid whole classes of error simply by reformulating the problem, eliminating aspects that would otherwise have caused trouble. This chapter does not attempt to be comprehensive in its treatment. The chapter is meant to provide the spirit of the enterprise: that error should be considered a normal part of operation, not an exception.

> *There already exists a system that accepts erroneous statements gracefully, usually managing to interpret actions correctly in spite of error, other times providing elegant correction procedures. The system?—human speech. Normal speech contains many errors and corrections, yet people have evolved such skillful procedures for correction that the listener seldom notices either the errors or the corrections. In a conversation, we do not complain of "syntax errors," even though there are many. Even simple errors of meaning can be tolerated if it is clear what was meant. When a speaker detects an error, the erroneous part of the utterance can be corrected while leaving the surrounding material unchanged,*

> *reducing effort and trauma by all concerned parties. When we fail to understand what a person has said we ask for correction. But we act as a cooperative team and minimize the effort required to deal with the error. Basically, we assume the person is trying to tell us something, so that even errors or local incoherence are still informative attempts. We treat understanding as a cooperative endeavor, requiring effort from both speaker and listener.*

If we think of interaction with a computer as a cooperative endeavor we see that each side has certain talents. The person is good at setting goals and constructing intentions. The computer is great on the details. Let the two work together, with the person in charge. And let the computer go out of its way to make things easy for the user, to make corrections easy, to go that extra step toward understanding that makes conversation with an intelligent colleague so fruitful.

CHAPTER **21**

Computer-Mediated Communication

LIAM J. BANNON

I noted in Chapter 19 that the focus of human–computer interaction tends to be restricted to the relationship between the individual user and the computer and I discussed a broader perspective on the field that would take into account the social context in which people use computers. In this chapter I expand on how the computer can augment our current communication facilities.

> *This approach is to be distinguished from another usage of the term "improved communication facilities" in the context of human–computer interaction, where the focus is on such features as natural language front-ends and better screen pointing devices to assist users in "communicating" with the* computer, *rather than with other* people, *which is the concern here.*

Once again, the focus is on improving the computer tools that are available for collaboration between people in an effort to exploit human capabilities more fully. This view extends the concept of human–computer interaction to that of human–computer–human interaction—namely, a perspective where the computer serves as a mediator between people. Within this framework, the computer takes its place as another piece of interactive communications technology that can be analyzed along with other electronic media such as the radio, television, and telephone.

One of the few people who foresaw the revolutionary potential of the computer as a medium for improving idea development and group communication was Douglas Engelbart, who conceived a project entitled "Augmenting the Human Intellect" at Stanford Research Institute in the 1960s. (See Bannon, 1985, for further discussion of Engelbart's work and other issues raised in this chapter.) His goal was to provide " a way of life in an integrated domain where hunches, cut-and-try, intangibles, and the human feel for a situation usefully coexist with powerful concepts, streamlined terminology and notation, sophisticated methods, and high-powered electronic aids" (Engelbart, 1962). Engelbart wanted to build a new kind of computerized working environment in which the emphasis was on how people could achieve significant gains in productivity as a result of the computerized support made available to them. Integral to Engelbart's scheme was the provision of computerized support to enhance communication between people. He wished to provide a complete new environment for "knowledge workers," an information space through which the worker navigates, and in which the worker can become totally "at home." This scenario is distinct from the more common perspective that attempts to provide the worker with a set of isolated computer tools. Having people work and live in this environment would, in Engelbart's view, ultimately lead to new insights by users of his system into the nature of problems and the evolution of a more capable society to deal with these problems.

One clear separation between Engelbart's viewpoint and that of many others working in the computing field was that he sought to develop a synergy between the computer and the human, a situation where, through the use of sophisticated tools, the human could gain new insight into problems. This was in contrast to many other system designers who focused more on the total automation of many human tasks. Another feature was that, although Engelbart's approach was criticized for a certain naivety concerning the difficulties involved in changing traditional modes of human behavior, he was aware of the need to look at a complete system—people working with computers in an organizational environment—in order to understand how technical developments might be used.

EXISTING SUPPORT FOR COMPUTER-MEDIATED COMMUNICATION

This section reviews existing support for collaboration activities, providing a rough taxonomy of the different kinds of support. It includes excerpts from responses to a query I had posted on several electronic networks concerning computer communication. I do not attempt to give a comprehensive survey of all systems available, but rather I show the range of computing facilities that can be used to facilitate person-to-person interaction and collaboration. In later sections I analyze some of the strengths and weaknesses of these facilities and relate them to different social situations.

Computer as Shared Facility

Simply having a computer system around, where people can prepare their papers and store their data, can increase collaboration. One person put it this way:

> Although we have only a few general-use data bases in the customary sense, there is a considerable amount of data-sharing in many forms—made practicably possible only by our super-mini-based computer system. And, programming packages for various applications (e.g., signal processing) often get developed for and by one group, then are applied by many of the others. Proposals and papers, more commonly than not, are prepared jointly with multiple versions of drafts being revised and edited back and forth between the various researchers in different fields. . . . people from a variety of different disciplines that, only a few years ago wouldn't have thought about sharing data or working cooperatively are, in fact, doing that now. This occurs despite the fact that we have only a rudimentary electronic mail system, and none of the nifty tools that supposedly provide for handy electronic interaction.

Electronic Mail Within and Across Systems

Many computer systems now have some form of "electronic mail" or computer messaging facility. All such systems allow people to send textual messages[1] via computer to another person on that system, or group of people, where the message is placed in a "mailbox" until it is read by the recipient. Obvious advantages of this medium over the physical mail system are the rapid delivery of the message, the ease of sending to groups of people, and the ease of editing and reviewing stored messages. Another reason for the popularity of such systems in both research and business operations is that there is no need for the recipient to be physically contactable at the time the message is sent, as there is if one wishes to establish a phone connection. Eliminating the game of "telephone tag" is often put forward as a key office productivity gain with electronic mail systems. Most of the systems allow for files to be included in messages, thus allowing several people to work on a paper, albeit clumsily, by passing mail back and forth.

Electronic mail really becomes interesting when it is not confined to mail between people on a single machine, but when it is possible to interconnect to other computers via computer networks, both local-area networks (LANs) and wide-area networks (WANs), some of them nation-wide and even world-wide, such as ARPANET, BITNET, USENET, MAILNET, etc. These national networks allow one to make contact via computer with a much wider circle of people than was possible previously. One person put it this way describing how important the ARPANET was in the development of the Ada language:

> Not only were the design and review processes mediated by the ARPANET, but the language design team was geographically distributed. Jean Ichbiah and several others in France, a group of key people in England, the administrative work and design of a test compiler in Minnesota, and several other key people in the United States, Germany, the Netherlands, etc.

Shared File System

Shared files are very useful in collaborative work, as they allow people to access and develop their work using the same set of files without having to coordinate the transfer of files or to use complex file transfer protocols. Safeguards to ensure that people are not updating the file simultaneously must be provided. An example:

---

[1] Systems are also available that support the delivery of spoken messages, for example the IBM Speech Filing System (Gould & Boies, 1983). Research is also underway to transmit voice and image information over long-haul networks.

> When I was at Xerox, we used electronic mail and shared electronic filing to great advantage—long-distance cooperation on documents was the rule, since about 2/3 of our people were in El Segundo and most of the rest in Palo Alto. I wrote a couple of papers this way, with coauthors down south; Star's Functional Description was maintained similarly . . . . The general approach was to keep the text in a public spot; give one author at a time the write access (often, divvy it up and let different folks be working on different parts. The "write access" was by consensus, rather than adjusting privileges—the privilege structure certainly would have supported it, but for papers we didn't bother. For code, it was another matter—that got formally checked out and back in again.) We'd comment and revise via electronic mail, with occasional check-ins where people got a new consistent version.

Computer Conferencing

The term computer conferencing is commonly used to refer to computer systems that provide extended facilities for keeping a record or transcript of all messages related to a topic, allowing one to set-up conferences and to browse through the topics and messages for each conference. There are usually facilities to send public and private messages, and to find out information on the conference participants. The use of the term "conference" is a bit misleading, as usually we refer to conferences as being in real-time, where all the participants are active at the same time. Although some of the systems provide a simultaneous or "real-time" mode, this is generally a less-used aspect of these computer conferencing systems.[2]

Computer "conferences" have the advantage that people can respond to topics in their own time and at whatever length they feel is appropriate rather than feeling pressured to "get in their say" before a regular meeting finishes. The quality of the conference is not determined by

---

[2] There are a number of well-known conferencing systems available, including the New Jersey Institute of Technology's EIES (Electronic Information Exchange System), designed principally by Turoff; the University of Michigan's CONFER system, developed by Parnes, the Institute for the Futures FORUM and PLANET system—marketed commercially now as the NOTEPAD system by Infomedia, Inc.—developed by Vallee, Johansen and others, and the Swedish COM and PortaCOM systems, developed by Palme and colleagues.

the ease of use or power of the technology, although that can play a minor role. More important is the quality of the participants and of the appointed moderator of the conference. It is the latter's task to ensure that the meeting stays on topic and to nudge people in the right directions without making their presence too visible or interrupting the flow of ideas. The technology of itself does not guarantee anything—the quality of the conference depends on how social roles get defined in conjunction with the technology. All of these systems have been the subject of testing and evaluations by various means—on- and off-line questionnaires given to the participants, analyses of usage statistics of the system, and participant observation (see Kerr & Hiltz, 1982, for a summary of these evaluations).

Terminal-to-Terminal Communication

Many systems have some simple means of "writing" or "sending" a note in "real-time" to another terminal—often used for short messages such as "want to go to dinner now?" They are generally not used for sustained discussion as the pressure of thinking "online" can be onerous and people get tired of typing conversational-style text; they would prefer to use the telephone. Early examples of such features include the TENEX Link facility that allows up to four people to be linked simultaneously, and the CDC Plato TALKOMATIC program that segments the screen into five windows, allowing up to five anonymous users to communicate, each in their separate, scrollable, window.

Another reason that people do not carry on any sort of sustained dialogue this way is because on some of these systems—TENEX **Link** and UNIX **write**, for example—the message on the screen can become garbled if the respondent starts to type as the initiator is still typing. A convention, or protocol, has to be established in order to cue the respondent when one has finished typing. Here, the opportunities for interruption are nonexistent. Some systems, e.g., the **talk** command in Berkeley UNIX, and Plato's TALKOMATIC, split the screen and allow simultaneous noninterfering typing, which makes for a significant difference in the perceived utility of the system, but only talking is allowed: You can't get access to another person's files, or see their screens, etc. One thing to note is how apparently minor changes to the system can significantly affect the utility of the system in the eyes of the user. Buxton (Chapter 15) notes how minor changes in the control mechanisms of input devices can significantly shift the complexity of certain operations. Engelbart, in reflecting on lessons learned from the development of the NLS system, has also commented on how important "lots of little things" are to the overall effectiveness of the system.

At quite a different level, the social utility of a **write** feature can be significantly affected by the social norms existing in the organization as to its use. I know of one case where the facility, although available, was rarely used, and discovered the reason behind this was not because of its ineffectiveness, but because an administrative person had strongly hinted that people did not like to have their screens messed up by people using the facility, and so it should not be used except in dire emergencies. Not surprisingly, the facility was therefore not used very much.

> *This example, although amusing, points out a standard trade-off. The* **talk** *facility is obtrusive, it can interrupt one's current activity unexpectedly, disturbing concentration. Alternative implementations of the facility might reserve a special screen area for such conversations that would reduce the scale of the interruption, although responding to the* **talk** *request necessarily entails a switching of attention away from the current task. There remains the issue of the subtleties available within a medium to express shades of meaning—the binary choice of having the message facility activated or not on your terminal does not come close to the subtle messages that it is possible to give in an office environment through varying the position of one's office door.*

Shared Screen Facilities

A powerful tool is a facility that allows one to link to another terminal in "real-time" and actually see what is on the other's screen.[3] There are several variants of the simple shared-screen capability. Some systems allow control to be passed back and forth between the participants, some allow more than two people to participate, some provide a chat facility together with the shared screen, some allow for access to files and programs from the remote terminal, etc. These facilities come close to implementing a comprehensive shared workspace for participants, where people can point to things, edit, and run programs jointly, in real-time, as discussed by Thompson (1984). Adding in an acoustic link could significantly enhance the utility of such features, as typing

[3] Several systems provide a version of such a facility: the ADVISE command on several DEC systems; TERMTALK on the CDC Plato system; CONFERENCE on the TYMSHARE AUGMENT office system (the descendant of NLS); CVIEW on IBM machines.

speed is a limiting factor in "real-time" terminal communications. The power of such facilities might be best demonstrated by the following example:

> We have a junior programmer working on the project I am "leading" (the programmer is located 2,000 miles away from the project leader who is reporting this incident!). We are working on a Tenex-like system. I wanted to find out how things were progressing . . . so instead of calling on the phone I found her on the system and linked to her. It came out that there was a problem with a piece of code she was working on. It was giving an error message that a certain record wasn't in the database, even though it was. She couldn't find anyone in the office who could tell her how to figure out what was wrong (it wasn't the obvious things). So I converted the link into an ADVISE link. (. . . Tenex ADVISE puts the "advisor" into a state where his keyboard input is put into the input stream of the advisee, as well as the shared output function of LINK.) Then I told her (via comment command) to show me the right source things. She jumped to the right code on the screen and I then gave some more commands to jump around to other procedures being called, etc. Didn't find anything obviously wrong. So I told her to get the thing running under our debugger and put a breakpoint at a certain place and do whatever it was you do to get that code to be executed. So she loaded the program and activated the debugger and went into the program and gave a command which resulted in the breakpoint being hit. Then I started giving debugging commands and scouted around in the executable image and eventually displayed the filename associated with a certain "statement identifier" (never mind what that is). As soon as I did that she said (typed) "Oh, I know, . . ." etc. And also that she didn't understand some of the debugger commands I gave. So I told her what I had done and why. And now she knows what needs to be fixed. . . and we broke off with gossip, goodbyes, etc.

With the capabilities of bit-mapped displays today, even further sophistication in workspace sharing is possible. Rather than a simple duplication of the bit stream to two terminals, possibilities for having only parts of the displays being shared are possible, allowing individual users to keep control of various parts of their screens.

Related Media

The computer can be used as the sole medium of communication, as in a computer conference between participants who are physically remote from each other, or it can be used to augment regular face-to-face meetings, or meetings that rely on other technology, such as audio or video. I would like to mention briefly how computer support might assist at a regular meeting, where all the participants are physically co-present in the same location, and also link in discussion of computer-mediated communication with the larger literature on communication media before refocusing on the computer medium. Consider this scenario:

> Imagine that each person at the meeting has an unobtrusive terminal available, allowing access to personal files. A large-screen display at the front of the room is used to display material from any of the participants screens. There is also an electronic blackboard, where people can write and make diagrams that are stored for later use. There is a project meeting being held about the deadline for a new product. Tom starts the meeting, and with a few keystrokes made on the standup terminal at the front of the room brings up on his display some graphical information which he wishes to discuss with the group. It is possible for Tom to link the terminals of the other people in the room to his, but in this case, he decides it is probably better to use the large display in the front of the room as a common reference point, so he simply enters a command that projects his screen display onto the large screen. Initially Tom has control of the pointer on the display—controlled by a simple ring worn on his index finger, or by the mouse attached to his keyboard, but he can give control over the pointer to anyone in the group, either by handing over the ring, or by slaving the person's terminal to his, thus making the other person's mouse active and visible on the large screen. While Tom and Mary are discussing something on the screen, another member at the meeting, Joe, has his memory jogged by something that was said, so he starts privately to search his files for a particular item. On finding the item, he interrupts the main conversation for a minute, is given control, and displays his current screen image on the main screen for a few seconds, marking certain objects. He then reorganizes the display by shifting his image and

> reducing it, and bringing up the earlier display that Tom had been referring to so as to compare aspects of the two figures. Tom likes the comparison, makes some connecting lines between the two diagrams with his finger on the touch sensitive large screen display (he could also make these marks and notes using the mouse and keyboard), and then "freezes" the display for later use in his project. As well as having a dynamic record of the large screen display, he also has a reference to the origin of the information—a pointer to the file from which Joe brought up the new piece of information, should Tom need to go back to the source. Tom takes back control and starts to run some simulation sequences showing how various factors would be affected by changes in the due date for the new product. These sequences can be frozen, stepped back and forth, and annotated in the course of this activity.

Although aspects of this scenario are not technically feasible today, much of this capability was actually available and demonstrated on the NLS project. (See Licklider, Taylor, & Herbert, 1968, for a brief description.) Of course, there are a lot of potential problems with the scenario, not only from a technical viewpoint but from a "social engineering" viewpoint as well. For instance, how do we ensure that people are paying attention if they can be continually distracted by their terminal screens at the meeting? It would appear that new forms of coordination and a new etiquette for holding such a meeting would have to be evolved. Some of these kinds of functionalities, and perhaps more importantly, the underlying computer tools necessary to support them, are currently being developed on the Colab project at the Xerox Palo Alto Research Center (Foster, 1984).

Returning to the more usual scenario, where one or more of the meeting participants is physically separated from the other participants, then a variety of technologies—audio, video, or computer—might be used to establish a connection between the participants. There have been a large number of studies done on these different "teleconferencing" media, and a very useful summary of this work is available in Johansen, Vallee, and Spangler (1979). They note five fundamental characteristics of all of these electronic conferencing media that fundamentally affect the nature of the interaction: physical separation of participants; access to remote resources; narrow communication channels; potential for control of group interaction; and dependence on technology.

Obviously, a key factor missing in all of these mediated interactions is the sense of *social presence* (Short, Williams, & Christie, 1976) possible at a face-to-face meeting. This impoverishment can have serious consequences under conditions where participants are not well known to each other and might have very different goals and cultural backgrounds. On the positive side, teleconferencing can allow a group to include experts at remote points in their deliberations, expanding the knowledge base of the group. Again, this can lead to difficulties if the expert is perceived to be too distant in attitude from the group, and the expert can miss the context provided by the rest of the group.

ANALYSIS OF COMPUTER-MEDIATED COMMUNICATION TOOLS

Here I discuss some basic distinctions among computer-mediated communication tools and show how different tools might be matched to office functions. I finish on a speculative vein with the topic of electronic communities. One central theme that needs to be stressed is that uses of the new media evolve from the interplay of social and technical factors.

Communication or Collaboration Support?

Two trends can be noted in work on computer-mediated communication. One tends to focus on the capabilities of the technology and shows how certain features, for example computer conferencing, affect group communication patterns (Freeman, 1980; Hiltz, 1984). Another focuses on actual work situations and attempts to show how work might be accomplished more effectively through use of the new media. The latter does not focus on the effects of the medium per se, but on what aspects of the medium might be utilized to produce more effective tools for collaboration and coordination. Here the focus is not simply on establishing a communication link between people, but on augmenting the possibilities for interaction by using the computer to help coordinate activities and support joint problem-solving by providing shared workspaces and tools for annotating and writing documents. In this context, even as simple a facility as the personal electronic calendar that is selectively accessible to others can be an important tool to assist in the coordination process. These two approaches selectively illuminate issues in the field of computer-mediated communication.

Basic Distinctions

If we attempt to categorize the communication facilities available on computing systems, probably the most common distinction drawn is that between synchronous facilities, where parties are connected in "real-time," and asynchronous facilities, where there is no such requirement for parties to be simultaneously present on the system. Attempting to come up with a proper taxonomy of forms of interaction can be a futile quest, as it is possible to argue endlessly about the correct distinctions (see Bretz et al., 1976). However, most analysts accept the "real-time"/"nonreal-time" dimension as fundamental.

> *Scollon (1982) argues that the division of systems into real and nonreal time facilities may not be the critical feature for providing insight into our activities. He imports a distinction made by Erickson (1980) between* **chronos** *and* **kairos** *with respect to time-related activities.* **Chronos** *-time is clock-governed time, whereas* **kairos** *-time is time "geared to appropriateness." The former emphasizes independence of events, the latter, interdependence. The interesting issue is whether the property of being* **chronos** *-timed or* **kairos** *-timed is inherent in the medium itself. Scollon argues that this is not so, citing how reading a book—quite definitely an asynchronous communication with respect to the communication between the reader and writer—can be viewed as being either* **chronos** *or* **kairos** *timed. He claims that some kind of real-time is inherent in each medium, while any medium can be geared to either* **chronos** *or* **kairos**. *This would imply that the traditional "real-time"/"nonreal time" distinction is misleading as it emphasizes a perspective that takes technical features as the distinguishing characteristic. From a personal/social perspective, we can separate activities as being geared to* **chronos** *(the timeclock) or* **kairos** *(appropriateness), and communication technologies can assist in both kinds of activities. On one dimension, reading a book is a nonreal time activity, yet reading it for an exam tomorrow is a* **chronos** *activity, and reading it as bedtime reading is a* **kairos** *activity. Many technical facilities are focused on* **chronos** *-type interactions, yet might also have aspects of relevance to* **kairos** *-type interactions. Specifically, computer conferencing can be utilized very effectively to support* **kairos** *-type activities, as control of when to interact and to what extent are usually up to the person, and are facilitated by aspects of this medium.*

This way of viewing new media shows that focusing on the point of view of the participant, rather than on characteristics of the technology, may lead to a better understanding of the relevant issues in computer-mediated communication.

Another key feature of communication facilities is to what extent a transcript plays a key role in the activity (see Carlstedt's message on page 32 of Bretz et al., 1976). Bulletin boards and computer conferencing systems rely heavily on the transcript, whereas for some synchronous facilities such as **talk** it plays a very fleeting role.

Asynchronous Interaction

Extensive use is already being made of available asynchronous facilities such as electronic mail and computer conferencing systems, and the benefits of being able to communicate with people separated in time and space are obvious—people can choose their own time for writing and reading messages, rather than being forced to respond instantly. Although electronic mail does in theory allow the user to respond as and when they desire, social forces push users into responding as soon as the message is received. In situations where it is known that users are on the system at least once every day, a delay of more than 12 hours in getting an answer to a message is viewed as being "bad manners." So here we have an interaction between the nature of the communication tool and expectations about how it should be used. The nature of the medium provides some constraints on its use, but other constraints are brought in from the social/work context.

Of course, one consequence of this flexibility in response is that it can be disconcerting for the initiator of a message to wait for a varying time length of anywhere from a minute to several weeks in order to obtain a response. This lag in feedback can be disruptive, especially in a group context, and individuals can become uncommunicative as a result. Such lags in response time are not possible in a face-to-face, or telephone encounter. Indeed Wilson (1985) notes how one role that had to be provided in order to ensure a successful computer conference was to designate an "absence coordinator" in order to know when people were away from their work, so as to ensure work was not held up while people awaited a reply from the absent person. Maude, Heaton, Gilbert, Wilson, and Marshall (1985) also note that "there is pressure on individuals participating to contribute some message no matter how trivial, each time they log into the system, to maintain a presence."

A problem in any computer conferencing system is how to provide the user with a means of keeping track of where they are in the various conference groups that get spawned, what messages have been read, etc. Facilities for rapid browsing of discussions and retrieval of relevant messages are still quite primitive. Certainly, having a written record of the message traffic can be an advantage in certain instances, but it can also be inhibiting, as Johansen, Vallee, and Spangler (1979) note, especially in situations where delicate issues are being discussed. Sometimes one does not want everything "on the record." That is why most systems also provide a private message facility as well. However, if the bulk of communication switches to this mode, then some of the key aspects of the conference concept are lost.

The "signal-to-noise" ratio of the conference can be another problem. The emergence of open electronic networks has reduced the effort involved in sending messages to a large number of people, and as a consequence, many people are deluged with "junk mail." Similarly, in many computer conferences, one can page through many screenfuls of text before arriving at any substantive discussion. This is particularly true of the more open conferences. Of course, this can be controlled through limiting participation in conferences when they are organized around a specific work-related project, and through active intervention of the conference organizer to keep the discussion focused.

In discussing the EIES conferencing system, Hiltz (1984) makes an important observation that social characteristics of the group can affect the evaluation of the system. In other words, the technology is mediated by the social process and any evaluation must take this into account. For instance, the role of the conference leader was discovered to be a crucial determinant of group effectiveness (as perceived by the group members). This person is responsible for two kinds of activities: an *administrative support* role, orienting new members, etc., and a *conference management* role, getting feedback from group members about various conference arrangements, summarizing discussions, etc. Over time, the role of the leader can change, and the need for a clear "leader" may decline, with various people in the group performing different "leadership" roles as the occasion warrants.

Wilson (1985) and Maude et al., (1985) report how a joint activity—in this case, writing a paper—can be conducted solely through the computer medium. Of interest here is the observation that such an "electronic mailbox" system is seen as being especially useful in focusing members of the group on the task at hand (Maude et al., 1985). Others argue that keeping computer conferences "on track" is exactly what one does not want, as one of the advantages of the electronic medium is that multiple threads of discourse can be present in the conversations. Whether this aspect is one that should be fostered or not will be determined, to my mind, by the nature of the task one is trying to accomplish (see Black, Levin, Mehan, & Quinn, 1983). Thus, within the computer medium, it is possible to focus discussion if required, but it also has the potential to support this "multiple thread discourse" pattern if desired, which is a feature that is virtually impossible to obtain in face-to-face interactions.

Synchronous Interaction

On systems that provide both asynchronous and synchronous facilities, the asynchronous facilities are much more heavily used (Palme, 1985). That is not too surprising in my view. A simple "talk" facility has limited usefulness, as most of the functions it serves are duplicated and improved upon by a phone connection when available. However, the more powerful synchronous linking facilities do have a special role that goes beyond simple exchanges of opinions on topics, and can be very helpful in consultation and tutorial sessions. Confirmation of this comes from several correspondents:

> The ability to link terminals seems to be of most use for remote demonstrations, instruction, or receiving expert assistance. I know of lots of people who are strongly in favor of it for these uses and I have seen it used effectively in these contexts.

> There were occasional problems, which mail and long-distance phone calls didn't really ease. Better was for one of us to log-in on the other's machine and run the offending code with the other one linked on, occasionally suggesting debugging probes to try or alternative strategies.

Being able to have a virtual "shared workspace" in which both remote parties can be active, sharing control, and commenting on their actions, approximates the feeling of collaboration that goes on when people are hunched side-by-side over a pad of paper, a blackboard, or a computer screen. The point is not that synchronous facilities will be utilized all the time but that they are invaluable at certain times.

The 'Ecological Niche' Metaphor

Each facility has an ecological "niche" in the space of system support, most of which have already been well-established and accepted by the user community. Within the working environment, electronic mail lies between the phone call and the office memo with respect to its degree of formality. It is useful to try and keep this medium in such a position, rather than try to shift it in a more formal direction, as this would endanger the unique aspects of the medium, making it simply a substitute for the formal office memo. Brown (1983) emphasizes this point by arguing for a "de-speller" for electronic messages to prevent this gradual melding of one facility into another. One might want to argue about the seriousness of this suggestion, but it at least focuses attention on the importance of preserving the uniqueness of each medium of communication.

A key feature is how computer-mediated facilities mesh with other facilities—person-to-person meetings, phone calls, and video meetings. The outmoded view, which emphasized direct productivity gains to be obtained from substitution of personal communication, by electronic communication, has been replaced with a more realistic view that stresses how overall effectiveness can be improved by selecting the appropriate media for the activity at hand. Palme (1985) notes that on the COM computer conferencing system, only about 13% of the time users spent on COM was replacing face-to-face communication; 6% replacing mail and circulars; and 14% replacing phone calls; 65% was new communication. This kind of pattern has been observed by others as well (Kerr & Hiltz, 1982). We should be more concerned about the linking of the various technologies together to provide a coherent working environment for people than with attempting to force all communication through a single medium.

The majority of the studies mentioned above have involved groups of users taking part in quasi-experimental studies to gauge the effects of the new technologies. An alternative approach would be to take a real-world environment, for example, the office, and see how current office functions could be supported by the technologies. The purpose of the following section is to give one perspective on what the different niches might be for the available computer-communication media.

An Example—The Office Environment

In this section I sketch how computer tools might be selected in order to fit into and serve the needs of a given organization. The framework adopted here is adopted from Barns (personal communication, January 18, 1985). The basic model (shown in Table 21.1) makes a three-tier

TABLE 21.1
MATCHING COMPUTER TOOLS TO OFFICE INTERACTIONS

	Type of Interaction		
	Operational		Directional
Computer Tools	Bounded	Unbounded	
Electronic Mail	++	+	+++
Synchronous Tools	+++	+	+
Computer Conferencing	++	+++	+

taxonomy of system types—electronic mail, synchronous tools, and computer conferencing. Organizational interactions are divided into two main classes, Directional and Operational, with the latter further subdivided into Bounded and Unbounded categories. The number of + signs reflects the degree of match between tools and interaction types: +++ indicates high match; + equals low match.

Directional communications in the working environment map well onto the organization chart view of communications, which shows the lines of official authority. These communications tend to be of a directing/controlling nature, where the authority figure outlines the tasks to be done and leaves subordinates some flexibility in actually carrying out the assignment. Communications between the two levels are usually quite brief, and of a question-and-answer clarificationary nature. Operational interactions, on the other hand, are normally not shown on the organizational chart as often they do not define a specific mission with respect to the organization, but are more pertinent to the needs of individuals, often involving sharing of information with others, sometimes on a reciprocal basis. Learning to use the computer would be an example of this kind of activity. This might be supported by organized activities, such as training classes, but often here is where the pickup of information from coworkers is noted (Chapter 19).

In Directional interactions, the volume of information being transmitted is often small, even though it can be important (e.g., a message from the manager "I must have this program online by 2 PM tomorrow") and does not usually involve a lot of "learning" by the subordinates, as they are supposed to know how to perform certain tasks. In Operational interactions the amount of information is variable, and much learning may go on between individuals or groups as there may be little shared understanding initially. The further subdivision into Bounded and Unbounded contexts relates to the fact that in some cases one needs to fill in information within some framework, whereas in the less-bounded situations one is actively seeking information but is unsure of exactly what all the relevant constraints are.

Given the taxonomy of organizational interactions and media types, how do they map on to each other? The major mappings are shown in Table 21.1. *Electronic mail* seems to be most relevant to interactions of the Directional sort as they handle small amounts of information well and users do not have much to learn in order to use the system. They can handle short question–answer interactions, if required (the "answer" or "reply" command available on many message systems supports this explicitly). They are less suited to Operational interactions, as it is hard to find out if similar questions had been asked and answered before (no historical record), and the separate spaces of mailboxes makes it hard to keep track of what might have been discussed by certain people before—as distinct from having a common database of information.

Synchronous conferencing tools seem most appropriate for bounded Operational interactions. The rapidity of exchange is fast, as one person can interrupt the other if a change in direction is required or the issue redefined. If the computer is itself being used to solve the problem posed in the interaction, then switching between the task itself and comments on the task (meta-description) are possible with minimal delays—as in face-to-face discussion. These tools are probably less good for unbounded Operational interactions as they only work well for limited contexts: They require more learning than electronic mail, and they do not support long background messages that might be required for certain interactions.

Computer conferencing tools seem well suited to unbounded Operational interactions because they have many facilities for cross-linking information, and they support enquiries that seek out background information on a topic. They are also more suited to handling large volumes of information and projects that exist over a long timespan that would be impossible with synchronous facilities. They are less suited to the other two interaction types as the overhead required for the cross-linkages is unnecessary in Directional interactions. Within a particular context, there is a danger of having much "useless" information within the conference format.

In sum, for relatively straightforward interactions electronic mail seems most appropriate. For more open, inquiring interactions, the much richer structures supported by conferencing systems seem worthwhile, even though the overhead in learning time is much greater. Further, more detailed, analyses of office interactions could provide the basis for developing aspects of this model and serve to show what kinds of system supports might be appropriate or need developing.

POSTSCRIPT—ELECTRONIC COMMUNITIES

To some analysts, new technological capabilities in the area of communications will have a profound impact on our society, creating the "Global Village" of McLuhan and the "Global Information Society" of Masuda (1982). These authors expect that the technologies will cause changes in personal and organizational work patterns. Debates on "telework" and opportunities for reductions in travel as a result of new technology often adopt this perspective. Such a position assumes that technology is the key force acting to change the nature of today's society, a position labeled *technological determinism.*

The alternative account, which I favor, places greater emphasis on the large number of factors that effect change in today's society, of which new technology is just one. The technology does not determine how society will evolve, but it does provide new possibilities. How society actually evolves will be determined by the complex interplay of social forces and technological opportunities. In the context of our current concerns with computer-mediated communication, the technology affords certain possibilities for affecting our current communication practices, and changes will occur, but they will evolve in the context of the decisions—not all of them rational—made by both individual, organizational, and societal actors. This does not thereby imply that we cannot attempt to use the technology in innovative ways, it simply warns of the limitations of Utopian thinking that is overly guided by technological possibilities.

Computer-mediated communication, in its many and varied forms, could under appropriate social conditions help to create new communities of people, bound together by a shared interest in a topic or a shared background. The grassroots development of community bulletin boards using privately owned personal computers is an example of such a nonwork-oriented community formation. Much has been made of the dangers of technology with respect to alienating individuals and reducing direct social interaction both in quality and quantity. The argument here is not to substitute for the richness inherent in face-to-face interaction, but to explore new ways and means of interaction through the computer medium as a supplement to other modes of interaction. Through international computer networks, it is possible to obtain information from a vast pool of human resources that would be difficult to tap in any other fashion.

At a more local level, many projects are concerned with providing both electronic access and production facilities for the local community. In some cases, due to time and distance constraints, face-to-face interactions may not be feasible, yet electronic interaction may still be possible. The "Community Memory" project in Berkeley, California involves a community bulletin board that is accessible from a number of terminals distributed in shopping areas and other community meeting places in the neighborhood. It is an attempt to refashion electronically aspects of the ancient marketplace which served many functions, including that of informal meeting place—the Greek *Agora*—where the development of community is strengthened by increased interactions at all levels of intensity and duration.

Thompson (1972) notes the importance placed by Jane Jacobs on the city sidewalk as being a place where people can pick up useful information in chance encounters, and wonders what would be the electronic equivalent—some electronic means of idly browsing information spaces and coming across potentially useful information, without actually searching explicitly for that information. Again, the intent is not to simply duplicate the kind of interactions that occur on sidewalks, but to provide some features of this kind of interaction in another medium. There is, of course, a limit to this enterprise: attempting to completely substitute one medium of communication, in all its variety, for another, is not likely to be successful. Rather, we need to develop a more comprehensive understanding of each medium, of the tasks we wish to accomplish, and the best match of the two for any particular problem. In this light, electronic communities do not replace but extend the notion of work community that I discussed in Chapter 19. The fundamental insight is the shift in perspective within the field of human-machine interaction to include the potential of the computer medium to significantly enhance the possibilities for communication and collaboration among people.

ACKNOWLEDGMENTS

I would like to thank Sondra Buffett and Donald Norman for their comments on an earlier draft of this paper. Also thanks to the "Asilomar Group" for their critiques. I am much indebted to the many people who responded to my query for information about software to support interaction over the USENET and ARPANET—especially the following people who have allowed me to include extracts from their responses in this paper: William Barns, Robert Cunningham, Robert Eachus, Jerry Farrell, and Tom Kaczmarek. A special debt is due to William Barns for his encouragement and his insightful comments. Other people at UCSD, from the Cognitive Science Laboratory, the Laboratory for Comparative Human Cognition, and the Interactive Technology Laboratory, have discussed many of the issues with me, especially Michael Cole and Jim Levin. These people are of course not responsible for any of the views expressed here.

SECTION VII

THE CONTEXT OF COMPUTING

This final section has only one chapter, one that in part functions as a sort of summary statement. The book is organized by "grain size," by the level at which the topics were approached. It is only fitting that the book concludes with the largest grain size, with a chapter that considers the way one might use computational tools to affect society, to change the tasks that we do and the way that we view them. This section, and the single chapter within it, addresses many of the issues raised in the book, but with the explicit attempt to exploit them by suggesting tools for society, tools that could change thinking, problem solving, writing, and communication.

In Chapter 22, Brown continues the discussion introduced by Hooper and Bannon (Chapters 1 and 2) on the role of the Bauhaus School of Architecture and the general role of the designer in society. Design should be responsible. Brown suggests that "We might begin to look for ways to design systems so as to engender social change, much as the Bauhaus School of design hoped to do through crafting of architectural artifacts." This is both an exciting challenge and a major problem. As the experience with the Pruitt Igoe Housing Project in St. Louis vividly points out (see Chapters 1 and 2), it is dangerous to make social policy statements prematurely. We certainly do not know enough about social factors to understand the full implications of the artifacts that we create. In an electronic mail interchange on this topic, Bannon summarized the issue like this:

> I am still struggling with the dilemma of how to reconcile the role of the designer as both creator of artifacts that can effect positive change in society and the obverse side of this, which is where the designer creates artifacts that, though designed with "good intentions," are totally unsuited to the needs of people.[1]

A recurring debate throughout the book is the view of the computer as a "tool." Norman (Chapter 3) makes this one of his themes, that the computer should be viewed as a tool, that it should serve in the interests of the user, neither dictating the work nor taking control, but rather acting together with the user to get better mastery of the task domain. Laurel, in Chapter 4, argues against the notion of computer as tool, but using "tool" in a different sense, as a intermediary that might stand in the way between a person and the task or experience. These two points of view are reconciled if the tool is invisible or transparent, so that the user notices the product and the task, not the tool being used to get there—a point of view consistent with the arguments for Direct Manipulation in Chapter 5. Brown comes down strongly in favor of the computer-as-tool view, and he shows that the proper tools can change one's perception of the task. Indeed, he argues that the best way to induce social change in a large design community is to provide tools carefully crafted to affect the way designers think of their task. (Brown uses as an example the design of copying machines in his own organization, but the clear intent is to generalize to design in general, in much larger contexts.)

An interesting commentary on the role of computer-based tools in society and how they might be crafted with concern for the social implications comes from the Scandinavian experience in constructing tools for newspaper composition,

[1] Those interested in pursuing these issues might start with the several chapters in this book that discuss social change, and then turn to Wolfe's (1981) treatment of the lessons from Architecture or to Giedion's general critique of architecture (1965).

using high-quality video displays. These tools are being designed by an innovative design institute, one partially funded by Labor Unions and Corporations: Norman, Chapter 3, reviews briefly some of their work. (See Ehn & Kyng, 1984.) A brief overview of the entire project can be found in Howard, 1985. It turns out that even the intense concern for social impact and explicit support by labor unions during development did not prevent difficulties when the systems were actually installed. This emphasizes the point that we should be very cautious in making social commentaries or in introducing technological artifacts designed for a particular social purpose: We risk tampering with the complexities of a delicate cultural process with insufficient knowledge and experience.

"User Understanding," the theme of Section III of this book, provides a major theme for Brown in this chapter. He argues that users form mental models of the systems with which they interact, and he echoes Norman's concern from Chapter 3 that the system itself provide the necessary information to support the development of an effective model (what Norman calls the *System Image*). Here, Brown builds upon the work in Cognitive Science on understanding (reviewed by Riley, Chapter 7), along with the discussion by Lewis (Chapter 8) of the way that people find explanations of their experiences. Brown argues that people need to construct explanations of the systems that they use, that they build upon a "naive theory of computation" (in the sense described by Owen, Chapter 9), and that these play major roles in the way that people recover from difficulties—"repair" the model and procedures being used. Here he borrows heavily from the experiences of natural language, where repair and communication take place continually and unobtrusively, points that serve as the themes for Chapter 14 by Reichman and 20 by Lewis and Norman.

The concept of "Information Flow"—the theme of Section VI—is central to Brown's chapter, so much so that we debated adding this chapter to that section. In many ways that is what the chapter is about: the flow of information between person and computer system, between people, and among social groups and organizations. Brown takes the strong position that most tasks are interactive, that we need more tools to support this interaction, and that technology now offers a way to provide these tools. Design, writing, communicating—all are areas that can benefit from enhanced tools; tools as artifact, not only of technology, but of society and culture. The theme that users must help one another is dominant here, reflecting the arguments made by Owen, O'Malley, and Bannon (Chapters 17, 18, 19, and 21). Brown points out the differences that occur when learning is done in isolation as opposed to with other people. "Eye contact" is critical for Brown, and he illustrates a case in which failure to be within sight of others learning a system was detrimental for the learner. As Bannon points out, in the design of computer (and other) systems, the social organization of work, and even the physical layout of the office may be as important as the manner by which the system was programmed.

Brown's chapter is provocative, raising important issues about the role of our field within the context of computing, indeed, within the context of society. It is a fitting way to close the book.

CHAPTER 22

From Cognitive to Social Ergonomics and Beyond

JOHN SEELY BROWN

Informational systems pose interesting new challenges to the traditional design goals, validation strategies and research methodologies of the industrial design and human factors communities. The purpose of the first section of this chapter is to consider the nature of these challenges and to explore some of the factors involved in meeting them. I suggest a new goal for interface design, namely to focus our efforts on developing not idiot-proof systems, but rather systems that facilitate *the management of trouble* by the casual user. I argue that in order to achieve major improvements in the human-computer interface, we must consider issues that typically arise not only in cognitive, but also in social, ergonomics—that is, how the overall social environment in which systems are embedded can serve as a powerful resource for helping users master their systems.

I believe that the theoretical and methodological "eyeglasses" used in the past to study human-machine communication have diverted us from recognizing and investigating fully the rich and subtle mechanisms of human communication, a point made throughout this book, especially in Chapter 14 by Reichman. The second section of the chapter discusses new methodological techniques and analytical frameworks aimed specifically at exploring human-machine interactions, and suggests theoretical and design directions that may emerge from this research. Some of these new directions have been explored in this book.

The third section focuses on the extraordinary power of information tools to change our perspective on the world and to transform our relationships, not only with technology, but also with other human beings: again, points made elsewhere in the book, especially Chapters 2 and 21 by Bannon. Properly crafted and disseminated, informational systems can act as powerful agents for social change—that is, computational artifacts are cultural artifacts that can pull social change in one direction or another. So, for example, if we believe that a powerful way to amplify human potential is by tapping the latent resources of a group, we might design informational tools that facilitate team brainstorming, effective sharing of community knowledge, and collaborative authoring and design. By fostering a collaborative spirit in schools and in the workplace, such tools can function as cultural drivers as well as intellectual and productivity amplifiers. I discuss how designers who take up this challenge might form a movement akin to the Bauhaus School of architecture, in which physical artifacts were designed explicitly to facilitate social change.

INFORMATIONAL SYSTEMS: NEW CHALLENGES TO USER CONTROL AND USER UNDERSTANDING

An implicit design goal in most discussions of human-computer interfaces is that system design should enable users, in particular casual users, to be in control of their technology. Unfortunately, in many instances, this is taken to mean no more than the self-evident proposition that people should be able easily to do the things they want with computers. The role of, and problems inherent in, understanding the system are usually glossed, perhaps with the maxims that the user interface should be intuitive, the procedures for operation clear and simple, the need for extensive documentation minimized, and so on.

We need to recognize how fully a user's sense of control rests on a robust understanding of how a given system functions, of why the procedures for operation are as they are, and of how the informational support system for the given system is organized. Learning procedures may enable users to accomplish some daily tasks adequately. However, without at least a common-sense understanding of how the procedures relate to the underlying system, users will be unable to adapt them to new situations, to deal with either system malfunctions or the consequences of their own errors, or to adapt to new or evolving systems. See Sections III (Users' Understandings) and VI (Information Flow) for further discussion.

Unfortunately, however, informational systems are inherently hard to understand. Let us first explore some reasons this appears to be so. We will then consider how, as designers or researchers in human-computer interaction, we might evolve means for tapping users' natural sense-making strategies—e.g., strategies that are used to understand ordinary conversation or to explain the embedding of physical and social worlds—as resources for making sense of informational systems.

System Opacity

Unlike mechanical systems generally, informational systems are largely *opaque*—that is, their function cannot be perceived from their structure. The user of physical systems is likely to develop a model of how the system works from the combination of structural evidence and a variety of implicit physical cues, interpreted in the context of use. In the case of informational systems, this natural sense-making process is short-circuited. The types of physical cues that in many technologies help show how something works, such as the spatial relationships, mechanical connections, mechanical movements, and differentiating sounds, are greatly decreased in informational systems. Compare, for example, the resources for understanding that confront a user peering at a computer screen and those that face a cyclist examining a bicycle's gears and derailleur. The computer user has far less evidence for what is transpiring in the system "beneath" the screen.

Informational systems do give off physical signs, in the form of sounds accompanying disk movement, console lights, and time delays in system response, but these are often confusing unless interpreted in light of a sophisticated mental model of how the system works. For example, a sudden delay in carrying out an operation during word processing makes sense only if one understands the relationship between one's own actions, the appearance of letters on the screen, and the functions being carried out as the system stores and accesses text in the main memory, on local disk, or on remote file servers, etc. As a result, new design strategies are needed to make evident to users the relationships between their activities in using the system and the system's underlying structural and functional characteristics.

Complexity From Multiple Processes

Complicating the issue of opacity is the fact that personal computers increasingly are designed to allow users to manipulate multiple processes and multiple windows at one time. Thus, users are presented with the riddle not of one, but of many, active "black boxes" in understanding a system or inferring its overall state.

> *Consider the case of a user who is running multiple processes (a message system, several text editing processes, etc.) on a system that is capable of buffering a series of user inputs. Such a system does not require the user to wait until a given process is completed before a new command is initiated. Perhaps the user initiates a hardcopy command in one window, tries to get mail in another, and issues a copyfile command at the top level. If the system then signals an error, the user faces a significant problem of interpretation, in addition to more straightforward recovery ones. Did the user cause the error, or did it result from some system-contingent problem, such as a "crashed" file server? If it were caused solely by the user, was it due to overloading the system with commands or to an illegal operation? And if the latter, which action was the offending one? Clearly a user requires no little sophistication even to know which diagnostic actions to initiate, or when the whole issue might be sidestepped without losing state.*

Users of an opaque single-process system gradually develop a set of input-output associations based on the system's behavior in response to single actions, which allow some predictions about the system. Users of a multiple-process system face the added difficulty of sorting out the relationships among multiple processes, their own triggering actions, and system responses.

Functional Complexity

A third source of difficulty in understanding informational systems is their *functional complexity*. Not only do users often encounter multiple kinds of functionality, such as editors, message systems, and electronic spreadsheets, but they also must cope with multiple layers of functionality within single applications. Advanced text editors like Emacs and Teco are examples of layered functionality in informational systems. As a result, users are likely to develop a number of limited and potentially conflicting models of the overall system as they encounter new functionalities. For example, the model that emerges from using a message system with intensionally specified distribution lists, and that can communicate with diverse operating systems, is likely to be quite different from that which emerges from a straightforward text-editing and -formatting system. At the same time, each is complex in its own right. An adequate model of a text-editing and formatting system must include specifications for distinguishing between how something appears on screen and how it will appear in hardcopy form—a crucial distinction even in those systems that are built around "what you see is what you get" philosophies. On the other hand, a message system introduces, in addition to notions about temporary information storage, the need to know social conventions and "legal" restrictions governing the use and updating of distribution lists (see the discussion by Bannon, Chapter 21).

Users are likely to vary widely in their knowledge of and facility with any given functionality. Designing interfaces for a range of expertise raises a host of difficult problems concerning the tradeoffs between learnablity and functionality, the need to encourage continuous learning in a variety of contexts, problems of establishing a useful explanatory framework, and so on. At the same time, even expert users often have the properties of casual users as they encounter new functionalities or as they attempt to carry out a new kind of task with their well-worn system. Thus, designing systems that minimize the trauma of "first encounters" is becoming increasingly important even for systems that are aimed at expert end users.

Lack of Causal Metaphors for Explanation

As several chapters have pointed out, one of the ways in which informational systems differ from physical systems is in the kinds of explanatory metaphors that capture how the system functions and provide a sound basis around which to "grow" a mental model (Norman, Chapter 3; Lewis, Chapter 8; Owen, Chapter 9). Mechanistic systems can often best be understood in terms of causal explanations; such explanations are able to draw on an arsenal of simple causal models and metaphors that most of us have abstracted from our experiences with the physical world and that we commonly use to construct plausible "stories" or "explanations" accounting for new phenomena. But what are the counterparts to these causal models for understanding informational systems? What in the information world corresponds to physical mechanisms represented as ideal springs, levers, wheels, pulleys, ramps, screws, etc.? We are just beginning to explore naive theories of computation (see Chapters 9 and 10 by Owen and diSessa) in order to provide better anchors for intuitively satisfying explanations of informational systems, much as we have investigated naive theories of physics in order to construct better instructional scenarios in the physical and engineering sciences.

> *It may be that people derive their best metaphors for system operation from their experience with social interactions and attributions. For example, teleological models that anthropomorphize simultaneously running processes as multiple (system) "intentions" can provide a useful and compelling way to describe some informational systems.*

It is important to realize (as Lewis points out in Chapter 8), that people naturally attempt to explain or make sense out of their experiences with systems by using naive empirical strategies and everyday reasoning. Moreover, as O'Malley and Bannon point out (Chapters 18 and 19) communities develop informal communication networks through which such explanations are propagated, forming a sort of community knowledge base around system use.

> *The power of community knowledge, and the fact that such knowledge is often crystallized in myths, is illustrated by the following anecdote. We recently discovered that members of a large research institute using* **Interlisp-D** *machines widely believe that one must simultaneously press the two "boot" buttons on the front panel of the machine before turning it off, in order to keep the disk head from crashing onto the surface of the disk and destroying it. This piece of folklore apparently has its origins in the properties of a very early prototpye of the current generation of* **D-machines**, *one that never made it beyond the designer's workbench some ten years ago. Nevertheless, the belief has spread throughout the subculture, apparently propagated through informal channels of communication and justified by an account that incorporates available explanatory "tokens."*

By recognizing the fact that users do engage in constructing explanatory stories and identifying the occasions, strategies, and reasoning processes that underlie these sense-making activities, we will be better prepared to help users build a sound understanding through improved cognitive and system resources (see Lewis, Chapter 8). In addition to strategies for increasing the intelligibility of informational systems, our design goals should include the creation of explanatory metaphors from which users can construct useful mental models—that is, models that provide a sound basis for reasoning and hypothesizing about new problems and tasks. We should recognize that the informal channels through which groups communicate constitute an important social resource for improving users' understanding. Tapping this natural tendency could be tremendously valuable for facilitating system understanding. (This point that has been made quite strongly in a number of chapters, especially those in Section VI, *Information Flow.*)

Interactivity and Ambiguity

Perhaps the most subtle and complex factors affecting the intelligibility of informational systems have to do with the problems inherent in making sense of *interactive systems.* While we generally recognize that inferring a "model" of the speaker is necessary for intelligibility in conversation, we are perhaps less aware that collaborative *action* also requires an interpretation of purpose against a set of assumptions, or "conversational postulates," about the nature of the interaction. This can be most easily seen in systems in which user and system act as coparticipants in getting a job done. Studies of system operability for complex copiers at the Xerox Palo Alto Research Center (PARC) have begun to identify the ambiguities that arise when the system's response not only is occasioned by the user's action, as in an online help system, but also serves as a request to the user to do the next thing that the system requires in order to proceed with the task. In such a case, it turns out, interpretation of the system response is not straightforward. The user must understand why the request is being made in order to understand what is being requested—that is, the user must infer the underlying intentionality of the system. In order to do so, the user brings a range of tacitly held conversational and pragmatic rules to bear.

> *The nature of the resulting ambiguities and the resources that users call on to help resolve them was initially discovered by Suchman (1985) as part of an operability project at PARC for analyzing users' "first encounters" with a sophisticated reprographics system. The system contains an electronic subsystem that provides instructions to the user reactively, that is, as occasioned by the user's latest action. In complex copying jobs, such as making 100 copies of a bound 200-page, two-sided document, the user must assist the copier by carrying out certain intermediate tasks (e.g., moving a stack of paper with one side already copied on it from an output paper tray to a second, input paper tray). Sometimes the system issued instructions that caused users to proceed down a "garden path" from which they had great difficulty extricating themselves. The causes of the garden path were unclear until Suchman's analysis revealed that part of the problem derived from the users' misinterpretation of the underlying purpose of*

> *the instruction. For example, when the system issued an instruction intended to allow the user to recover from an error in carrying out the most recent step in the procedure, the user interpreted it as specifying the next step in the procedure, and therefore as confirming that the previous step was correct. The simple fact of believing the current action to be the next step in the procedure, when in fact it was providing the opportunity for repair, caused a chain of problems with subsequent instructions.*

Successful human-machine collaboration rests to a large degree on the ability to infer the underlying intentionality of a given action or imperative. From the user's point of view, understanding the intentionality of interactive systems depends on having a model of the functions and mechanisms of the underlying system; of the system of procedures for carrying out a given task; and of the system's assumptions about, and interpretation of, the user's own actions. Moreover, users base their interpretations of interactive behavior in part on certain tacitly held assumptions about the conventions that govern an interaction. For example, a response to a collaborator calling for an action is assumed to be relevant both to accomplishing the overall task and to the preceding action. These bear some resemblance to the Cooperative Principle derived from Grice's (1975) analysis of conversational implicatures, suggesting that the interactional basis for cooperative action is much the same as that for conversation. I will return later to contributions that studies of human interaction and communication, in particular everyday conversation, may make toward achieving natural communication between human and machine. For the moment, let it suffice to say that understanding the nature of tacit interactional rules can help us identify likely sources of interpretative difficulty.

DESIGN FOR THE MANAGEMENT OF TROUBLE

The previous section identified five properties of informational systems: *system opacity; complexity from multiple agency; functional complexity; lack of causal metaphors for explanation; and interpretative problems arising from system interactivity and ambiguity.* Taken together, these properties challenge users' natural resources for making sense of a system behavior, increased cognitive demands in dealing with system complexity, and increase the need for new research and design strategies for creating of effective interactive artifacts. What is increasingly clear is that the classic design goal of designing "idiot-proof" systems—including not only informational, but also procedural, instructional, and physical systems—is profoundly wrong, a point made strongly by Bannon in Chapter 2. This is not to say that such design goals as searching out clarity in the user interface or consistency in the underlying organization of system functionality, are not desirable. Meeting these goals does not seem to prevent interpretative and other troubles in the use of informational systems. One can never anticipate and "design away" the exigencies, misunderstandings, and problems that will arise in people's use of systems; instead, we need to recognize and develop resources for dealing with the fact that the unexpected always happens. Said differently, we must widen our focus from designing for the avoidance of trouble to design for the *management of trouble*— of both a communicative and an operative nature. From this perspective, the notion of *repair*, as a fundamental mechanism of natural communication, becomes a crucial construct for interface design; that is, we need to design user-system interaction that supports human participants in using the particular context of the interaction, including the larger social setting, to aid in the recognition and repair of misunderstandings.

The notion of designing for the management of trouble has cognitive, social and instructional implications. One important aspect of rethinking our design philosophy is the possibility of creating self-explicating informational systems—that is, systems that overcome the cognitive problems of penetrating system opacity and complexity by providing various sorts of "grist" or resources for user understanding. The basic idea is simple: We need to construct the overall system, including its supporting subsystems (e.g., instructional/help facilities, "undo" facilities, etc.) so that the user is encouraged both to construct mental models of the system and to use those models to "guess" at how to handle unanticipated problems. Furthermore, we need to understand that the social system in which informational systems are embedded may be seen as a "subsystem" of the informational system, in that it significantly affects the way people learn about and use informational systems (see Bannon, Chapter 19 and O'Malley, Chapter 18). It is possible to engineer social infrastructures so that the social setting both functions as an information resource and encourages experimentation, collaborative learning, and sharing of knowledge around informational systems. Finally, we need to devise new methods for training people to use complex systems. These should reflect the need for continuous learning by helping students develop strategies for using mental models as a basis for extending their knowledge, for coping with incomplete knowledge, for framing useful questions, and for identifying the gaps in their understanding.

Let me now expand these issues in order to see how, taken together, they form an approach for designing for the management of trouble. The phrase "taken together" is serious. One of my claims, and the claim of this book, is that design of tomorrow's informational systems must transcend the system per se to take into consideration the design of the social infrastructure in which these systems are to be embedded. It is the interplay between informational systems and social systems that will make the difference in effective individual and social use of computational systems. Indeed, the final section of this chapter goes one step further and suggests that the informational systems themselves can help engender appropriate social infrastructures.

Mental Models

A precondition of being able to adapt procedures for operating informational systems is that users understand the reasons for, and interrelationships among, the steps of the procedure—that is, the procedure needs to be *semantically rationalized.* Mental models of how the system functions relative to both its constituent parts and a given task provide the most stable and robust basis for such an understanding. Such models are also a crucial resource for facilitating informal discovery learning through a sort of task-oriented empiricism, as they form a cognitive structure about which hypotheses can be formed and tested. And finally, an understanding of system functionalities in terms of mental models and explanatory metaphors provides a basis for resolving ambiguities in system behavior relative to specific aspects of system operations and therefore for gradually deepening one's knowledge through experience.

> *An example will help illustrate this fact. Since I am writing this paper at home using* **Tedit**, *a Lisp-based editing system, I use a floppy disk as the connected file system for holding the sources of this paper. Two events that occurred as I was working are noteworthy. First, I remembered a procedural rule that I shouldn't remove the floppy from the drive even after I have accessed the file stored on the floppy. Second, I was surprised as I scrolled backwards over the document that my floppy started spinning. When I heard the spinning, I was relieved that I hadn't removed the floppy. But I was also confused about why it should matter; the file was already loaded into* **Tedit** *and I was not in the process of storing it on the floppy.*
>
> *I quickly realized that my initial confusion lay in the fact that my mental model of the system did not include the notion that*

> *the edit system buffers a limited number of pages from the source file, pages that then get replaced as I move either forward or backward in the file. I had assumed that the editor either simply read the entire file into its virtual memory or at least read each portion into its virtual memory as I accessed it. Had either of these assumptions been true, going backward in the file would cause the system to read text into edit buffer from the internal hard disk, not the floppy. When using the same system with a remote file server at work, I had been unable to hear the various sounds that accompany reading from the storage device, and so had no occasion to think of this interpretation. Now I understood the source of the various mysterious delays I had previously experienced at work, delays caused by network and remote file server contentions, but that I had attributed to strange happenings in the prereleased version of the editor.*

My incident with the text editor gave me a better mental model of the flow of information inside my computer. But note some of the details of this experience. First and most importantly, when this event transpired, I had both a substantial mental model and an arsenal of metaphors containing the relevant constructs of buffers, piece tables, virtual memory, internal disks, floppies, etc. It was in terms of these cognitive resources that I was able to make sense out of the evidence of the spinning floppy. Second, even though it left unspecified many inferences and causal relations, I had constructed an "explanation" for the earlier delays that was within the realm of possibility and that might have accounted for the system behavior of which I was aware.

Third, both physical signals and an element of a procedure served as cues that prompted me to extend my model, because they required interpretation. The procedural constraint, not to remove the floppy after accessing the file, constituted evidence both that I did not fully understand some aspect of the system's functioning and confirming evidence that my later account was probably correct. (This view of the use of procedures is in sharp contrast to the usual notions about the role of procedures—that they should tell us what to do, not why to do it. However, note that procedures, like any other aspect of experience, require explanation; they serve either as an element in the fabric of a satisfying explanation or as a clue to aid in constructing one.) Likewise, physical signs—delays in the earlier cases and the spinning of the floppy in the latter—had served as both the occasion for and a resource in constructing an explanatory account.

And finally, even within the context of a fairly sophisticated model of the system, the signs that the system presented were not always unambiguous. Further, they relied for interpretation on pieces of information deriving from the actual or *situated* context of carrying out a specific task. Without the disambiguating evidence of information actually being read from a source file (in this case the floppy) at a particular moment (when I issued a "scroll" command), the delays in the performance of the editor would have a wide variety of possible explanations.

> *Note also that being able to interpret cues in terms of an explanatory account is important to my sense of control over the system. My first reaction to the unexpected sounds of the spinning floppy was anxiety that I might lose all my work, followed shortly by relief that I had not removed the floppy, followed by a spate of mental work in which I resolved the mystery so that I might not be caught by surprise again. The desire to feel in control is a strong one. It is likely to motivate a great deal of cognitive work in interpreting system behavior. From a motivational point of view, it is probably less important for the system to provide a completely "bullet-proof" explanation or account of its operation than to provide the resources with which people can construct their own explanations, repair their misunderstandings, and so on. An important motivational factor in the development and use of mental models is the existence of the right situation with the right kind of evidence, together with an occasion in which the interpretation becomes important.*

Design implications. Computers are multimedia devices; their power and flexibility in handling text, graphics, animation, and sound provide an opportunity for representing the internal structure and processes of the system. However, it is important to be aware of the subtleties that are involved in constructing representations that are cognitively transparent. That is, it is not enough simply to try to show the user how the system is functioning beneath its opaque surfaces; a useful representation must be cognitively transparent in the sense of facilitating the user's ability to "grow" a productive mental model of relevant aspects of the system. We must be careful to separate physical fidelity from cognitive fidelity, recognizing that an "accurate" rendition of the system's inner workings does not necessarily provide the best resource for constructing a clear mental picture of its central abstractions. Chapter 9 on naive theories of computation discusses ways of explaining computational constructs so as to tap the intuitions that are apt already to exist from exposure to procedural systems, albeit not necessarily computer-based ones. This kind of work is aimed at helping specify not only what the ideal cognitive "target structures" are for understanding a given system, but also how they might be used to rationalize particular procedures for using that system (see Brown, Moran, & Williams, 1983, for more discussion of this topic).

Another possibility is to find ways of representing and animating the steps in a complex procedure, so that users can see the direction and goals of the procedure in terms of their relationship to the underlying system. Many researchers are beginning to construct representation schemes for capturing procedural abstractions and related procedural epistemologies, including the plans underlying the fully specified procedure and the key ideas or tricks comprising the intensionality of an algorithm. At a minimum, our aim for constructing self-explicating systems is to make visible the flow of control of a procedural system by animating the choice points, the choices and the computational state that necessitates the particular choices.

Finally, we need to recognize that visual representations, like actions and language, require interpretation in terms of the user's existing understanding and within the situated context of use. As in all cognitive work to construct meanings, understanding—including the inevitable processes of backtracking and repair—uses resources from the environment as mutually relevant means of interpretation and disambiguation. Thus, an instruction not to remove the floppy while text editing not only will be interpreted in terms of delays on the screen and sounds from the disk, but also serves as a context for interpretation. One heuristic for designing and animating representations that will be interpretable is to relate them to implicit system cues such as sounds, delays, etc. To do so we must become more aware of the existence and nature of these cues and develop strategies for making them relevant to system understanding. In this way we can help reconstruct in ways understandable by casual users the interrelationships between structure, function and use that help us in understanding simpler mechanical systems.

New Learning Strategies

It is becoming increasingly clear that in order to deal effectively with rapidly evolving technologies and the "explosion" of information, we require new strategies for ongoing learning, as well as new cultural attitudes about what constitutes expertise and about the role of errors and uncertainty in learning. Informational systems, as a result of their rapid evolution and multiple functionality, are beginning to present us with problems of cognitive overload and the need to engage comfortably in "cognitive bootstrapping" in order to feel in control. As a result, users need to develop strategies for guessing or hypothesizing about how a system is likely to respond in a given situation based on partial models or understanding; for forming maximally informative tests of a given hypothesis; and for folding the knowledge gained from experimenting back into their current mental model, thereby developing a more accurate and powerful cognitive tool for future reasoning and problem solving.

Consider how a simple structural model can help extend a user's knowledge by suggesting interesting experiments to run or questions to pose. The strategy for hypothesis formation represented in this example might be called "stress the boundary conditions." One representation of the Hewlett-Packard stack calculator consists of a picture of a four-element pushdown stack that represents the calculator's memory buffer (see Figure 22.1). This model is meant to clarify the structural characteristics of the buffer and to provide the basis for running the system in the mind's eye to see how it works. When the top element is "popped" or called off the stack, each remaining element moves to the next highest memory location. But the model is ambiguous at one of its boundary conditions: What happens to the bottom position of the stack when its contents are popped up a level? Experimenting with the actual calculator quickly reveals the interesting property that the element leaves a copy of itself in the bottom cell; that is, popping the stack results in the bottom two cells having the same number. This effect is repeated next time the stack is popped, so that now the bottom three cells have the same number, and so on. This suggests a new use for the stack, in which a constant can be pushed to the bottom of the stack and then popped to the top, with the result that the constant can now be popped off the top of the stack as many times as necessary.

Design implications. A crucial factor in learning through experimentation and discovery is that these activities should never lead to disaster. As far as possible, systems should be constructed with a complete undo facility, precisely to facilitate productive guessing.

> *Some years ago, I was teaching an introductory computer science course. One of my students was a 35-year-old Navy jet pilot. He came into the computing center one day to start an assignment. He picked up the instruction sheet, got somewhat confused, and asked for help. I suggested he try typing "---------" and walked away. Fifteen minutes later I looked back in his direction and saw him sitting there with a puzzled*

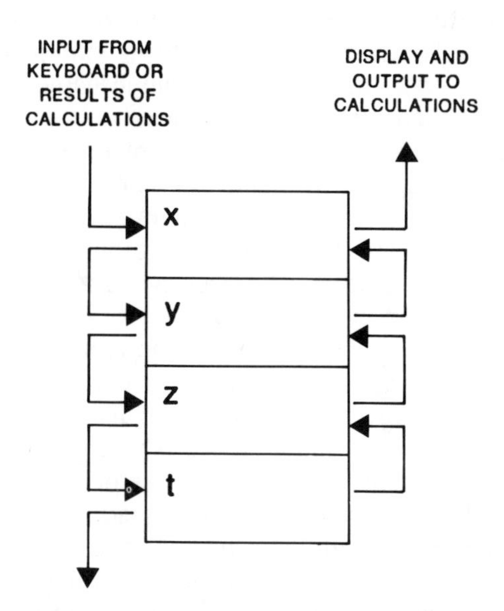

FIGURE 22.1. This model of the HP stack calculator shows the operation of a pushdown stack memory. As each new value is entered at the top of the stack, existing values are "pushed down" to the lower cells. When the stack is popped, the values migrate up the cells again, so that values can be recovered in the opposite order from the their order of entry. However, the model is ambiguous at one of its boundary conditions. Suppose that the stack is full. What happens to the value in the bottom cell when the stack is popped? By testing the boundary conditions, a user can discover that the calculator leaves a copy of value **t** in the bottom cell when the full stack is popped. Repeated popping of the stack creates a stream of constants that can be popped off the stack as required.

> *expression on his face. I walked over to see what the problem was. Lo and behold, there sitting on the screen was the line I had asked him to type. I had forgotten to tell him to hit "return." What made his behavior even more striking was that he had typed the return key after every previous command. Shocked, I asked him why he hadn't hit the return. "But you didn't tell me to do that." "But," I stammered, "after you saw that nothing was happening, didn't you think about trying something similar to what you had done in the past?" "Of course not," he said. "That could be dangerous. I'm not sure I would know how to recover."*

My student's experience with a complex technology had taught him a very important lesson: Experimentation is dangerous because it may lead you into a situation from which there is no recovery. In the case of fighter jets, we compensate for the unforgiving qualities of the system by extensive training of experts, who become the sole qualified operators. The challenge for a cognitive and social ergonomics of informational systems is to find ways to encourage users to take an active role in learning through experimentation: to guess and to experiment. Adequate mental models and the use of implicit system cues to infer underlying processes are key cognitive resources for the ability to extend one's learning through experimentation and guessing. In succeeding sections, I discuss a variety of social and instructional resources needed to provide a coherent and supportive context for the development of strategies and skill in extending learning through experimentation. However, the first prerequisite for developing these flexible learning behaviors and attitudes is that guessing be "safe," that is, that it not lead to informational disaster or even major inconveniences. In fact, through its ability to change attitudes toward errors and toward experimentation with known facts and procedures, a flexible undo facility is a small, but potentially powerful example of how the design of a computational artifact can have impact on the surrounding culture.

Social Setting

The effective use of tomorrow's computers will be a function of the interplay between social and system factors. As earlier chapters point out, the social infrastructure in which systems are embedded plays a crucial role in determining styles of use and in supporting or discouraging various types of learning behaviors. For example, the use of system facilities and resources for cognitive bootstrapping depends on a shift in perspective away from perfectionism and toward the realization that even experts ask for help, share knowledge, and engage in guessing. The development of social strategies in learning to use complex systems can be particularly powerful aids to developing skills for ongoing learning in informal settings.

Recognizing the need of others for cognitive repair, for guessing, and for ongoing learning is likely to encourage users to seek help from peers rather than struggling alone. The willingness to ask for help enables users to make use of the full range of communicative resources for achieving intelligibility that are available in face-to-face human communication and to relate more general information about system functioning to a particular and personally meaningful context. Collaboration in problem solving taps the shared knowledge resources of the collaborators while encouraging the development of metacognitive knowledge through the process of articulating and reasoning about a problem. It may also be that the opportunity to work out learning problems with peers provides a strong motivational context for moving away from perfectionism.

> *Consider this example based on the experience of an administrative assistant that we studied as she was trying to learn how to use a set of complex office systems. She was an ambitious and capable young woman whose accomplishments rested at least in part on her high performance expectations for herself. She was deeply frustrated at her inability to master quickly these systems, but was unwilling in general to turn to the numerous programmers and researchers for help, though they were quite willing to provide it. We now realize that the problem would have been significantly diminished if she had been able to work in a setting in which she was in easy eye contact both with other learners and with experts. For then she would have been able to see that the experts also ran into problems. In addition, she would have been in a position to collaborate with other learners, both providing and receiving help. Her sense of helplessness and ineptitude would have been lessened by seeing that she did have knowledge that was valuable in helping others, while her willingness to receive help would have been bolstered by feeling herself to be part of a community that regularly gave and received help.*

Design implications. Design for the management of trouble includes recognizing the role of the social infrastructure in encouraging awareness of the kind of work involved in informal learning, and in facilitating the dissemination of both systems knowledge and strategies for extending knowledge. Some changes in social attitudes are required, which in turn might require special arrangement of offices, creation of appropriate settings for informal interaction among office members, etc.

> *The design of the physical and social matrix can have a significant impact on social behavior and attitudes. For example, in order to encourage a social support system for teaching the computer language Logo to Humanities graduate students, we exercised great care in laying out the computing lab. In contrast to traditional labs, one of our goals was to be sure that a*

> *student at a terminal could be in eye contact with other members doing the same problem set, so that each could see that others also struggled, experiencing periods of frustration followed by periods of "ahas."*

Systems design itself can have a significant impact on the modes of behavior within an organization. One example of using computational artifacts to influence the social infrastructure is the creation of distribution lists on electronic mail systems that are dedicated to sharing knowledge, seeking help, etc. (as discussed by Bannon, Chapters 19 and 21). Community bulletin boards might be used to give credit to helpers, or to recognize those who come up with the best questions, or to record the best suggestions for new ways of using (or designing) a given system. Another possibility is the creation of tools for building community knowledge bases, much like Owen's "Did You Know" program (Chapter 17), in which knowledge sharing and a process of dialectical exchange become the vehicles for learning among community members. Such systems enable users to contribute and access explanations and discussions of system features as a means of developing a fuller understanding of, and a fuller participation in, the conditions affecting community life. In the third section of this chapter I look at ways in which systems design can bring about social change, not only by changing the ways in which information is exchanged or disseminated, but also by restructuring the information itself to bring about changes in the kinds of information used in problem solving, communicating, and collaborating.

Training and Training Design

In addition to supporting new attitudes toward, and resources for, informal learning, we need new approaches to formal training. First, we should move away from training individuals to use rote procedures. As Riley (Chapter 7) points out, rote training lessens the users' ability to adapt procedures, handle unforeseen consequences, and reconstruct forgotten procedures from an understanding of their structure and purpose. Aside from enabling more flexible control of systems, training that includes the development of coherent mental models and the rationalization of procedures provides a better basis for transferring knowledge to new or modified systems.

Perhaps more important, we need to consider designing instruction explicitly to help users develop strategies for ongoing learning—strategies for drawing on themselves and others as resources, and for making use of informational resources such as documentation in ways that help them extend their understanding. For example, users might be given exercises aimed at improving their ability to formulate good questions based on their current understanding. In order to help users become comfortable with guessing as a mode of extending their knowledge, we might purposely teach incomplete submodels of a system and then provide students with problems that force them to use those models as a basis for deriving solution strategies. Practice in identifying and articulating the implications of the boundary conditions of one's current model should lead to hypotheses and tests to extend the model.

SITUATED HELP AND MUTUAL INTELLIGIBILITY: ISSUES AND METHODOLOGIES

I have discussed some of the issues and design strategies relevant to helping users gain an understanding of systems: animating internal processes and procedural structures; encouraging experimentation by providing extensive undo facilities; encouraging an awareness of the importance of the embedding social infrastructure for the effective use of informational systems; and developing new training strategies to teach relevant new skills. However, the interactivity of the computer, along with its capability to represent, record, and manipulate information both about itself and the user's actions make it possible to consider building systems that provide situated help—that is, assistance and feedback related specifically to users' actions. In order to realize this possibility, it is necessary to develop systems that are able to communicate naturally with users. The observable inadequacies and unnaturalness of current systems and new research on the nature of their failures, are leading us to recognize some of the issues that attend interactive communication in general, and human-machine interaction in particular. Two dimensions are particularly salient: (a) the issues surrounding the need for the user to infer the intentionality of the design as a basis for interpreting system behavior; and (b) the issues surrounding design approaches that enable the system to infer a model of the user.

Like human interaction, the interaction between human and computer takes place under real-time constraints. And like human interaction, successful human-computer interaction requires the participants to interpret the actions and responses of the other. However, human-computer interaction lacks a crucial set of resources that implicitly guide human conversation and allow users to achieve mutual intelligibility. These include a shared model of the communication, including the complete history of the conversation, shared access to an immediate physical and social context, and a verifiable set of assumptions and inferences about the intentionality of one's conversational partner. Moreover, studies of human interaction show that even with a rich set of shared resources, the process of achieving mutual intelligibility in face-to-face human communication rests on detection and repair of misunderstandings through the use of a variety of linguistic, contextual, and cognitive resources—a capability that current interactive systems crucially lack.

Our studies at PARC on the operability of complex reprographics systems have provided a new analytic framework and set of methodologies for considering the problems of human-machine interaction. One of our studies examined (Suchman, 1985) an instructional system designed to aid casual users in carrying out complex procedures on a sophisticated reprographics system. The system, called Bluebonnet, was intended to provide prompts and guidance to users so that they could accomplish any one of a large number of different tasks. One goal of the system was to have each piece of procedural advice appear to result from the previous action of the user. A second goal was to provide the user with the maximum amount of freedom in carrying out the procedures, and in the event of an impasse, to provide help that would make use of the steps already carried out. The technical problems were solved by using advanced AI tools, having the system compute a plan on the basis of the user's initial task selection and from it determine the set of prerequisites for the steps of the particular copying task. If all went smoothly, the user would be presented, one at a time, with a set of steps, each appearing to result from the previous operation.

In her studies of Bluebonnet, Suchman gathered naturalistic "action protocols" by videotaping users' encounters. She was able to analyze nonverbal behaviors, such as visual focus of attention, searches for components, etc., that are particularly relevant to interaction between human and machine, but that are lost in other data-gathering methodologies. In order to have access not only to users' actions, but also to their reasoning about what they encountered, users were observed carrying out tasks in teams of two, a social situation that leads to natural conversation about the task. The data were then put into an *interaction framework* (see Figure 22.2) that shows (a) the user and machine actions that are available to both, (b) the user actions that are inaccessible to the machine, and (c) the design rationale or assumptions underlying the machine's presentation of each new instruction (unavailable to the user).

Suchman's format for her data highlights those features of the data

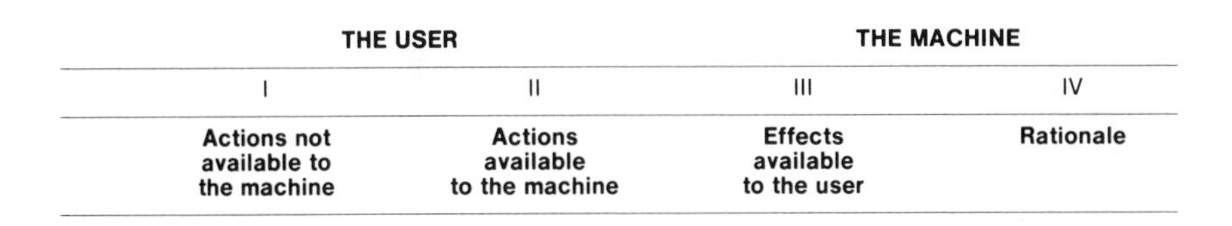

FIGURE 22.2. Suchman's four-category framework for analyzing interactions between novice users and Bluebonnet, an intelligent advice system interfaced with a complex copier. The framework helps highlight the distinction between the varieties of evidence by which a human observer understands the user's actions, on one hand, and the narrowly construed acts available to the system, on the other.

that are interactionally relevant. What is particularly interesting about the analysis is that it shows how problems of mutual intelligibility affect the progress of human–machine communication, even in constrained domains. Bluebonnet attempts to provide situated, interactionally coherent help to users carrying out tasks in which (a) the goal of the interaction is known; (b) the system's communication to the user is based on a logical decomposition of the task; and (c) the communication consists of a set of straightforward procedural instructions. Nevertheless, many subjects experienced considerable difficulty and frustration in attempting to carry out their intended tasks. Their troubles, when viewed from the perspective of an interactional framework, highlight the Gulf of Ambiguity that exists in human–computer interactions and some of the reasons that it does. (The Gulf of Ambiguity is a tributary of the Gulf of Evaluation of Chapters 3 and 5.)

Suchman's analysis graphically illustrates the severe limitations on the bandwidth of communication between user and system (see Figure 22.3). Much of the users' behavior is unavailable to the system. The system cannot see the subtle signs and actions by which people reveal their intentions and their interpretations. As a result, the system is unable to recognize, and to initiate repair of, many communicative failures.

The study of human interaction shows that the processes by which people make sense of interactions consist largely in *local* interpretations

	THE USER		THE MACHINE	
	I	II	III	IV
	Actions not available to the machine	**Actions available to the machine**	**Effects available to the user**	**Rationale**
S1	[places first page in document handler]	Original in document handler		
	"Press START." Where's the Start button? [looks around machine, then to display.]		INSTRUCTION 1: Push START (to produce 4 copies in output tray)	Ready to print (Assumption: document is in document handler)
S2	[points to display] Start? Right there it is			
S1	There, okay	Selects START	STARTS	Document is being copied
			DELIVERS COPIES	
	So it made four of the first?		INSTRUCTION 2: Remove originals from document handler	**Job complete (Assumption: copies have been made)**
	[looks at display]			User removing documents from document handler
	[takes first page out of document handler]	Removes original		
			INSTRUCTION 3: Remove copies from output tray	**(Assumption: copies have been made)**
				Removing copies
	[holding second page over document handler, looks to display] Does it say to put it in yet? [puts second page into document handler]			
		Original in document handler		
			INSTRUCTION 2: Remove originals from document handler	**Job complete (Assumption: copies have been made)**
	"Remove the original..." Okay, I've re...I've moved the original. And put in the second copy.			

FIGURE 22.3. In this selection from Suchman's protocol, the subjects are attempting to make four one-sided copies of an unbound document, carrying out procedural steps in response to the promptings of Bluebonnet. While the instructions are based on the assumption that the originals will be fed to the copier all at once, the subjects expect to feed the originals to the copier one at a time. The system's design understands the user's actions in terms of a detailed, top down plan computed from the initial goal of the user (in this case, to make four one-sided copies). But as becomes clear in the last two interactions, the user's actions, while affecting the machine's state in ways anticipated by the design, are in fact open to multiple interpretations. It is this mismatch between the design's assumptions about the meaning of the user's actions, and the user's actual intentions at the time of the interaction, that causes the communicative impasse. (The boldfaced text shows the user's behavior that serves as evidence of a different intention than that anticipated by the design, on the one hand, and the related assumptions by the design, on the other.)

of interactive events in the context of the particular situation at the time of the interaction. As such, they are only weakly governed or influenced by top-level plans, even though the interaction can be neatly analyzed, postfacto, as if it had flowed from a highly detailed plan. Thus, Suchman's work sets a flag that the descriptive accounts that we, as cognitive scientists, tend to give to an extended interaction probably bear little resemblance to how that interaction was in fact produced. A descriptive account is not necessarily the same as an explanatory account, that is, as an account of the mechanisms and processes by which the interaction actually took place.

The failure to distinguish between these kinds of analyses is at the root of the tendency in AI to account for user behavior in interactive systems solely in terms of well-formulated plans. In a system like Bluebonnet, for example, changes in machine state occasioned by users' actions are compared to a plan for accomplishing a particular task that is based on a logical decomposition of the task into a set of subgoals and procedural steps. As a result, each action has but a single possible interpretation within the context of the plan computed for that task. However, Suchman's analysis shows that many communicative problems arise precisely from the fact that users' actions, rather than flowing smoothly according to a fully instantiated plan, often require interpretation in terms of a variety of situated factors which are, in turn, under interpretation by the user.

Examination of users' interactions with informational systems using methodologies and analytic frameworks such as these can help us to recognize a number of crucial factors affecting human-machine interactions. We can come to understand the sources of ambiguities in users' interpretations of system behavior, as well as the distance between the rational decomposition of the task within the system and the actual actions of the user. We can begin to understand users' strategies in attempting to make sense of system behavior, including the story-making strategies that are used to explain both expected and unexpected behavior. We can begin to recognize the degree to which mutual intelligibility is necessary for successful communication, and as a corollary, the sophistication that will be required by intelligent systems in order to achieve successful communication with users. In particular, the possibility of constructing systems capable of conversational interactions with users requires the capability for constant, ongoing conversational repair on both sides. In order to detect misunderstandings as well as to provide the "grist" or evidence on which the partner can carry out a repair, both partners must infer the underlying intentionalities of the other, while revealing the assumptions that each uses in making those inferences. Access of user and system to evidence for the underlying intentionality that both bring to the interaction, and explicit strategies for interpreting that evidence, can provide an improved basis for building mutual intelligibility.

BAUHAUS REVISITED

Thus far, I have been concerned with the general problems of designing informational systems so that users can understand and be in control of them and with the cognitive, computational, and social implications of such design goals. Now I take a somewhat different tack. The purpose of this section is to discuss the extraordinary power of information tools to change our perspective on the world, and to bring about changes in social and organizational infrastructures. In other words, I raise the question of how we might design computational artifacts to optimize their function as cultural artifacts—that is, to create artifacts that can act as the seeds of new perspectives or "cultures." As the title of this section suggests, we might begin to view design as a proactive as well as a reactive process. We might begin to look for ways to design systems so as to engender social change, much as the Bauhaus School of design hoped to do through crafting of architectural artifacts (e.g., public housing). In fact, because of the computer's ability to change what is difficult about a task, and its potential influence on the organization and structure of information, computational artifacts probably have a greater ability to modify points of view than did the physical artifacts stemming from the Bauhaus School (Whitford, 1984).

My interest in using design as a proactive tool for social change emerged, in part, from the need to transfer the perspectives on interfaces to the designers of copiers at Xerox. It is easy to give talks about abstract ideas and even to get folks to understand them "intellectually," but it is quite a different thing to have ideas actually affect people's beliefs, actions and ways of thinking about a given problem. Perhaps, I reasoned, we could transfer the new design methodology directly by embedding it in a set of computer-based design tools; the aim would be to craft tools to act as carriers of a new paradigm. Indeed, might it be possible to go a few steps further? Might we build these tools with the aim of facilitating a sense of collaboration within the design community, increasing the likelihood of building on the effective design abstractions of others by making it among the easiest and most natural things to do with the system?

However, the design of tools that require a paradigm shift, or new set of "eyeglasses," in order to attain their full power raises the question of how to induce users to accept or become aware of the implicit new perspective that the tool brings. In general, accepting a new perspective is hampered by cognitive and social investments in current ways of working, and by the "costs," in terms of lowered productivity and effectiveness of changing existing systems of doing things. Perhaps to be effective, we need only take this fact into consideration and construct tools that, on the one hand, have a perceived utility with respect to the current perspectives and methods of work and, on the other, induce a strong pull or gradient toward another perspective, a perspective that emerges from the user's gradual exploration of the system's power.

Several years ago, Henderson (1983) constructed a knowledge-based design system, called *Trillium*, that was aimed at influencing the conceptualization of copier interfaces. Formerly, design practice relied heavily on the use of flow charts to lay out the possible sequences and choices in specifying copying tasks. *Trillium* affects the conceptualization of the interface by employing a design medium that leads one to think in terms of constraints between the task options. Our belief is that interfaces that can be simply represented in this constraint "language" will make it easier for users to form useful mental models of system use. By providing a graphics-based, rapid prototyping tool, the system pulls designers toward a more exploratory and flexible attitude toward their own designs. Rather than requiring a month or more of hard work, designers could progress from a design idea to a runable artifact in a day or two. As a result, a designer is likely to have less ego involvement in the particular design and be more willing to critique it and change it.

Trillium keeps track of the history of the design process. This makes it possible to ship both individual designs and the history of the design process among geographically distributed design groups for criticism and comments, thereby improving the sense of coherence and collaboration within the design community as a whole. Moreover, by maintaining the history of design ideas and revisions in shareable form, the system promotes group learning, not only about design problems and solutions, but perhaps more importantly, about the design process. The role of revisions, "garden paths," and breakthroughs becomes more apparent and, thus, more accepted. In these ways, *Trillium* can influence the formation of new design communities, the establishment of a corporate design memory, and the consequent acceleration of the corporate learning curve.

Systems such as *Trillium* can also affect social infrastructures through their impact on the relationship among working groups. To some extent, and in some communities, the possession of sophisticated tools define status within the culture. Our experience with *Trillium* suggests that computational artifacts can act as means of changing power structures by imparting new standing to a subculture of users and improving their perceived value within the larger community. Indeed, there is ample evidence of similar developments in history, such as the development and appropriation of forceps by the emergent medical profession during the 18th century and its unfortunate influence on the subsequent decline of midwifery (Jordan, 1983).

Trillium shows how proactively designed informational systems can affect a specialized community's status, as well as its attitudes toward experimentation and revision, toward communication and collaboration within the community, and toward learning from others. I now give some broader examples that illustrate the ways in which computational artifacts act as cultural artifacts by changing the ways people formulate problems and the methods and strategies by which they attempt to solve them.

Process vs. Product in Writing: From Chaos to Order

Current writing technology focuses almost exclusively on the crafting and production of final documents, making easier the editing, formatting, and printing of finished papers. However, word processing systems do little to support writing during the early stages of idea creation and revision and finding an organization, although these are arguably the most difficult and important parts of the writing process—these are the points at which the underlying ideas and presentation are gotten right, reformulated, modified, and discarded. Indeed, it is likely that current systems militate against the user considering fundamental revisions in writing, by focusing attention on the surface details of the text and making it difficult to see its overall structure. Moreover, current tools do very little to support collaborative writing, as they do not take into account the types of communication that are required for successful collaboration. At best, it is possible to compare complete alternate versions of a text; even normal editorial markings cannot be represented, nor, more importantly, can annotations that comment directly on a particular passage.

Consider tools that reify all stages of the process of authoring and help users move through the stages from chaos to order. In order to be effective, such tools must match the level of chaos appropriate to the particular stage of writing. At the earliest stage, when a writer might normally make notes on the backs of envelopes or on pads of paper, we must allow for a certain imprecision of thought, for multiple organizations, and for sketches of relationships among ideas. While pencil and paper are perfectly adequate for creating notes and sketches, electronic tools make it possible: (a) to manage an indefinite number; (b) to annotate and expand and, where possible, to use those expansions directly in a draft; and (c) to try out different spatial organizations in order to find the relationships.

At PARC, we are experimenting with a system, *IdeaSketch*, that approximates an indefinitely large sketch pad or white board. Users can type notes and sketch diagrams, shrink them to facilitate looking at relationships among groups of ideas, move them, expand them again to look at the components of an idea group, draw boxes and arrows between collections of ideas, and so on.

As the formation of ideas progresses, the user might wish to move to a system that allows for the imposition of more structure, at the same time that it supports the expansion of individual ideas into whole pieces of text. We have developed an experimental system, *NoteCards*, that provides, among others, tools for creating pieces of text in individual "notecards," storing them in "fileboxes," and embedding links representing different kinds of relationships between ideas in a piece of text. By activating the links in a succession of linked notecards, it is possible to follow a progression of ideas to determine the flow of an argument. It is also possible to view a graph (a " browser") of a particular set of the relationships among cards, thus obtaining an overview of the emerging organization. The contents of cards can be copied into a single text window, or an electronic "make-document" capability can be invoked to create a linear text from a specified series of cards.

The point of these examples is to demonstrate the power of appropriate computational tools to pull the user toward a different way of conceptualizing a task—in this case of coming to view writing as a difficult and complex *process* with different levels of formality and precision. Writers are often left to flounder through the early stages, with expectations loaded onto the final stages. The provision of new kinds of tools provides a fundamentally different perspective on what constitutes appropriate products at various stages of writing by allowing idea creation and structuring processes to be externalized and manipulated in new or more flexible ways. Clearly, in order to develop good versions of such tools, designers must find new abstractions based on a clearer understanding of the idea creation and organizing processes of writers and of how those processes are related to producing finished documents.

Making Implicit Knowledge Explicit: Argument Structures for Problem-Solving

In a business and social environment characterized by rapid change, we need to improve our ability to solve novel problems and to find creative solutions to old ones at both the organizational and the individual level. Unfortunately, while rigid, hierarchical infrastructures are effective mechanisms for execution of policies, they are often poor communication and accountability structures for problem solving. In part, this is the result of the way in which information is passed around and up the organization; hierarchies encourage formation of conclusions at the lowest organizational level possible. All too often, the dialectical process of decision making is suppressed, lost, or hidden within the final presentation. This process is the means by which assumptions are defined, pros and cons are weighed, alternatives considered, and lines of reasoning are laid out. The loss is important at the local level of the team of people working on the particular problem, who would benefit from being able easily to reassess assumptions and arguments in light of the evolving realities of implementation. But it is also important at the organizational levels at which corporate-wide policy is made. Another result is that crucial assumptions affecting a variety of decisions may not be considered from the broader strategic perspective of the responsible executive.

Current communications tools and methods force the crafting of complex arguments into linear form for presentation, so that the web-like connections among ideas is hidden from view, making it difficult to see alternate interpretations and points of view. Moreover, the cognitive difficulty of maintaining all the arguments and their interrelationships while writing or speaking, or of inducing them while reading or listening, is frequently overwhelming. As a result, many of the underlying ideas, arguments and assumptions either remain implicit or are lost altogether. But consider the possibility of crafting new informational tools to capture not just conclusions and the view of matters that supports them, but to allow the explicit representation of underlying assumptions and argument structures.

Such a tool would allow the representation and management of a level of complexity not hitherto possible. However, representation is clearly not enough; in addition to facilities for crafting argument structures, we also need facilities to aid in comprehending them. Rather than, or in addition to, browsers that provide views onto pieces of data based on keyword search or other intrinsic information, we require browsers that allow us to look at the various relationships among ideas. We might, for example, wish to examine all the links that have been established between ideas in the hope of discovering forgotten or hidden connections or of establishing new ones. On the other hand, we might wish to take a more focussed view by considering all the evidence in support of an argument, or evaluating an argument at varying levels of detail, or pursuing all the comments of a particular individual.

To accomplish these goals, we need a taxonomy of epistemological links for relating ideas, as well as link-related filters. That is, we must now think about giving users access to and utilization of not just undifferentiated links, but links with appropriate kinds of labels.

SUMMARY

As informational systems permeate our society we, as designers, bear a responsibility to ensure that users retain a sense of control. In order to do so, we must seek means to make informational systems more intelligible, using the flexibility of the computer in representing information about itself to overcome problems of opacity and complexity. Research on formulating and presenting effective metaphors and models of informational systems can aid in developing more effective instructional strategies, while influencing our approach to building "self-explicating" machines.

At the same time, users must assume, and feel safe in assuming, *a more active experimental role* in learning about informational systems. Through the crafting of systems and through new directions in research, we need to enhance recognition of the ways that social attitudes and the organization of social infrastructures affect users' willingness and ability to engage in active, ongoing learning in order to understand and master complex systems. Tools that encourage collaboration in learning to use and understand systems and that enable creation of community knowledge bases may function to pull organizations toward more open sharing of information. The approach to training is equally important. The development of skills in forming and testing hypotheses, formulating useful questions, making use of a variety of information resources, and developing and extending mental models should occupy a more central position in instruction.

The difficulty of understanding informational systems and the inherent ambiguities of interactive communication require us to focus our research and design efforts on understanding and enhancing the properties of systems—computational, social and instructional—that facilitate the management of trouble and that acknowledge the inevitable need for repair. Our design strategies should reflect the understanding that repair functions not only as a minor "error-correction routine," but as a crucial part of an ongoing process by which users infer and adjust their models of system design in order to make sense of their interaction with the system. Seen as such, repair becomes part of a creative learning process by which users broaden their understanding or extend their models of complex systems.

Studies of human interaction and the processes of conversational repair provide an interesting new paradigm for understanding the "everyday" strategies and resources—including repair—by which people infer or attribute intentionality as a partial basis for interpreting the meaning of an interaction. Application of this paradigm to human–computer interaction suggests that the capability for repair in interactive communication between user and system, as between individuals, rests on success in achieving mutual intelligibility. If repair in human–computer interaction is to approximate the same collaborative, natural process that it is in human interaction (as Lewis and Norman urge in Chapter 20), we must devise strategies for improving the ability of systems to infer the intentionality of users, as well as the other way around.

Finally, while the design of informational systems entails responsibility, it also presents new opportunities to influence the attitudes, beliefs and behavior of the surrounding culture. The organization of information and of the processes of problem solving in informational systems can have a powerful impact on the way that users perceive their work, approach problem solving, and relate to a community. Crafting tools that make implicit knowledge explicit, that reify processes as appropriate products for different stages of an activity, and that change the abstractions through which people view their tasks can alter the focus of attention within an entire community. Building tools that enhance collaboration can bring about the development of community support structures that more effectively tap individual resources and increase group learning. In order to realize this potential, we need to be aware of the subtle ways in which computational artifacts function as cultural artifacts.

ACKNOWLEDGMENTS

I am extraordinarily indebted to Susan Newman who ingeniously transformed a set of very rough, initial notes plus dozens of dictated pages into this paper which would not have come about without her skill, patience and ceaseless quest for coherence. I am also grateful to Lucy Suchman and Austin Henderson for their long-term collaboration on the Operability Project. It was Lucy who first convinced me to look beyond current paradigms of cognitive science in order to understand some of the subtleties of human–computer interactions.

Glossary

DONALD A. NORMAN, STEPHEN W. DRAPER,
and LIAM J. BANNON

Bauhaus. An artistic movement active in Germany between 1919 and 1933 that attracted many notable artists and designers of the period and has had an influence to this day. The Bauhaus was founded by Walter Gropius. Leading members included Paul Klee, Wassily Kandinsky, and Laszlo Moholy-Nagy. The Bauhaus emphasized the fusion of the fine arts with the crafts and the building of forms that were appropriate to the kind of material being used. The movement has had an enormous impact on 20th Century architecture.

Bluebonnet. An experimental artificial intelligence interface to a complex copier manufactured by the Xerox Corporation. Bluebonnet attempted to assist users in accomplishing their tasks by determining what the users' intentions and plans were and providing guidance for each step of the plan. Although useful in many respects, Bluebonnet occasionally failed dramatically when its plan failed to match the plan of the person. (See Suchman, 1985 for an extensive critique.) As a result, the program provided an illustration of the difficulties of discovering the intentions of users.

Bootstrapping. Possibly derived from the phrase "to pull one's self up by one's own bootstrap," bootstrapping now has come to refer to almost any selfstarting approach where a system uses its own action to get itself into action. The problem in starting up a computer is: How does the first program or operating system get into the machine's memory and started up? (It can't read from disk, say, unless it already has a running "reading" program.) The general solution involves a cascade of "bootstrap" programs, each of which knows how to copy the next (progressively larger and more complex) program in from some secondary storage (e.g., a disk), and then start it running. The cascade begins with some very short and simple version (which may only be able to copy a limited-size program in from some fixed location). Formerly this might have been keyed-in to memory from switches, but nowadays is typically permanently stored in a special Read Only Memory (ROM) chip, having been "keyed in" once and for all in the factory. This process is often called "booting" the computer. See the section introduction to Users' Understandings for a discussion of its application (by analogy) to Users' problems of learning.

Clicking. A jargon term used to refer an action that typically results in the selection of a particular screen object or item on a menu. Thus, selecting a displayed object by moving the mouse-controlled cursor to it and pressing the mouse button is called "clicking on the object." The jargon continues. Operations that require two or three rapid clicks are called "double clicking" and "triple clicking." Operations done by pushing the mouse button down can be called "clicking on" and those by releasing the mouse button are called "clicking off."

Computer Conferencing. Computer conferencing systems provide extended facilities for keeping a record or transcript of all messages related to a topic, allowing one to hold conferences with groups of people, and to browse through the topics and messages for each conference. There are usually facilities to send public and private messages, and to find out information on the conference participants. The use of the term "conference" is a bit misleading, as we usually we think of a conference as having simultaneous interaction among participants. Although some of the systems provide a simultaneous or "real-time" mode, this is generally a less-used aspect of these computer conferencing systems.

Conceptual Model. There really are two different concepts to be considered: the conceptualization of the system held by designer and the conceptual model constructed by the user. Both are forms of "mental models," the first being the *Design Model*, the second the *User's Model.*

Design Model. The Design Model is the conceptual model of the system to be built, held by the designer. It is to be distinguished from the User's Model of the system.

Direct Engagement. In Chapter 4, Laurel introduced the concept of *Engagement*, arguing that "an interface, like a play, must represent a comprehensible world comprehensibly. That representation must have qualities which enable a person to become engaged, rationally and emotionally, in its unique context." In Chapter 5, Hutchins, Hollan, and Norman expanded this concept by adding "directness": *Direct Engagement* occurs when one has the feeling of directly manipulating the objects of the task domain themselves. The concept of *Direct Engagement* plays a major role in their chapter on *Direct Manipulation*.

Direct Manipulation. Originally coined by Shneiderman (1982), the term "Direct Manipulation" refers to systems in which one manipulates the objects on the screen usually with some kind of pointing device. Key words associated with the notion are "icons," "windows," "pop-up menus," "easy," "explorable," "learnable." Chapter 5 by Hutchins, Hollan, and Norman shows that the term "direct" actually has many components. Direct Manipulation is much mor complex than is usually realized.

Displacement Activity. Descended from the Freudian concept of displacement, the term "displacement activity" originated in the field of Ethology: when an animal substitutes one form of behavior for what would normally be expected, especially when the substitute behavior is irrelevant to the goal, it is called a displacement activity. Thus, if a bird is under attack and, indecisive about whether to attack or flee, the bird instead pecks at the ground, the pecking is called displacement activity. We use the term to refer to any activity used to displace the desired activity. Thus, if one is supposed to be debugging a program or writing a long-overdue letter, displacement activity often takes the form of listing the directory, finding out who else is on the system, or, in general, doing anything but the one activity that should be done.

DWIM: Do What I Mean. A system developed by Warren Teitelman, for the Interlisp programming system (Teitelman & Masinter, 1981). DWIM is quite successful in recognizing typing errors and translating them into the item that the typist actually meant. DWIM has been expanded in the Interlisp environment beyond a simple spelling corrector, mainly by using context (including the syntax of the language, the set of identifyers declared so far, etc.) to determine what the target set of words is against which the input is to be matched. See discussion in Lewis and Norman.

Electronic Mail (E-Mail) A facility on computer systems that allows people to send textual messages via computer to other people on that system, with the message placed in a "mailbox" until it is read. Obvious advantages of this medium over the conventional mail system of transporting pieces of paper are the rapid delivery of the message, the ease of sending to groups of people, and the ease of editing and reviewing stored messages. Another reason for the popularity of such systems, in both research and business operations, is that there is no need for the recipient to be physically contactable at the time the message is sent, as there is if one wishes to establish a phone connection. Electronic mail is especially useful in conjunction with electronic networks, allowing messages to be passed between different machines over vast distances (e.g., from Japan to California to England during the writing of this book).

The Empire Strikes Back. Second in the "Star Wars" film trilogy.

First Personness. The term, introduced by Laurel in Chapter 4, represents the experience of directly living and acting within the world established by the computer. One can have this first-person experience in many contexts, most obviously in arcade-style games or simulated vehicles, but also in spreadsheet manipulation, drawing programs, and even text editing. Laurel describes it this way: "Most movies and novels, for example, are third-person experiences; the viewer or reader is 'outside' the action, and would describe what goes on using third-person pronouns: 'First he did this, then they did that.' Most instructional documents are second-person affairs: 'Place the diskette in drive B.' Operating a computer program with an intermediary interface is a second-person experience: The user makes imperative statements to the system and asks it questions; the system tells the user what to do and what it has done ('File access denied, please try again'). Walking through the woods is a first-person experience; so is playing cowboys and Indians, writing a letter, or wielding a hammer. . . . The underlying principle here is *mimetic*; that is, first-personness is enhanced by an interface that enables inputs and outputs that are more nearly like their real-world referents, in all relevant sensory modalities."

History List. A record of previous inputs or activities on a computer system: a list of the history of activities. History lists are often made available to the user so that earlier commands can be duplicated without having to retype the entire command.

Hypertext. The notion of documents with optional pointers and flags from words or points in the text to other texts. Thus, if this book was being read on such a system, this glossary would not need to have a separate existence. Upon finding a term (such as "hypertext") in the text that was not understood, one could simply point at the word and have the book page expand into an explanation of the term. One could point at the words "Engelbart" or "Nelson" and have the name expand into the appropriate reference for hypertext. The essential idea is to escape from the one- or two- dimensional approach of ordinary text, which forces a single fixed order on to the expression of ideas regardless of the structure of the ideas or the needs of individual readers. In theory, one can imagine hypertext referring to other sections of the same document, to expansions upon the document, to comments made by others. In practice, hypertext systems barely exist. See Engelbart (1962, 1984) and Nelson (1981).

Inter-Referential I/O. A term invented by Draper (Chapter 16). The concept that the item one sees as an output (O) on the computer screen also serves as part of the input (I) command sequence, or, in general, that inputs and outputs may refer to each other directly. Inter-Referential I/O is a critical component of Laurel's "first personness" and Hutchins, Hollan, and Norman's "direct engagement, the integration of input and output being an important feature." When some part of the form of a previous output expression is incorporated in the form of a new input expression the input and output are said to be inter-referential. (See Draper, Chapter 16.)

Logo. A programming language and facility, originally developed by Seymour Papert, Wallace Feurzeig, and Daniel Bobrow in the late 1960s. Logo is based on Lisp, but is much more accessible to beginning programmers. Perhaps the best description is given by Papert (1980, page 217): "Logo is the name of a philosophy of education in a growing family of computer languages that goes with it. Characteristic features of the Logo family of languages include procedural definitions with local variables to permit recursion. Thus, in Logo it is possible to define new commands and functions which then can be used exactly like primitive ones. Logo is an interpretive language. This means that it can be used interactively. The modern Logo systems have full list structure, that is to say, the language can operate on lists whose members can themselves be lists, lists of lists, and so forth."

A major component of Logo is heavy use of graphics, coupled with an easy-to-use command for manipulating a "turtle"—first a mechanical object that crawled on the floor, leaving a trail of ball point line behind, now most often a screen display object that draws lines on the screen. The turtle uses an egocentric geometry, and this, coupled with heavy use of recursion, leads to a unique and special "turtle geometry" (see Abelson & diSessa, 1981). The Boxer programming language is meant to be the modern successor to Logo (see diSessa, Chapter 6).

Mental Model. The notion is that a person forms an internal, mental model of themselves and of the things and people with whom they interact. These models provide predictive and explanatory power for understanding the interaction. The term *Mental Model* is often used to mean different things, however, which leads to obvious confusion. In particular, the term sometimes refers to the model a user has of the system, sometimes the model the designer has of the system, and sometimes even the model the designer or the system has of the user. In this book, the term refers to the model that the user has of the system. We introduce the terms *User Model, Design Model,* and *Conceptual Model* to make the distinctions clear (see relevant entries in this **Glossary**).

Mouse. A mechanical, hand sized input device with buttons on top usually with a tail consisting of a wire connecting it to the computer. Moving the mouse on a horizontal surface sends the computer information about its position changes along the *x* and *y* axes. Tailless mice exist. The mouse was developed in 1964 at the Stanford Research Institute by a group headed by Doug Engelbart and has now become an important member of the class of "pointing devices," a way of pointing at a screen object to signify what is to be done. The mouse has the virtue of being fairly easy to learn and to use. It signals relative motion not absolute position, which is often an advantage, sometimes a problem. The mouse is not the best pointing device for great precision. Some people complain that moving the hand on a horizontal plane is an indirect way of specifying a point on the vertical screen. And finally, the mouse, along with other pointing devices, offers difficulties when frequent transitions between the mouse and other input devices, (usually the keyboard) are required. Nonetheless it is currently the best general purpose pointing device. See Buxton, Chapter 15.

NoteCards. NoteCards is a research tool developed at the Xerox Palo Alto Research Center. NoteCards was developed as an electronic NoteCard file system, where the note is an idea-sized unit of text, graphics, images, or whatever. NoteCards can be used as a fairly general database system for loosely structured information. Different kinds of note cards are defined in an inheritance hierarchy of note card types (e.g., text cards, sketch cards, query cards, etc.). On the screen, multiple cards can be simultaneously displayed, each one in a separate window having an underlying editor appropriate to the card type. Individual note cards can be connected to other note cards through links, thus forming networks of related cards and allowing the user to view the same information through different perspectives.

Pole Position. A coin-operated game developed by Atari in 1982; it became a home computer game in 1983. A low-resolution, 3-D animated simulation of driving a race car on a track.

Pop-Up Menu. The menu that, when called, "pops-up" on the screen, covering whatever was there before. When the user then makes a selection from the menu it disappears, revealing the original screen display.

Rabbit. An innovative experimental retrieval system developed by Williams and Tou, then at Xerox Palo Alto Research Center (see Tou, 1982 & Tou et al., 1982). Rabbit was an iterative scheme in which users were first given an example of an item from the database which they could then refine, stating which components were of interest, which were desirable, which were undesirable, and which they did not care about. By iteratively critiquing examples from the database, the user would eventually be led to the desired target item. Rabbit is especially noteworthy because the user need not have any prior understanding of the way that items are organized in the database or the type of query language required.

Rumplestiltskin. An interactive fairy tale developed by CyberVision, Inc. in 1978, combined low-resolution graphic still frames with simple 2-D animation, and featured a synchronized cassette sound track. Interaction, confined to a few bracing nodes, was limited by the constraints of the cassette medium.

Scroll Bar. An area or bar on the sides of a window or display screen that allows one to control just what part of a large document is actually displayed on the screen at any time. Moving a marker up and down the scroll bar moves the window up or down over the document, revealing whatever portion of the document fits in the area of the window.

Shells of Competency. The concept that a person learns by developing a series of shells or incomplete models: sort of an onion-skin theory of learning—highly over simplified, but not a bad approximation. Each Shell serves to protect a person from complexity and engender a feeling of compentency. The idea of a shell is meant to capture the dual nature of the tradeoff: Initially a Shell provides protection but subsequently can restrict further growth and development. This notion is intended to lead designers to consider how to support the development of an effective series of Shells as users expand their knowledge of a system. Originally formulated in this fashion by Jim Hollan, based on ideas of Vygotsky.

Simula. A computer programming language especially developed for simulation purposes, with many interesting features, such as classes. Developed by Kristen Nygaard in the 1960s, it was the first "object-oriented" language.

Space Invaders. Originally a coin-operated game developed by Taito Corporation in the late 1970s, it became a home game for the Atari VCS and home computer systems in the early 80s. It featured a two-dimensional animated display in which wave after wave of alien spacecraft descended upon the user, whose defense consisted of moving and firing a cannon at the bottom of the screen. One of the first skill and action games.

Star Raiders. The first computer game developed for the Atari Home Computer in 1979. A 3-D simulation of operating a starship, including control panel display, maps and charts, and animated forward and aft views from the ship. Voted best computer game of the year for several years.

Star: The Xerox 1110 Workstation. Perhaps the first attempt at a modern workstation. The Star used a high resolution, bit-map display, a mouse with a pointing device, introduced windows to the workstation, and was designed with careful attention to the needs of the user (including continual test with sample user populations). The user interface is notable, among other things, for its pioneering use of the desk-top metaphor, and a thorough application of the principle of consistency. A good description is provided in the article by Smith, Irby, Kimball, Verplank, and Harslem (1982). Although it has its problems, the Star is a major landmark in user interface design, and many of the ideas behind its design have been used in later systems, e.g., the Apple Macintosh.

Steamer. An instructional system for learning about complex steam plant operation. Developed by the Navy Personnel Research and Development Center and Bolt, Beranek, and Newman, Steamer is the major example of the philosophy of "making visible the invisible," making it easier for the student to understand the underlying conceptual model of the steam plant. (See Hollan, Hutchins, & Weitzman, 1984.)

System Image. Users develop their conceptualization of the system—the *User's Model*—through their experiences with it, coupled with their prior knowledge and expectations. The experience includes all aspects of the system, including the physical appearance, the documentation and instruction, and the nature of the interaction. The image transmitted to the user by these physical aspects of the system is the *System Image.* In many ways, the primary task of the designer is to construct an appropriate System Image, realizing that everything the user interacts with helps to form that image. The designer should want the User's Model to be compatible with the underlying conceptual model, the Design Model. And this can only happen through interaction with the System Image.

User Friendly. This term (and its companion "idiot proof system") has been banished from this book. It is still extremely common, particularly in marketing literature. One reason for avoiding it is its devaluation (like the phrase "to be frank," it tends nowadays to be applied only to things to which its application is in fact dubious). Another is that whether the most desirable quality for the interface experience is "friendliness" is an interesting but debatable issue (would you characterize the best interfaces you use—pen and paper, books, automobiles—as "friendly"?) In practice it most often refers to verbose or chatty interfaces with many long prompts, messages, and menus. While this may help some users, its benefits must be traded against the penalties of slow display times, distracting displays, lack of flexibility, and degrading assumptions about the user. Since the metaphor does not suggest attention to this central tradeoff, it is of dubious value to designers.

User Interface Management System (UIMS). A UIMS provides a way for a designer to specify the interface in a high-level language. The UIMS then translates that specification into a working interface, managing both the details of the display and its associated input and output and also the interaction with the rest of the program. UIMS systems allow the generation of high-quality interfaces with much less effort than programming the interfaces directly. The disadvantages are that one is restricted to the type of interface supported by the particular UIMS, which may not always match the desired application well. Current UIMS systems are in an early stage of development and the high-level language is not always very usable, very complete, or particularly at high-level. In addition, there are often penalties in performance.

User Model. The user's Mental Model of the system, formed through interacting with it and from its *System Image*. Often simply called a *Mental Model*. Unfortunately, the term *User's Model* can refer to three different things: the individual user's own personal, idiosyncratic model; the generalized "typical user" model that the designer develops to help in the formulation of the *Design Model*; or the model that an intelligent program constructs of the person with which it is interacting. Alas, all three meanings are used in this book (the former primarily in Chapter 3 by Norman and Chapter 7 by Riley, the latter by Mark in Chapter 11).

Windows. A term referring to a subdivision of the display screen where one set of output is displayed. A window is usually a rectangular area outlined with distinctive boundaries, marked by a visible border. In a window-based system, many windows may be displayed simultaneously. Some window systems carefully "tile" the screen so that windows are placed neatly and do not overlap. As new windows are added, previously displayed windows are made smaller to allow room for the new ones. In other systems, windows may overlap and so can be plopped haphazardly upon the screen, the user adjusting the location and size of each new window at will. This means there may be large numbers of windows, some being completely obscured by windows placed on top of them. An important point is that each window can represent an independent process, allowing a reasonably effective way of presenting multiple activities at the same time on one display terminal.

WYSIWIG ("What You See Is What You Get": pronounced "Whiz-ee-wig") WYSIWIG refers to a text editor whose display on the screen is exactly what will be printed on the paper. The theory is that this enables better composition because the user need not do elaborate mental computation to imagine what the final output will look like. In fact, this is only approximated because even the best of the high-resolution screen displays cannot hold a full-page image of a printed page with the same resolution that can actually be printed on paper. Moreover, many WYSIWIG systems fall very short of the mark.

Zork. One of the "originals" of the text adventure game genre, developed on a large computer at MIT in the early 70s and translated to leading microcomputers in the early 80s.

References

Abelson, H., & diSessa, A. A. (1981). *Turtle geometry: The computer as a medium for exploring mathematics.* Cambridge, MA: MIT Press.

Alexander, C. (1964). *Notes on the synthesis of form.* Cambridge, MA: Harvard.

Alexander, C., Ishikawa, S., Silverstein, M., Jacobson, M., Fiksdahl-King, I., & Angel, S. (1977). *A pattern language: Towns, buildings, construction.* New York: Oxford University Press.

Alty, J. L., & Coombs, M. J. (1980). Face-to-face guidance of university computer users: Part I. A study of advisory services. *International Journal of Man-Machine Studies, 12,* 389-405.

Aristotle, (1954). *Poetics* (I. Bywater, Trans.). New York: The Modern Library.

Backer, D. S., & Gano, S. (1983). *Movie manuals* (Videotape). Cambridge: Massachusetts Institute of Technology.

Baecker, R. (1980). Towards an effective characterization of graphical interaction. In R. A. Guedj, P. ten Hagen, F. R. Hopgood, H. Tucker, & D. A. Duce (Eds.), *Methodology of interaction* (pp. 127-148). Amsterdam: North-Holland.

Bannon, L. (1985). *Extending the design boundaries of human-computer interaction* (Tech. Rep. No. 8505). La Jolla: University of California, San Diego, Institute for Cognitive Science.

Bannon, L., Cypher, A., Greenspan, S., & Monty, M. L. (1983). Evaluation and analysis of users' activity organization. In A. Janda (Ed.), *Proceedings of the CHI '83 Conference on Human Factors in Computing Systems* (pp. 54-57). New York: ACM.

Bikson, T. K., & Gutek, B. A. (1983). Advanced office systems: An empirical look at use and satisfaction. *Proceedings of the AFIPS National Computer Conference*, *52*, 319-328.

Black, S. D., Levin, J. A., Mehan, H., & Quinn, C. N. (1983). Real and non-real time interaction: Unravelling multiple threads of discourse. *Discourse Processes*, *6*, 59-75.

Blomberg, J. L. (in preparation). *Social interaction and office communication.* Palo Alto, CA: Xerox Palo Alto Research Center.

Bly, S. (1982). *Sound and computer information presentation* (Rep. No. UCRL-53282). Livermore, CA: Lawrence Livermore Laboratory.

Bobrow, D. G., & Stefik, S. (1982). *The Loops manual* (Tech. Rep.). Palo Alto, CA: Xerox Corp., Knowledge-Based VLSI Design Group.

Boddy, D., & Buchanan, D. (1982). Information technology and the experience of work. In L. Bannon, U. Barry, & O. Holst (Eds.), *Information technology: Impact on the way of life* (pp. 144-157). Dun Laoghaire, Dublin: Tycooly International.

Boden, M. (1977). *Artificial intelligence and natural man.* New York: Basic Books.

Bolt, R. A. (1985, February/March). Conversing with computers. *Technology Review*, pp. 34-43.

Borning, A. (1979). *Thinglab: A constraint-oriented simulation laboratory* (Tech. Rep. SSL-79-3). Palo Alto, CA: Xerox PARC.

Bott, R. A. (1979). *A study of complex learning, theory, and methodologies* (Tech. Rep. No. 82). La Jolla, CA: University of California, San Diego, Center for Human Information Processing.

Brachman, R. (1978). *A structural paradigm for representing knowledge* (Tech. Rep. 3605). Cambridge, MA: Bolt, Beranek & Newman.

Brachman, R., Fikes, R., & Levesque, H. (1983). KRYPTON: A functional approach to knowledge representation. *IEEE Computer*, 67-73.

Brennan, S. (1984). *Interface agents.* Manuscript submitted for publication.

Bretz, R., Carlisle, J. H., Carlstedt, J., Crocker, D. H., Levin, J. A., & Press, L. (1976). *A teleconference on teleconferencing* (Working Paper ISI/WP-4). Marina del Rey, CA: University of Southern California, Information Sciences Institute.

Brown, J. S. (1983, December). *When user hits machine, or when is artificial ignorance better than artificial intelligence?* Paper presented at ACM SIGCHI/HFS CHI '83 Conference on Human Factors in Computing Systems, Boston.

Brown, J. S. (1985). Idea amplifiers: New kinds of electronic learning. *Educational Horizons, 63 (3).*

Brown, J. S., Burton, R. R., & de Kleer, J. (1983). Pedagogical, natural language, and knowledge engineering techniques in SOPHIE I, II, and III. In D. Sleeman & J. S. Brown (Eds.), *Intelligent tutoring systems.* New York: Academic Press.

Brown, J. S., Moran, T. P., & Williams, M. D. (1983). *The semantics of procedures: A cognitive basis for training procedural skills for complex maintenance* (CIS Working Paper). Palo Alto, CA: Xerox PARC.

Budge, B. (1983). *Pinball construction set* (Computer program). San Mateo, CA: Electronic Arts.

Buxton, W. (1982). An informal study of selection-positioning tasks. *Proceedings of Graphics Interface '82*, 323-328.

Buxton, W. (1983). Lexical and pragmatic issues of input structures. *Computer Graphics, 17(1)*, 31-37.

Buxton, W., Fiume, E., Hill, R., Lee, A., & Woo, C. (1983). Continuous hand-gesture driven input. *Proceedings of Graphics Interface '83*, 191-195.

Buxton, W., Lamb, M., Sherman, D., & Smith, K. C. (1983). Towards a comprehensive user interface management system. *Computer Graphics, 17(3)*, 35-42.

Card, S. K., English, W., & Burr, B. (1978). Evaluation of mouse, rate-controlled isometric joystick, step keys, and text keys. *Ergonomics, 8*, 601-613.

Card, S. K., Moran, T., & Newell, A. (1980). The keystroke-level model for user performance time with interactive systems. *Communications of the ACM, 23*, 396-410.

Card, S. K., Moran, T. P., & Newell, A. (1983). *The psychology of human-computer interaction.* Hillsdale, NJ: Lawrence Erlbaum Associates

Carroll, J. M. (1985). Minimalist design for active users. In B. Shackel (Ed.), *INTERACT '84: First conference on human-computer interaction.* Amsterdam: North-Holland.

Ciccarelli, E. C. (1984). *Presentation based user interfaces* (Tech. Rep. AI-TR-794). Cambridge: Massachusetts Institute of Technology, Artificial Intelligence Laboratory.

Clement, J. (1983). A conceptual model discussed by Galileo and used intuitively by physics students. In D. Gentner & A. L. Stevens (Eds.), *Mental models* (pp. 325-340). Hillsdale, NJ: Lawrence Erlbaum Associates.

Compton, M. (1984, November). Don Norman speaks out: An interview with the UNIX system's best known critic. *UNIX Review*, pp. 38-49, 94-96.

Coombs, M. J., & Alty, J. L. (1980). Face-to-face guidance of university computer users: Part II. Characterizing advisory interactions. *International Journal of Man-Machine Studies, 12*, 407-429.

Coombs, M. J., & Alty, J. L. (1984). Expert systems: An alternative paradigm. *International Journal of Man-Machine Studies, 20*, 21-43.

de Kleer, J., & Brown, J. S. (1983). Assumptions and ambiguities in mechanistic mental models. In D. Gentner & A. L. Stevens (Eds.), *Mental models* (pp. 155-191). Hillsdale, NJ: Lawrence Erlbaum Associates.

diSessa, A. A. (1982). Unlearning aristotelian physics: A study of knowledge-based learning. *Cognitive Science*, *6*, 37-75.

diSessa, A. A. (1983). Phenomemology and the evolution of intuition. In D. Gentner & A. L. Stevens (Eds.), *Mental models* (pp. 15-33). Hillsdale, NJ: Lawrence Erlbaum Associates.

diSessa, A. A. (1985). A principled design for an integrated computational environment. *Human-Computer Interaction*, *1*, 1-47.

Douglas, S. A., & Moran, T. P. (1983). Learning text editor semantics by analogy. In A. Janda (Ed.), *Proceedings of the CHI '83 Conference on Human Factors in Computing Systems* (pp. 207-211). New York: ACM.

Draper, S. W. (1985). The nature of expertise in UNIX. In B. Shackel (Ed.), *INTERACT '84: First conference on human computer-interaction*.

Draper, S. W., & Norman, D. A. (1984). Software engineering for user interfaces. In *Proceedings of the 7th International Conference on Software Engineering* (pp. 214-220). Silver Spring, MD: IEEE.

du Boulay, B., & Matthew, I. (1984). Fatal error in pass zero: How not to confuse novices. *Behaviour & Information Technology*, *3*, 109-118.

du Boulay, B., O'Shea, T., & Monk, J. (1981). The black box inside the glass box: Presenting computing concepts to novices. *International Journal of Man-Machine Studies*, *14*, 237-250.

Duffy, T. M., Curran, T. E., & Sass, D. (1983). Document design for technical job tasks: An evaluation. *Human Factors*, *25*, 143-160.

Dunn, R. M. (1984, June). The importance of interaction. *Computer Graphics World*, pp. 71-76.

Ehn, P., & Kyng, M. (1984). *A tool perspective on design of interactive computer support for skilled workers.* Unpublished manuscript, Swedish Center for Working Life, Stockholm.

Ehrlich, K., & Soloway, E. (1983). An empirical investigation of the tacit plan knowledge in programming. In J. Thomas & M. L. Schneider (Eds.), *Human factors in computer systems* (pp. 113-133). Norwood, NJ: Ablex.

Eisenstadt, M. (1983). A user-friendly software environment for the novice programmer. *Communications of the ACM*, *26*, 1058-1064.

Engelbart, D. C. (1962). *Augmenting human intellect: A conceptual framework* (AFOSR-3223). Menlo Park, CA: Stanford Research Institute

Engelbart, D. C. (1984). Collaboration support provisions in AUGMENT. In C. U. Greaser (Ed.), *AFIPS Office Automation Conference Digest* (pp. 51-58). Los Angeles: AFIPS Press.

Ennals, R. (1983). *Beginning Micro-Prolog.* West Sussex, England: Ellis Horwood.

Erickson, F. (1980). *Timing and context in everyday discourse: Implications for the study of referential and social meaning* (Sociolinguistic Working Paper No. 67). Austin, TX: Southwest Educational Development Laboratory.

Evans, K.,Tanner, P., & Wein, M. (1981). Tablet-based valuators that produce one, two or three degrees of freedom. *Computer Graphics*, *15(3)*, 91-97.

Feiner, S., Nagy, S., & Van Dam, A. (1982). An experimental system for creating and presenting interactive graphical documents. *ACM Transactions on Graphics*, *1*, 59-77.

Fikes, R. E., & Nilsson, N. J. (1971). STRIPS: A new approach to the application of theorem proving to problem solving. *Artificial Intelligence*, *3*, 251-288.

Foley, J. D., & Wallace, V. L. (1974). The art of graphic man-machine conversation. *Proceedings of IEEE*, *62*, 462-470.

Foley, J. D., Wallace, V. L., & Chan, P. (1984). The human factors of computer graphics interaction techniques. *IEEE Computer Graphics and Applications*, *4(11)*, 13-48.

Foster, G. (1984). *CoLab: Tools for computer-based cooperation* (UCB/CSD 84/215). Berkeley: University of California, Computer Science Division.

Freeman, L. C. (1980). Q-analysis and the structure of friendship networks. *International Journal of Man-Machine Studies*, *12*, 367-378.

Furnas, G. W. (1983). *The FISHEYE view: A new look at structured files* (Tech. Rep.). Murray Hill, NJ: Bell Laboratories.

Galton, F. (1894). Arithmetic by smell. *Psychological Review*, *1*, 61-62.

Gentner, D., & Stevens, A. L. (Eds.) (1983). *Mental models*. Hillsdale, NJ: Lawrence Erlbaum Associates.

Giedion, S. (1965). *Space, time, and architecture* (4th ed.). Cambridge, MA: Harvard University Press.

Goldberg, A. (1984). *Smalltalk-80: The interactive programming environment*. Reading, MA: Addison-Wesley.

Gould, J. D., & Boies, S. J. (1983). Human factors challenges in creating a principal support office system: The speech filing approach. *ACM Transactions on Office Information Systems*, *1*, 273-298.

Gould, J. D., & Lewis, C. (1985). Designing for usability: Key principles and what designers think. In B. Shackel (Ed.), *INTERACT '84: First conference on human-computer interaction*. Amsterdam: North-Holland.

Gould, L., & Finzer, W. (1984, June). Programming by rehearsal. *Byte*, pp. 187-210.

Greeno, J. G. (1977). Process of understanding in problem solving. In N. J. Castellan, D. B. Pisoni, & G. R. Potts (Eds.), *Cognitive theory* (Vol. 2, pp. 43-83). Hillsdale, NJ: Lawrence Erlbaum Associates.

Greeno, J. G. (1978). Understanding and procedural knowledge in mathematics instruction. *Educational Psychologist*, *12*, 262-283.

Greeno, J. G. (1983). Conceptual entities. In D. Gentner & A. L. Stevens (Eds.), *Mental models* (pp. 227-252). Hillsdale, NJ: Lawrence Erlbaum Associates.

Grice, H. (1975). Logic and conversation. In P. Cole & J. L. Morgan (Eds.), *Syntax and semantics*. New York: Academic Press.

Gross, L., & Payne, H. (1984). *Intelligent correlation agent: An environment for development of ocean surveillance tactical data fusion technologies* (R-030-84).

San Diego, CA: Verac.

Guedj, R. A., ten Hagen, P., Hopgood, F. R., Tucker, H., & Duce, D. A. (Eds.) (1980). *Methodology of interaction.* Amsterdam: North-Holland.

Halasz, F. G., & Moran, T. P. (1982). Analogy considered harmful. In *Proceedings of the Conference on Human Factors in Computer Systems* (pp. 383-386). New York: ACM.

Halasz, F. G., & Moran, T. P. (1983). Mental models and problem solving in using a calculator. In A. Janda (Ed.), *Proceedings of the CHI '83 Conference on Human Factors in Computing Systems* (pp. 212-216). New York: ACM.

Halbert, D. (1984). *Programming by example.* Unpublished doctoral dissertation, Stanford University.

Hayes, P. J. (1978). The naive physics manifesto. In D. Michie (Ed.), *Expert systems in the microelectronic age* (pp. 242-270). Edinburgh, Scotland: Edinburgh University Press.

Hayes, P. J., & Szekeley, P. A. (1983). Graceful interaction through the COUSIN command interface. *International Journal of Man-Machine Studies, 19*, 285-306.

Henderson, A. (1983). *Trillium: A design environment for copier interfaces* (Videotape). Palo Alto, CA: Xerox PARC.

Hiltz, S. R. (1984). *Online communities: A case study of the office of the future.* Norwood, NJ: Ablex.

Hollan, J. D., Hutchins, E. L., & Weitzman, L. (1984, Summer). STEAMER: An interactive inspectable simulation-based training system. *AI Magazine*, pp. 15-27.

Hollan, J. D., Stevens, A. L., & Williams, M. D. (1980). STEAMER: An advanced computer-assisted instruction system for propulsion engineering. *Proceedings of Summer Computer Simulation Conference*, 400-404.

Howard, R. (1985, April). UTOPIA: Where workers craft new technology. *Technology Review*, pp. 42-49.

Hulteen, E. A. (1984). *Comprehensibility as a criterion for interface design.* Manuscript submitted for publication.

Illich, I. (1973). *Tools for conviviality.* New York: Harper & Row.

Jencks, C. (1984). *The language of post-modern architecture.* New York: Rizzoli.

Johansen, R., Vallee, J., & Spangler, K. (1979). *Electronic meetings.* Reading, MA: Addison-Wesley.

Johnson-Laird, P. N. (1983). *Mental models.* Cambridge, MA: Harvard University Press.

Jordan, B. (1983). *Birth in four cultures.* Montreal: Eden Press.

Joyce, J. (1984, August). Interview with Bill Joy. *UNIX Review*, 58-65.

Kaczmarek, T., Mark, W., & Sondheimer, N. (1983). The Consul/CUE interface: An integrated interactive environment. In A. Janda (Ed.), *Proceedings of the CHI '83 Conference on Human Factors in Computing Systems* (pp. 98-102). New York: ACM.

Kahney, H., & Eisenstadt, M. (1982). Programmers' mental models of their programming tasks. *Proceedings on the Fourth Annual Meeting of the Cognitive Science Society*, 143-145.

Kay, A. (1984, March). Computer software. *Scientific American*, pp. 52-59.

Kerr, E. B., & Hiltz, S. R. (1982). *Computer-mediated communication systems.* New York: Academic Press.

Kieras, D. E., & Bovair, S. (1984). The role of a mental model in learning to operate a device. *Cognitive Science*, *8*, 255-273.

Kieras, D. E., & Polson, P. G. (1983). A generalized transition network representation for interactive systems. In A. Janda (Ed.), *Proceedings of the CHI '83 Conference on Human Factors in Computing Systems* (pp. 103-106). New York: ACM.

Kieras, D. E., & Polson, P. G. (1985). An approach to the formal analysis of user complexity. *International Journal of Man-Machine Studies, 22*, 365-394.

Kling, R. (1980). Social analyses of computing: Theoretical perspectives in recent empirical research. *ACM Computing Surveys*, *12*, 61-110.

Kling, R. (1984, June). Assimilating social values in computer-based technologies. *Telecommunications Policy*, pp. 127-147.

Kling, R., & Scacchi, W. (1982). The web of computing: Computer technology as social organization. *Advances in computers*, *21*, 1-90.

Kowalski, R. A. (1979). *Logic for problem solving.* New York: North-Holland.

Kurland, D. M., & Pea, R. D. (1983). Mental models of recursive Logo programs. *Proceedings of the Fifth Annual Meeting of the Cognitive Science Society.*

Lang, K. N., Auld, R., & Lang, T. (1982). The goals and methods of computer users. *International Journal of Man-Machine Studies*, *17*, 375-399.

Lang, T., Lang, K. N., & Auld, R. (1981). Support for users of operating systems and applications software. *International Journal of Man-Machine Studies*, *14*, 269-282.

Laurel, B. K. (1985a). *The design of a computer-based interactive fantasy system.* Unpublished doctoral dissertation, Ohio State University.

Laurel, B. K. (1985b). Interactive fantasy in a multimodal interface environment. *Proceedings of AI '85: Artificial Intelligence and Advanced Computer Technology Conference.*

Lebowitz, M. (1984). *Creating characters in a story telling universe.* Unpublished manuscript, Columbia University, Department of Computer Science, New York.

Lewis, C. H., & Mack, R. L. (1982). Learning to use a text processing system: Evidence from "thinking aloud" protocols. In *Proceedings of the Conference on Human Factors in Computer Systems* (pp. 387-392). New York: ACM.

Licklider, J. C., Taylor, R. W., & Herbert, E. (1968). The computer as a communication device. *International Science and Technology*, *76*, 21-31.

Lieberman, H. (1984). Seeing what your programs are doing. *International*

Journal of Man-Machine Studies, *21*, 311-331.

Luchins, A. S., & Luchins, E. H. (1959). *Rigidity of behavior: A variational approach to the effects of einstellung.* Eugene: University of Oregon Press.

Lynch, K. (1964). *The image of the city.* Cambridge, MA: MIT Press.

Mack, R. L. (1984). *Understanding and learning text editing skills: Evidence from predictions and descriptions given by naive people* (Rep. No. 103330). Yorktown, NY: IBM.

Mack, R. L., Lewis, C. H., & Carroll, J. M. (1983). Learning to use word processors: Problems and prospects. *ACM Transactions on Office Information Systems*, *1*, 254-271.

Malone, T. W. (1981). Toward a theory of intrinsically motivating instruction. *Cognitive Science*, *5*, 333-368. Unpublished doctoral dissertation, Stanford University

Malone, T. W. (1983). How do people organize their desks: Implications for designing office automation systems. *ACM Transactions on Office Information Systems*, *1*, 99-112.

Mark, W. (1984). *The Consul project* (Annual Tech. Rep.). Marina del Rey, CA: University of Southern California, Information Sciences Institute.

Masuda, Y. (1982). Vision of the global information society. In L. Bannon, U. Barry, & O. Holst (Eds.), *Information technology: Impact on the way of life* (pp. 55-58). Dun Laoghaire, Dublin: Tycooly International.

Maude, T. I., Heaton, N. O., Gilbert, G. N., Wilson, P. A., & Marshall, C. J. (1985). An experiment in group working on mailbox systems. In B. Shackel (Ed.), *INTERACT '84: First conference on human-computer interaction.* Amsterdam: North-Holland.

May, R. (1975). *The courage to create.* New York: Norton.

Mayer, R. E. (1981). The psychology of how novices learn computer programming. *ACM Computing Surveys*, *13*, 121-141.

McClelland, J. L., Rumelhart, D. E., & the PDP Research Group (1986). *Parallel distributing processing: Explorations in the microstructure of cognition: Vol. 2. Psychological and Biological Models.* Cambridge, MA: MIT Press/Bradford.

McCloskey, M., Washburn, A., & Felch, L. (1983). Intuitive physics: The straight-down belief and its origin. *Journal of Experimental Psychology: Learning, Memory, & Cognition*, *9*, 636-649.

Mehan, H. (1985). *Computers in the classroom* (Final Report to the Department of Education, California). La Jolla, CA: University of California, San Diego, Teacher Education Program.

Meyrowitz, N., & Van Dam, A. (1982). Interactive editing systems. In J. Nievergelt, G. Coray, J. D. Nicoud, & A. C. Shaw (Eds.), *Document preparation systems* (pp. 21-123). Amsterdam: North-Holland.

Miller, G. A., Galanter, E., & Pribram, K. (1960). *Plans and the structure of behavior.* New York: Holt, Rinehart, & Winston.

Minsky, M. R. (1967). Why programming is a good medium for expressing poorly-understood and sloppily-formulated ideas. In M. Krampen & P. Seitz (Eds.), *Design and planning: Vol. 2. Computers in design and communication* (pp. 117-121). New York: Visual Committee Books, Hastings House.

Minsky, M. R. (1984). Manipulating simulated objects with real-world gestures using a force card position sensitive screen. *Computer Graphics, 18 (3)*, 195-203.

Mishra, P., Trojan, B., Burke, R., & Douglas, S. A. (1984). A quasi-natural language interface for UNIX. In G. Salvendy (Ed.), *Human-computer interaction* (pp. 403-406). Amsterdam: Elsevier.

Miyake, N. (1986). Constructive interaction. *Cognitive Science, 10 (2).*

Miyake, N., & Norman, D. A. (1979). To ask a question, one must know enough to know what is not known. *Journal of Verbal Learning and Verbal Behavior, 18*, 357-364.

Moran, T. P. (1983). Getting into a system: External-internal task mapping analysis. In A. Janda (Ed.), *Proceedings of the CHI '83 Conference on Human Factors in Computing Systems* (pp. 45-49). New York: ACM.

Nelson, T. H. (1967). Getting it out of our system. In G. Schecter (Ed.), *Information retrieval: A critical review* (pp. 191-210). Washington, DC: Thompson.

Nelson, T. H. (1981). *Literary machines.* Swarthmore, PA: Ted Nelson

Newell, A. (1980). Physical symbol systems. *Cognitive Science, 4*, 135-183.

Nievergelt, J., & Weydert, J. (1980). Sites, modes, and trails: Telling the user of an interactive system where he is, what he can do, and how to get places. In R. A. Guedj, P. ten Hagen, F. R. Hopgood, H. Tucker, & D. A. Duce (Eds.), *Methodology of interaction* (pp. 327-338). Amsterdam: North-Holland.

Norman, D. A. (1976). *Memory and attention: An introduction to human information processing* (2nd ed.). New York: Wiley.

Norman, D. A. (1981a). Categorization of action slips. *Psychological Review, 88*, 1-15.

Norman, D. A. (1981b, November). The trouble with UNIX: The user interface is horrid. *Datamation*, pp. 139-150.

Norman, D. A. (1983a). Design principles for human-computer interfaces. In A. Janda (Ed.), *Proceedings of the CHI '83 Conference on Human Factors in Computing Systems.* New York: ACM. (Also in *Applications of Cognitive Psychology: Computing and Education,* edited by D. E. Berger, K. Pezdek, & W. Banks. Hillsdale, NJ: Lawrence Erlbaum Associates, 1986.)

Norman, D. A. (1983b). Design rules based on analyses of human error. *Communications of the ACM, 4*, 254-258.

Norman, D. A. (1983c). Some observations on mental models. In D. Gentner & A. L. Stevens (Eds.), *Mental models* (pp. 7-14). Hillsdale, NJ: Lawrence Erlbaum Associates.

Norman, D. A. (1984a). Stages and levels in human-machine interaction. *International Journal of Man-Machine Studies, 21*, 365-375.

Norman, D. A. (1984b). *Working papers on errors and error detection.* Unpublished manuscript, University of California, San Diego, Institute for Cognitive Science, La Jolla, CA..

Norman, D. A. (1985). Four stages of user activities. In B. Shackel (Ed.), *INTERACT '84: First conference on human-computer interaction.* Amsterdam: North-Holland.

Norman, D. A. (1986). New views of information processing: Implications for intelligent decision support systems. In E. Hollnagel, G. Mancini & D. Woods (Eds.), *Intelligent decision aids in process environments.* New York: Springer-Verlag

Norman, D. A., & Bobrow, D. G. (1979). Descriptions: An intermediate stage in memory retrieval. *Cognitive Psychology, 11,* 107-123.

Norman, D. A., & Shallice, T. (1986). Attention to action: Willed and automatic control of behavior. In R. J. Davidson, G. E. Schwartz, & D. Shapiro (Eds.), *Advances in research: Vol. 4. Consciousness and self regulation..* New York: Plenum Press.

Olsen, D. R., Buxton, W., Ehrich, R., Kasik, D., Rhyne, J., & Sibert, J. (1984). A context for user interface management. *IEEE Computer Graphics and Applications, 4(12),* 33-42.

Palme, J. (1985). Experience with the COM computer conferencing system. In B. Shackel (Ed.), *INTERACT '84: First conference on human-computer interaction.* Amsterdam: North-Holland.

Papert, S. A. (1980). *Mindstorms: Children computers and powerful ideas.* New York: Basic Books.

Papert, S., diSessa, A., Watt, W., & Weir, S. (1980). *Final technical report to the National Science Foundation: Documentation and assessment of a children's computer laboratory* (Tech. Rep. 345). Cambridge: Massachusetts Institute of Technology, Artificial Intelligence Laboratory.

Pavel, M., Card, S., & Farrell, J. (1983). *Cognitive and perceptual principles of window-based computer dialogues* (Grant No. NAG 2-269). NASA-AMES.

Payne, S. J., & Green, T. R. G. (1983). The user's perception of the interaction language: A two-level model. In A. Janda (Ed.), *Proceedings of the CHI '83 Conference on Human Factors in Computing Systems* (pp. 202-206). New York: ACM.

Perlis, A. J. (1982, September). Epigrams on programming. *SIGPLAN Notices,* pp. 7-13.

Polanyi, L. (1978). False starts can be true. *Proceedings of the Berkeley Linguistics Society.*

Polson, P. G., & Kieras, D. E. (1984). A formal description of users' knowledge of how to operate a device and user complexity. *Behavioral Research Methods, Instruments, and Computers, 16,* 249-255.

Powers, W. T. (1973). *Behavior: The control of perception.* Chicago: Aldine.

Price, L. A. (1982). THUMB: An interactive tool for accessing and maintaining text. *IEEE Transactions on Systems, Man, and Cybernetics, 12,* 155-161.

Raskin, A. J. (1974). The flow-language for computer programming. *Computers and the Humanities*, *8*, 231-237.

Rasmussen, J. (in press). The role of hierarchical knowledge representation in decision making and system management. *IEEE Transactions on Systems, Man and Cybernetics.*

Rasmussen, J., & Lind, M. (1981). *Coping with complexity* (Rep. No. M-2293). Roskilde, Denmark: Riso National Laboratory.

Reason, J., & Mycielska, K. (1982). *Absent minded? The psychology of mental lapses and everyday errors.* Englewood Cliffs, NJ: Prentice-Hall.

Reichman, R. (1978). Conversational coherency. *Cognitive Science*, *2*, 283-327.

Reichman, R. (1985). *Getting computers to talk like you and me.* Cambridge, MA: MIT Press/Bradford.

Reisner, P. (1981). Formal grammar and human factors design of an interactive graphics system. *IEEE Transactions on Software Engineering*, *5*, 229-240.

Rich, E. (1982). Programs as data for their help systems. *Proceedings of the AFIPS National Computer Conference*, 481-485.

Riley, M. S. (1984). *Structural understanding in performance and learning.* Unpublished doctoral dissertation, University of Pittsburgh.

Riley, M. S., & O'Malley, C. (1985). Planning nets: A framework for analyzing user-computer interactions. In B. Shackel (Ed.), *INTERACT '84: First conference on human-computer interaction.* Amsterdam: North-Holland.

Ritchie, D. M., & Thompson, K. (1974). The UNIX time-sharing system. *Communications of the ACM*, *17*, 365-375.

Roberts, T. L., & Moran, T. P. (1983). The evaluation of text editors: Methodology and empirical results. *Communications of the ACM*, *26*, 265-283.

Robertson, S. P., & Black, J. B. (1983). Planning units in text editing behavior. In A. Janda (Ed.), *Proceedings of the CHI '83 Conference on Human Factors in Computing Systems* (pp. 217-221). New York: ACM.

Rumelhart, D. E., McClelland, J. L., & the PDP Research Group (1986). *Parallel distributing processing: Explorations in the microstructure of cognition: Vol. 1. Foundations.* Cambridge, MA: MIT Press/Bradford.

Rumelhart, D. E., & Norman, D. A. (1981). Analogical processes in learning. In J. R. Anderson (Ed.), *Cognitive skills and their acquisition* (pp. 335-359). Hillsdale, NJ: Lawrence Erlbaum Associates.

Rumelhart, D. E., & Norman, D. A. (1982). Simulating a skilled typist: A study of skilled cognitive-motor performance. *Cognitive Science*, *6*, 1-36.

Sacerdoti, E. D. (1974). Planning in a hierarchy of abstraction spaces. *Artificial Intelligence*, *5*, 115-135.

Sacerdoti, E. D. (1975). The non-linear nature of plans. *Proceedings of the Fourth International Joint Conference on Artificial Intelligence*, 206-214.

Sacerdoti, E. D. (1977). *A structure for plans and behavior.* New York: Elsevier.

Sacks, H., Schegloff, E., & Jefferson, G. (1974). A Simplest systematics for the organization of turn-taking for conversation. *Language*, *50*, 696-735.

Schank, R. C. (1982). *Dynamic memory.* New York: Cambridge University Press.

Scharer, L. L. (1983, July). User training: Less is more. *Datamation*, pp. 175-182.

Schmandt, C. M., & Arons, B. (1984). A conversational telephone messaging system. *IEEE Transactions on Consumer Electronics*, *30*, xxi-xxv.

Schwamberger, J. A. (1980). *The nature of dramatic character.* Unpublished doctoral dissertation, Ohio State University.

Scollon, R. (1982). Computer conferencing: A medium for appropriate time. *Quarterly Newsletter of the Laboratory of Comparative Human Cognition*, *5 (3)*, 67-68.

Sheil, B. A. (1981). *Coping with complexity* (CIS-15). Palo Alto, CA: Xerox PARC.

Shiffrin, R. M. (1986). Attention. In R. C. Atkinson, R. J. Herrnstein, G. Lindzey, & R. D. Luce (Eds.), *Stevens' handbook of experimental psychology* (2nd ed.). New York: Wiley.

Shneiderman, B. (1974, February). A computer graphics system for polynomials. *The Mathematics Teacher*, pp. 111-113.

Shneiderman, B. (1980). *Software psychology: Human factors in computer and information systems.* Cambridge, MA: Winthrop.

Shneiderman, B. (1982). The future of interactive systems and the emergence of direct manipulation. *Behavior and Information Technology*, *1*, 237-256.

Shneiderman, B. (1983). Direct manipulation: A step beyond programming languages. *IEEE Computer*, *16(8)*, 57-69.

Short, J. A., Williams, E., & Christie, B. (1976). *The social psychology of telecommunications.* London: Wiley.

Smith, D. C. (1975). *Pygmalion.* Boston: Birkhauser.

Smith, D. C., Irby, C., Kimball, R., Verplank, W., & Harslem, E. (1982). Designing the Star user interface. *Byte*, *7 (4)*, 242-282.

Soloway, E., Bonar, J., & Ehrlich, K. (1983). Cognitive strategies and looping constructs: An empirical study. *Communications of the ACM*, *26*, 573-560.

Soloway, E., Ehrlich, K., Bonar, J., & Greenspan, J. (1982). What do novices know about programming? In A. Badre & B. Shneiderman (Eds.), *Directions in human-computer interactions* (pp. 27-54). Norwood, NJ: Ablex.

Sommer, R. (1971). *Design awareness.* San Francisco: Rinehart Press.

Streitz, N. A. (1985). Cognitive ergonomics: An approach for the design of user-oriented interactive systems. In F. Klix (Ed.), *Man-computer interaction research.* Amsterdam: North-Holland.

Suchman, L. A. (1983). Office procedures as practical action: Models of work and system design. *ACM Transactions on Office Information Systems*, *1*, 320-328.

Suchman, L. A. (1985). *Plans and situated actions: The problem of human-*

machine communication (ISL-6). Palo Alto, CA: Xerox PARC.

Sutherland, I. E. (1963). Sketchpad: A man-machine graphical communication system. *Proceedings of the Spring Joint Computer Conference*, 329-346.

Sutherland, W. R. (1966). *The on-line graphical specification of computer procedures.* Unpublished doctoral dissertation, Massachusetts Institute of Technology.

Tanner, P., & Buxton, W. (1985). Some issues in future interface management system (UIMS) development. In G. Pfaff (Ed.), *User interface management systems* (pp. 67-69). Berlin: Springer-Vertag.

Teitelman, W., & Masinter, L. (1981, April). The Interlisp programming environment. *Computer*, pp. 25-33.

Tessler, L. (1981, August). The Smalltalk environment. *Byte*, pp. 90-147.

Thomas, J., & Hamlin, G. (Eds.) (1983). Graphical Input Interaction Technique (GIIT) workshop summary. *Computer Graphics*, *17(1)*, 5-30.

Thompson, G. (1972). Three characterizations of communications revolutions. In S. Winkler (Ed.), *Computer communication: Impacts and implications* (pp. 36-37). New York: ACM.

Thompson, G. (1984). Information technology: A question of perception. *Telesis*, *11 (2)*, 2-7.

Tou, F. N. (1982). *RABBIT: An interface for information retrieval by reformulation.* Unpublished doctoral dissertation, Massachusetts Institute of Technology.

Tou, F. N., Williams, M. D., Fikes, R., Henderson, A., & Malone, T. (1982). RABBIT: An intelligent database assistant. *Proceedings of the Conference of the American Association for Artificial Intelligence*, 314-318.

VanLehn, K., & Brown, J. S. (1980). Planning nets: A representation for formalizing analogies and semantic models of procedural skills. In R. E. Snow, P. A. Federico, & W. E. Montague (Eds.), *Aptitude, learning, and instruction: Vol. 2. Cognitive process analyses of learning and problem solving* (pp. 95-137). Hillsdale, NJ: Lawrence Erlbaum Associates.

Vygotsky, L. S. (1978). *Mind in society: The development of higher psychological processes* . Cambridge, MA: Harvard University Press.

Wason, P. C., & Johnson-Laird, P. N. (1972). *Psychology of reasoning: Structure and content.* Cambridge, MA: Harvard University Press.

Waters, R. C. (1984). The programmer's apprentice: Knowledge based program editing. In D. R. Barstow, H. E. Shrobe, & E. Sandewall (Eds.), *Interactive programming environments* (pp. 464-486). New York: McGraw-Hill.

Weinberg, G. M. (1971). *Psychology of computer programming.* New York: Van Nostrand Reinholt.

Whitford, F. (1984). *Bauhaus.* London: Thames and Hudson.

Wilczynski, D. (1981). Knowledge acquisition in the Consul system. *Proceedings of the Seventh International Joint Conference on Artificial Intelligence*, 135-140.

Wilson, P. (1985). Structures for group working in mailbox systems. In B. Shackel (Ed.), *INTERACT '84: First conference on human-computer interaction.* Amsterdam: North-Holland.

Winograd, T. (1984a). Beyond programming languages. In D. R. Barstow, H. E. Shrobe, & E. Sandewall (Eds.), *Interactive programming environments* (pp. 517-534). New York: McGraw-Hill.

Winograd, T. (1984b). Breaking the complexity barrier (again). In D. R. Barstow, H. E. Shrobe, & E. Sandewall (Eds.), *Interactive programming environments* (pp. 3-18). New York: McGraw-Hill.

Winston, P. H. (1975). Learning structural descriptions from examples. In P. H. Winston (Ed.), *The psychology of computer vision* (pp. 157-209). New York: McGraw-Hill.

Wolfe, T. (1981). *From Bauhaus to our house.* New York: Washington Square Press.

Wynn, E. H. (1979). *Office conversation as an information medium.* Unpublished doctoral dissertation, University of California, Berkeley .

Young, R. M. (1981). The machine inside the machine: Users' models of pocket calculators. *International Journal of Man-Machine Studies, 15,* 51-85.

Young, R. M. (1983). Surrogates and mappings: Two kinds of conceptual models for interactive devices. In D. Gentner & A. L. Stevens (Eds.), *Mental models* (pp. 35-52). Hillsdale, NJ: Lawrence Erlbaum Associates.

Zipf, G. F. (1965). *Human behavior and the principle of least effort: An introduction to human ecology* (2nd ed.). New York: Hafner.

Index